Pinta
(Abingdon)
Darwin
Wolf
Roca
Redonda
Volcán Wolf
Cabo
Berkeley
Volcán Ecuador
Cabo Marshall
Volcán
Darwin
Punta Garcia
James Bay
Santiago
(San Salvador,
James)
Punta
Espinosa
Cabo Douglas
Fernandina
(Narborough)
Canal Bolivar
Tagus
Cove
Urbina
Bay
Volcán
Alcedo
Rábida
(Jervis)
Cabo Hammond
Elizabeth Bay
Punta Moreno
Pinzón
(Duncan)
Isabela
(Albemarle)
Volcán Sierra Negra
Volcán
Chico
Punta Cristóbal
Volcán
Cerro Azul
Puerto Villamil
Tortuga
Ibarra
Floreana
45 km
28 mi

Islands

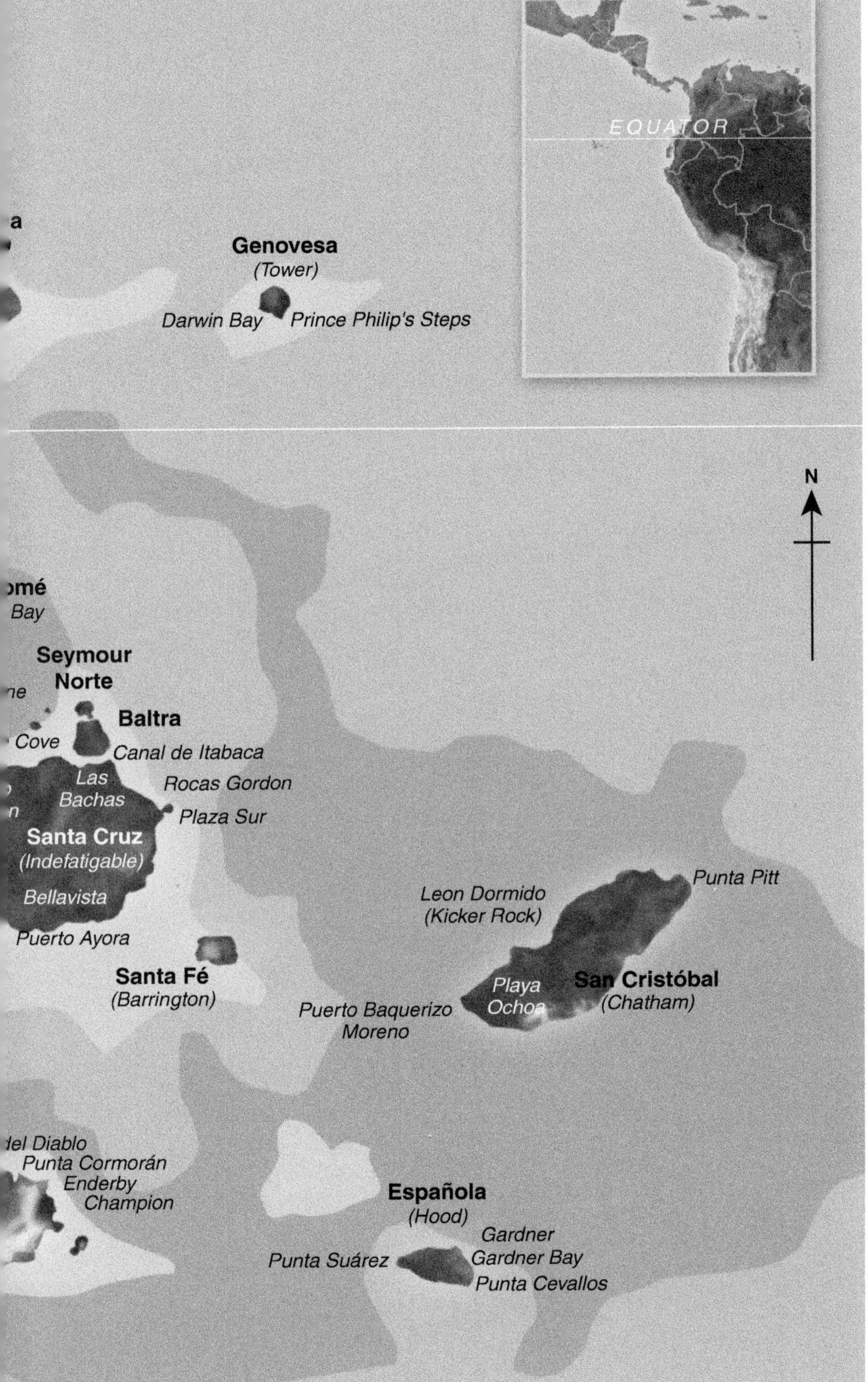

EQUATOR
N
a
Genovesa
(Tower)
Darwin Bay
Prince Philip's Steps
omé
Bay
Seymour
Norte
ne
Baltra
Cove
Canal de Itabaca
Las
Bachas
Rocas Gordon
Plaza Sur
Santa Cruz
(Indefatigable)
Bellavista
Puerto Ayora
Santa Fé
(Barrington)
Leon Dormido
(Kicker Rock)
Punta Pitt
Playa
Ochoa
San Cristóbal
(Chatham)
Puerto Baquerizo
Moreno
del Diablo
Punta Cormorán
Enderby
Champion
Española
(Hood)
Gardner
Gardner Bay
Punta Suárez
Punta Cevallos

Understanding Galápagos: What You'll See and What It Means

Randy Moore and Sehoya Cotner

The natural history of these islands is eminently curious, and well deserves attention.
—Charles Darwin, *Journal of Researches*, 1845

Connect Learn Succeed™

The McGraw-Hill Companies

Mc Graw Hill — Connect Learn Succeed™

UNDERSTANDING GALÁPAGOS: WHAT YOU'LL SEE AND WHAT IT MEANS
Published by McGraw-Hill, a business unit of The McGraw-Hill Companies, Inc., 1221 Avenue of the Americas, New York, NY, 10020.

This book is printed on acid-free paper.

2 3 4 5 6 7 QVS/QVS 23 22 21 20 19

ISBN 978-0-07-353228-8
MHID 0-07-353228-2

Senior Vice President, Products & Markets: *Kurt L. Strand*
Vice President, General Manager: *Marty Lange*
Vice President, Content Production & Technology Services: *Kimberly Meriwether David*
Publisher: *Michael Hackett*
Director of Development: *Elizabeth Sievers*
Brand Manager: *Rebecca Olson*
Marketing Manager: *Patrick E. Reidy*
Director, Content Production: *Terri Schiesl*
Lead Project Manager: *Jane Mohr*
Buyer: *Nichole Birkenholz*
Cover Designer: *Studio Montage, St. Louis, MO.*
Front and back cover images: *Daniel Fitter*
Typeface: *9.5/13 ITC Stone Serif*
Compositor: *Laserwords Private Limited*
Printer: *R. R. Donnelley*

Library of Congress Cataloging-in-Publication Data

Moore, Randy.
Understanding Galapagos: what you'll see and what it means/Randy Moore and Sehoya Cotner.
pages cm
Includes bibliographical references and index.
ISBN 978-0-07-353228-8 (alk. paper)
1. Natural history—Galapagos Islands—Guidebooks. 2. Galapagos Islands—Guidebooks. I. Cotner, Sehoya. II. Title.
QH198.G3M66 2014
508.866′5—dc23 2012041209

www.mhhe.com

DEDICATION

To Flannery and Lachlan, who showed me a new Galápagos
—Sehoya Cotner

To my Mother and Father, the heroes of my life
—Randy Moore

TABLE OF CONTENTS

PREFACE

I look forward to the Galápagos with more interest than any other part of this voyage.

—Charles Darwin, in a letter sent from Lima, Perú, to his cousin W. D. Fox, 1835

Although early visitors dismissed the Galápagos Islands as "useless," "hell," "so parch'd that [the islands] will not bear a man," and "the most dreary, barren, and desolate [place] I have ever beheld," today the islands are among the world's top destinations. Indeed, more than 170,000 tourists now visit Galápagos each year (Figure 1), despite the high cost of getting there and the archipelago's remote location, desolate landscapes, and occasional lack of amenities such as cell phone coverage, reliable Internet connections, hot showers, and fancy coffee shops. Long lists of tourism-related awards—not to mention ever-increasing numbers of tourists—testify to the appeal and lure of Galápagos. How did islands once described as "worthless" and "one of the most discouraging spots in the world" become one of the most famous places in all of science and one of the world's top destinations? Why is Galápagos—an archipelago barely visible on a world map—so popular?

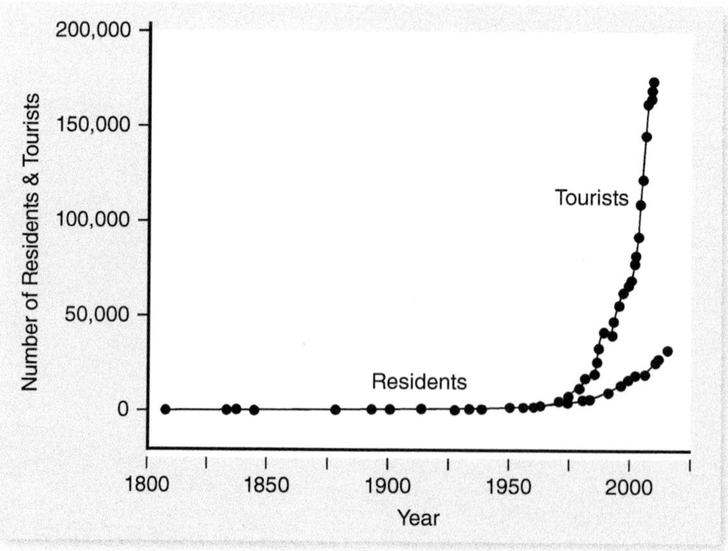

FIGURE 1 Although early visitors such as Herman Melville wrote of the islands' "emphatic uninhabitableness" and likened them to an "inhospitable inferno," Galápagos is now flooded with tourists, and its population is skyrocketing. Galápagos—the islands that guided the thinking of Charles Darwin—is one of the world's top once-in-a-lifetime destinations.

People go to Galápagos for a variety of reasons. The islands are unique: cattle in the highlands eat avocados, strange lizards swim in the surf, penguins mingle with cacti and tropical birds, and several landscapes make visitors feel as if they are on the Moon. No place is more closely associated with biology and evolution than Galápagos, and many scientists visit the archipelago to retrace the footsteps of Charles Darwin and experience one of his inspirations for conceiving of evolution by natural selection. Other visitors are lured by the islands' volcanic landscapes, world-class scuba diving, and unsolved mysteries. But the main attractions that lure most people to Galápagos are the famously unafraid and curious animals that visitors can watch from a few meters away. Many of these animals are *endemic*, meaning that they are found nowhere else except in Galápagos—and in some instances, only on particular islands in Galápagos. As Darwin noted with understatement, "the natural history of this archipelago is very remarkable; it seems to be a little world within itself."

But why another book about Galápagos? Most of the current books about Galápagos are field guides that stress traditional natural history; a few others focus on specific issues such as conservation and history. We have used and enjoyed many of these books. Although these books provide basic information about the identities of islands and their organisms, Galápagos is famous for its association with evolution, and we have often wanted more information about evolution in Galápagos as we learned about the islands. All books about Galápagos mention "Darwin's finches," despite the fact that it was mockingbirds, and not finches, that impressed Darwin during his visit to Galápagos in 1835 (Chapters 2, 5). But what about the other organisms there? They evolved, too, but this is seldom discussed. In this book, you will learn their stories.

Several books about Galápagos are strikingly Lamarckian, while others gloss over evolution without linking it to what visitors actually see in the archipelago. Granted, the islands' marine iguanas, flightless cormorants, and other iconic plants and animals (see color images) are interesting in and of themselves, but why are they in Galápagos and nowhere else? Why are there so few endemic mammals in Galápagos? Why don't the animals run when approached by humans? Answering these questions helps you understand what you see, but requires an understanding of evolution. Chapter 1 summarizes what evolution is and how populations evolve.

The Truth About Darwin's Finches

Throughout this book you will encounter stories about evolution in Galápagos, and how those stories have been influenced by geology, climate, and human history. The most famous of these stories involves finches, which are small birds ubiquitous in the islands.

Given their common name—that is, "Darwin's finches"—many visitors assume that Darwin's observations of these birds while he was in Galápagos somehow inspired his theory of evolution by natural selection. Many biologists have made similar assumptions. For example, famed evolutionary biologist V. C. Wynne-Edwards claimed that Darwin was "deeply stimulated by his own observations of the Galápagos finches during the voyage of the 'Beagle.'" And in his best-selling biology textbook, John Kimball claimed "probably no evidence was more important for [Darwin's evolutionary] thinking than the example of his finches." Similarly, many students believe that Darwin's observations of finches in Galápagos convinced him that species are mutable and were a foundation for his monumental *On the Origin of Species* (1859).

Not true.

Darwin's finches *are* an iconic piece of evidence for evolution by natural selection; that is why you will find them described in almost *every* introductory biology textbook. However, their fame came long after Darwin's visit to Galápagos, and their legend does not match the historical facts. Indeed,

While in Galápagos, Darwin paid little attention to finches; he did not even recognize them as different species. Instead, Darwin believed they were "only varieties."

Darwin did not mention finches in any of his major works. For example, you will not find the word *finch* in any of the six editions of *On the Origin of Species*, in his *Red Notebook*, nor even in any of his famed *Transmutation of Species* notebooks. After briefly mentioning them in the second edition of his *Journal of Researches* (1845), Darwin never again published anything else about finches.

Darwin never publicly used finches as evidence of evolution by natural selection.

Darwin brought home only 31 finches, all of which were poorly labeled (e.g., Darwin did not record the islands on which they were collected). For comparison, the 1905–1906 expedition by the California Academy of Science brought back 8,691 finches.

The iconic name "Darwin's finches" was not ascribed to the birds by Darwin, but instead by ornithologist Percy Lowe in 1936. The name was made famous by David Lack's book titled *Darwin's Finches* (1947) more than 100 years after Darwin visited Galápagos.

Although recent work by Rosemary and Peter Grant and their colleagues has documented evolution in Darwin's finches (Chapter 5), the fame of these iconic birds far exceeds their role in Darwin's formulation of evolution by natural selection.

We wanted a book that includes up-to-date information about evolution in Galápagos, yet is accessible by professional biologists, laypeople, students, armchair travelers, and the merely curious. Naming the endemic—and often curious—organisms in Galápagos from checklists can be rewarding; that's why you will find checklists of the islands' iconic species in this book (Chapter 12). But *understanding* what you see is a different adventure, and that is one thing that we hope distinguishes this book from others. In short, we wanted to write a book that would help you not just identify what you see, but *understand* what you see. Such an understanding goes beyond identification guides and traditional natural history, and requires a context that includes the islands' biology, geology, climate, and history. For example,

Why do Nazca booby chicks kill their siblings?

Why do land iguanas look and behave so differently from marine iguanas?

Where should I go to see marine iguanas feeding, and why do they feed there?

What are lava tunnels, and how did they form?

What is the significance of foot color in blue-footed boobies?

How can parts of islands such as Santa Cruz be arid desert, while other parts of the same islands are humid and covered with lush vegetation?

What are the seasons in Galápagos, and how do they affect what you will see?

Why is the ocean so cold in Galápagos?

When and where in Galápagos is sailing the roughest? Where is it most dangerous?

Why do some islands have almost no introduced species, while on other islands the introduced species are abundant?

Although Charles Darwin's visit in 1835 later helped make Galápagos famous, other people—with names such as Cobos, Wittmer, Angermeyer, Gil, and Villamil—profoundly shaped the islands. These people are legends in the archipelago; you will see their names throughout the inhabited islands. Who were these people, and what did they do?

Other questions are more general:

Where did the islands come from, and why is this important for understanding what you will see there?

Why are the plants and animals in Galápagos so unusual? Why are different plants and animals often found on specific islands?

Why are animals on some islands less skittish than on other islands?

Why doesn't Galápagos have any indigenous people?

After being discovered in 1535, why were the islands ignored for 200 years?

How many of the invasive species have been eradicated, and how? At what cost?

When did tourism begin in Galápagos, and how has it grown? Is this good or bad?

Why did scientists lament the disappearance of iconic species such as tortoises, yet kill tens of thousands of animals themselves?

What do landmark events such as the *Jessica* oil spill, repatriations of endemic species, and dramatic increases in tourism and population growth mean for the islands?

How have tensions between tourists, scientists, conservationists, politicians, and local residents been managed in Galápagos?

What are the future prospects for the islands?

In this book, you will find discussions (and checklists) of the native and endemic organisms you'll see, especially the iconic ones such as penguins, land and marine iguanas, finches, giant tortoises, and albatrosses (see color images). However, we have not described *all* of the organisms on the islands—after all, there are more than 600 species of lichens and 2,900 species of marine organisms, not to mention the more than 50 families, 300 genera, and 370 species of endemic and native beetles. Similarly, we discuss many, but not all, of the most troublesome introduced species (at last count, there were more than 1,400 such species). We have focused on the native and endemic "stars" of Galápagos that fill tourists' must-see lists.

Although our book emphasizes evolution, we have also addressed several other topics, for evolution cannot be understood without a corresponding understanding of its context: namely, the geology, climate, and historical impact of humans on the ecosystem. This is why you will find discussions of the islands' geology, weather, trails, and Visitor Sites in this book.[1] However, we have not tried to discuss

[1] We have dedicated relatively little space in this book to hotels, restaurants, shops, and similar places. The names and locations of these businesses often change, and there is a growing number of new places that would soon render a discussion obsolete. The best way to find good hotels, restaurants, and shops is to ask your travel agent, tour guide, boat crew, and local residents for recommendations. You might also check the Internet, but beware—there is much inaccurate information on commercial pages and blogs about the islands.

all of the more than 150 Visitor Sites in Galápagos, some of which (e.g., Darwin and Wolf Islands) are remote and visited only by dive boats. Instead, we have described sites that you will actually see during a typical visit; these discussions are supplemented by descriptions of the trails and organisms that you will find there. Virtually everyone wants to see giant tortoises, marine and land iguanas, albatrosses, and penguins, and we will tell you where and when to find them and other organisms, as well as their unique adaptations. We have also included discussions about several other aspects of your trip, including how to prepare for your trip, what to take, how to take good photographs in the islands, how to avoid common problems, and how you will travel in the islands (see Appendices 1–8).

We also wrote this book to emphasize the importance of preserving Galápagos, as well as to help readers understand the challenges facing the islands. These issues are multifaceted and complicated, and include many historical and human interest issues associated with the islands—not just the disappearances, murders, and mysteries that have contributed to the mystique of Galápagos, but also the impacts and political decisions of humans in the archipelago. Understanding the islands' history, from the earliest visits by hungry pirates and whalers to today's political tensions and floods of tourists, will help you to better understand what you see in Galápagos, from Visitor Sites to towns' street names, and will enrich your visit to the islands.

Finding Accurate Information about Galápagos

Those who have read much about Galápagos appreciate the honesty and insights of Carmen Angermeyer, a famous Galápagos resident who noted, "a lot has been written about Galápagos that is wrong. It is quite a problem. Sometimes people come down here for a short time, go around collecting little bits here and there. Sometimes you just have a conversation with them, and they go home and write down whatever they heard. What they think they heard. And the problem is that writers come later, and read what somebody else wrote, and the same errors go on and on for quite a time." She's right; for example, the geographic data available online and in many books about Galápagos often vary widely; some sources list heights of volcanoes that differ by hundreds of meters, others misidentify islands and Visitor Sites, and some even describe nonexistent places (e.g., the "Bachas Islands").[2] Similarly, there is much inaccurate information about the archipelago's wildlife and Visitor Sites, as well as about some of the most famous and influential events in the islands' history, including Darwin's visit in 1835 (Chapter 2).

Misconceptions about Galápagos extend all the way to the name of the archipelago's most famous residents: Galápagos tortoises. Many books and articles, arguing that *galápago* is an old Spanish word meaning *saddle,* claim that Spanish explorers named Galápagos tortoises *galápagos* because the tortoises' shells resembled Spanish saddles. Galápago *is* an old Spanish word for *tortoise.* However, as Galápagos researcher John Woram has noted, Tomás de Berlanga, the discoverer of Galápagos (Chapter 2), wrote in 1535 of *"muchos lobos marinos, tortugas, higuanas, galápagos . . . ,"* but mentioned neither saddles nor similarities of the *galápagos* to saddles. Similarly, Ambrose Cowley's map of Galápagos included a footnote stating, "these islands derive their name from the resort of Tortoises to them in order to lay their eggs; for in Spanish, Gallapagos signifies a Tortoise," but did not mention saddles.

[2] There are no "Bachas Islands" in Galápagos. The writers were presumably referring to "Las Bachas," a popular Terrestrial Visitor Site on the north shore of Santa Cruz (Chapter 4).

In 1815, Captain David Porter described the shells of Española tortoises as "turning up forward, in the manner of a Spanish saddle." Although this was the first known mention of the similarity of the carapace of a Galápagos tortoise and a Spanish saddle, Porter said nothing about the saddlelike shape being called a *galápago*. In 1845, Darwin referenced Porter's description of Española tortoises, but did not say anything about Spanish saddles being called *galápagos*, possibly because the dome-shaped tortoises that Darwin saw in Galápagos (on Santiago and San Cristóbal) do not resemble Spanish saddles at all.

The first printed link between *galápago* and a saddle did not appear until 1859, when Spanish writer José Coello y Quesada described a particular type of saddle called a *silla-galápago*;[3] and the first dictionary to link *galápago* with a saddle (the 11th edition of *Diccionario de la Lengua Española*) did not appear until 1869. Thus, although particular kinds of saddles were named in honor of tortoises (e.g., *silla-galápago*), there is scant evidence to support the legend that Galápagos tortoises were named with an old Spanish word for saddles.

Visitors to Galápagos also soon learn that the spelling of several locations varies significantly, even on official signs and documents. For example, most books mention "Sullivan Bay" near Santiago (the correct spelling is Sulivan), and "Margaret Wittmer" of Floreana Island (the correct spelling is Margret). Herman Melville is memorialized by San Cristóbal's Avenida "Hernan Malville," and signs in San Cristóbal direct tourists to "Playa Man," "Playaman," and "Playa Mann" (only the last of which is correct[4]). When you're in Galápagos, expect to see alternate spellings.

To ensure accuracy and consistency, the information in this book was derived from visiting the islands more than 30 times over many years; talking with numerous Galápagos National Park Service guides, boat captains, boat crews, school teachers and other residents of Galápagos, and former prisoners at the Wall of Tears penal colony; documenting the GPS coordinates of more than 350 sites, usually with two independent GNPS units (Appendix 1); studying documents at a variety of libraries, research stations, and related locations such as GNPS, Charles Darwin Research Station, National Geospatial Intelligence Agency, and Instituto Oceanográfico de la Armada; being on the bridge with boat captains in the most challenging conditions and regions of the islands (Appendix 2); and, when necessary, hiking, scuba diving, and interviewing local residents to resolve controversies (e.g., the exact location of *Jessica*; see Chapter 11).

In short, we have tried to write the book that we looked for when we came to Galápagos almost 20 years ago—a book that would help us understand what we were seeing in Galápagos. We hope that you will enjoy this book, and that it will enhance your visit to the islands.[5]

Randy Moore & Sehoya Cotner

December, 2012

---

[3] *Silla* is Spanish for *chair* or *seat*.

[4] Playa Mann is named for Nathaniel and Betty Mann, who operated the nearby refrigeration plant at the original La Predial (Chapter 2).

[5] For updates and further information about understanding Galápagos, please visit http://www.cbs.umn.edu/lab/galapagos

ABOUT THE AUTHORS

RANDY MOORE earned a PhD in biology from UCLA, after which he worked as a biology professor at several large universities. He edited *The American Biology Teacher* for 20 years, teaches courses about evolution and creationism, and has authored and coauthored several books about evolution and the evolution-creationism controversy, including *Evolution 101* (2006), *More Than Darwin: The People and Places of the Evolution-Creationism Controversy* (2008), *Biology Laboratory Manual* (2013), *Chronology of the Evolution-Creationism Controversy* (2009), and *Arguing for Evolution* (2011). Randy is H. T. Morse-Alumni Distinguished Teaching Professor of Biology at the University of Minnesota.

SEHOYA COTNER earned a PhD in conservation biology from the University of Minnesota and is currently a professor of biology at the University of Minnesota. Sehoya's research focuses on science education, specifically, effective strategies for teaching about evolution and the nature of science. She coauthored *Chronology of the Evolution-Creationism Controversy* (2009) and *Arguing for Evolution* (2011). Sehoya and Randy teach classes in Galápagos and take students to the archipelago every year.

ACKNOWLEDGEMENTS

We are grateful to many people for their help with this book and our visits to Galápagos over the years. Abrazos to our many Galápagos National Park Service (GNPS) guides, especially Rafael Rubio and his family, Victor Hugo Mendia and his family, Valerio, Martha Chica, Andrés Moreano, Christian Rueda, and Julio Pachay.

We also thank the crews of our boats for their courtesy and safe travels, especially Ricardo Andrés Bermúdez "Colombia" Escobar, Carlos Alberto "Alfredo" Bangera, Javier Alfredo Lavayen Lara and his family, Andrea Adelaida Velez Macias, Gastón, Danielito, José Luís, Edwardo, Augustin, Willi, Gonzalo Quiroga, Hermano, Maryuri Raquel Acosta Conza and her family, Ivan, and Byron Herrera "Canela" Santana. We appreciate the help of Captains Roberto Alvarado Zavala and Freddy Peñaherrera Andrade of *Yate Floreana*, and divemaster and underwater photographer José Bravo and his family. We are particularly grateful to Javier, Maryuri, Freddy, Andrea, and Rafael and Cecilia for inviting us into their homes during our many stays in Guayaquil, Puerto Ayora, Puerto Baquerizo Moreno, and Puerto Villamil.

We thank Fernando, Mafita, José, and Blanquita Icaza, who helped us countless times in Guayaquil; Jorge Alvarado, who helped us in Quito; and the hundreds of students we have taken to Galápagos, who have consistently identified new things for us to study. A special thanks and a hug to Debbie Sturdivant and Laurie McLaughlin of Holbrook Travel, who did much more than just help coordinate so many of our trips.

We also thank Alton Biggs, Robert Dennison, Darrell Vodopich, Dennis Geist, C. Josh Donlan, Fausto Rodriguez, Daniel Fitter, Jim Cotner, Betty and William Harris, Mariosi, Sarah Darling, Mark Decker, Bette Englund, Jane Super, Molly Adams, Margaret Johnson, Pat MacDonell, Lorraine Gabbert, Tammy Beward, Geri Grosinger, Nancy Becker, Sarah Knutie, Tom Alfoldi, Jon Degenhardt, Julia Ponder, and many officials at GNPS and the Charles Darwin Research Station (CDRS) for their help. We are especially grateful to Roslyn Cameron of CDRS for her expert help and gracious companionship. We also thank Lori Bradshaw, Kevin Campbell, Kala Ramachandran, Jane Mohr, and Becky Olson for their patience and helpful guidance through the production of this book. We appreciate the many other people who have helped us, but whom we can thank only collectively and anonymously.

Most of all we thank Maryuri, who helped start it all.

Randy Moore and Sehoya Cotner

December, 2012

GALÁPAGOS AT A GLANCE...

Official name: Archipiélago de Colón

Location: 620 miles west of Ecuador, the country to which Galápagos belongs. Although Galápagos straddles the equator, only Isabela is crossed by the equator.

GPS Coordinates: 1°40'N to 1°25'S, 89°16' W to 92°01'W (Appendix 1)

Number of islands: There are approximately 128 named islands and islets, with 13 principal islands and more than 60 other named reefs and rocks. These islands have 95% of their pre-human diversity of species.

Number of inhabited islands: 5 (Santa Cruz, San Cristóbal, Floreana, Isabela, and Baltra)

Main ports: Puerto Ayora (Santa Cruz), Puerto Baquerizo Moreno (San Cristóbal), Puerto Villamil (Isabela), and Puerto Velasco Ibarra (Floreana). Three islands have airports: Baltra, San Cristóbal, and Isabela.

Economy: Most people in Galápagos make their living from tourism, fishing, farming, and ranching. Per capita income is higher in Galápagos than anywhere else in Ecuador.

Population: 25,124 (legally registered permanent residents in 2010)[1]

Area code: 593-52

Government: Galápagos is administered by a provincial government headquartered in Puerto Baquerizo Moreno (San Cristóbal). On a per-capita basis, Galápagos receives more governmental money than any of Ecuador's provinces.

Busiest months, in decreasing order: July (10.6% of year's visitors), August, March, June, May, February, December, April, January, November, October, September (5.8% of year's visitors)

Official public holidays: January 1 (New Year's Day), February (Carnival), March or April (Easter Week), May 1 (Labor Day), May 24 (Battle of Pichincha), August 10 (Independence Day), October 9 (Independence of Guayaquil), November 2 (All Souls' Day), November 3 (Independence of Cuenca), and December 24–25 (Christmas)[2]

[1] For comparison, Ecuador has a population of 14,483,499; its three largest cities are Guayaquil (population = 2,291,158), Quito (population = 1,619,146), and Cuenca (331,888). The official population of Galápagos, which does not include an estimated 1,800 temporary residents or 5,000 illegal "irregular" immigrants, comprises approximately 0.17% of the population of Ecuador. Approximately 75% of the people living in Galápagos migrated from outside the archipelago, and more than 80% of the population lives in coastal villages (e.g., Puerto Ayora, Puerto Baquerizo Moreno, Puerto Villamil).

[2] Other widely observed holidays that are not official public holidays include December 6 (Foundation of Quito) and December 31 (New Year's Eve). Other nonofficial public holidays are observed locally.

Flights and baggage: All flights to Galápagos come from Quito or Guayaquil; international flights to Galápagos are not allowed. Flights to Galápagos allow a maximum of 20 kilograms (44 lbs) of luggage, plus a 6.8 kilograms (15 lbs) limit for carry-on bags.

Tipping: Tipping and gratuities are a personal matter, but on tour boats they usually average at least $60 per person for the guide and $110 for the crew.

Currency: U.S. dollar (US$). Bring small bills; you will need $1 and $5 bills for tips, and many shops and restaurants do not accept $100 bills.

Taxis: If you take a taxi anywhere in Quito or Guayaquil, make sure that it is yellow, has a running meter, and has an identification number on the doors. If the taxi does not have a meter, negotiate the fare before you get into the cab (most trips will be less than $10). In Galápagos, most taxis are small, white trucks; fares to local destinations are $1–2.

Time: GMT minus 6 hours; this is equivalent to Central Standard Time in the United States. Mainland Ecuador is GMT minus 5 hours, which is Eastern Standard Time (i.e., the same as Miami and New York City).

Vaccinations: No vaccinations are required for entry into Ecuador or Galápagos unless you are arriving from an endemic yellow fever area within 6 weeks of entry.

Language: Spanish and Quechua, but many people in Galápagos speak English.

Religion: More than 90% of Ecuador's population is Roman Catholic.

Passport: You will need a passport valid for at least 6 months from the date you enter Ecuador. You will not need a visa if you are staying 90 or fewer days. If you are not traveling with a U.S. passport, check with the Ecuadorian embassy for requirements accompanying your nationality.

Electricity: 110 V; if you visit Galápagos from the United States, you will not need any adapters.

Park entry fee: $110 ($100 Park Entrance Fee, $10 Transit Control Card), payable upon arrival at the airport; nonresidents of Ecuador under 12 years of age pay an entrance fee of only $50.[3] You will need your Transit Control Card when you leave Galápagos, so keep it in a safe place. All visitors except those on private yachts are required to go through mainland Ecuador.

Airport departure tax: In 2011, all airlines began including the airport departure tax in the cost of their tickets. You will not have to pay this tax at the airport.

Major designations: Marine Reserve (1986), UNESCO World Heritage Site (1978), Whale and Shark Sanctuary (1990) (for others, see Chapter 2)

---

[3] Residents of Ecuador pay entrance fees of $6 (adults) or $3 (children).

CHAPTER 1

Evolution and the Galápagos Archipelago

> *[O]n oceanic islands radiations have occurred not only among the land birds, but also among insects, land mollusks, and land plants, [suggesting] that the evolutionary factors involved in these other groups are fundamentally the same as they are in Darwin's finches. This is one of the chief reasons why I hope* [Darwin's Finches] *may have been worth writing.*
>
> —David Lack, *Darwin's Finches,* 1947

THROUGHOUT RECENT HISTORY, THE GALÁPAGOS ARCHIPELAGO has meant different things to different people. The islands were a convenient location for pirates to stash their booty, for whalers to stockpile fresh water for continuing oceanic voyages, and for prospectors of all sorts—Norwegian fishermen, tyrannical plantation owners, and even an ambitious baroness—to try to strike it rich. Yet, in recent decades, Galápagos has become a famous destination aligned with Darwin's theory of evolution by natural selection, and the local economy feeds off the tourism associated with the islands' connection with evolution. Thus, if you want to understand what you see in Galápagos, it is important to clarify what evolution is, and what natural selection is. But to fully appreciate the scientific power of evolutionary thought, we must first define what *science* is.

WHAT IS SCIENCE?

Science is a logical, empirical, evidence-based way to understand the natural world. The process of science involves making observations, generating logical explanations (or *hypotheses*) for these observations, and developing and testing predictions by gathering data. For example, you will see blue-footed boobies on several of the islands in Galápagos. You may wonder about the significance of these birds' blue feet, and you may generate your own hypotheses about whether blue feet have a specific meaning—and if so, what is it? One suggested hypothesis is that the color of booby feet indicates the health of individual birds; specifically, that birds choose mates whose feet are a particular shade of blue because that shade signals something about the mate's quality. Two testable *predictions* related to this hypothesis are:

1. Birds choose mates whose feet are a particular shade of blue.
2. Blue coloration is an indicator of a mate's overall health.

These predictions can be tested by observation and experimentation:

For Prediction 1: Observations of blue-footed boobies have demonstrated that female boobies prefer to mate with males having greenish-blue feet rather than deeper, duller shades of blue. Experimentation, in which the feet of some males (the *experimental* group) were painted different shades of blue, has confirmed these preferences.

For Prediction 2: Experimental work with blue-footed boobies, in which scientists manipulated the amount of food and carotenoids (such as beta-carotene) boobies received, revealed that (1) hungry birds have duller-colored feet, (2) birds deprived of dietary carotenoids have duller foot color as well, and (3) birds with bright, greenish-blue feet have stronger immune systems.

In summary, both of the predictions listed above can be supported through observation and experimentation. These results indicate that foot color is an honest signal of male health, and that female boobies choose mates accordingly. However, science is open to revision; further observation and experimentation could cast doubt on the foot-color-as-an-honest-signal hypothesis.

A *scientific theory* is a related set of hypotheses that, collectively, explain fundamental aspects of nature. In other words, a *theory* is a well-supported, broad idea about how things work. A biological example of a theory is the theory of evolution. Although scientific knowledge is provisional, many theories are supported by so much evidence that they are not questioned by serious scientists. Evolution is one such theory.

EVOLUTION

Evolution is a broad, theoretical framework for understanding life and life's history. As a result, it is important and useful for biologists because it underlies life's many facets, including speciation, physiology, genetics, ecological interactions, and behavior. Simply, evolution refers to changes in the frequencies of units of inheritance called *alleles,* which are variants of a *gene.* Key to this definition is the fact that evolution is a population-level phenomenon: Populations evolve; individuals do not evolve. If there is *any* change in the frequency of *any* gene—even a change that appears insignificant—then evolution has occurred.

Populations can evolve in different ways (Figure 1.1):

1. Random genetic changes, or *mutations,* occur in all natural populations. Mutations can have a positive effect, a negative effect, or no effect on the bearer. Most mutations have no discernable impact on the individual or the population. That is, while an organism's *genotype* (i.e., DNA) is affected, the *phenotype* (i.e., the expression of the genes as a trait) is not altered. Yet, in generating diversity, mutations provide the raw material for evolutionary change. Mutations change genes, but by themselves are not adaptive.
2. *Genetic drift* occurs as a result of random fluctuations in the frequency of alleles. Some alleles may be eliminated from the population, while others are perpetuated; however, by definition, the alleles that survive genetic drift do so without any obvious cause. Like mutation, genetic drift is not adaptive. Genetic drift can be especially influential with small populations, such as those that colonize isolated islands (e.g., Galápagos) and those that remain after predation, habitat loss, and climatic disruptions (all of which have been common in Galápagos for at least two centuries). The "genetic bottleneck" of small populations reduces genetic diversity and the ability of the species to respond to future challenges, thus rendering the population especially vulnerable to drift. For example, the random loss of 20 iguanas from a continental population of 500,000 iguanas produces a much smaller change in genetic diversity than does a random loss of 20 iguanas from an island population of 30 individuals. However, even in these situations, drift is not the only possible cause of low genetic diversity. Indeed, most populations in Galápagos were started by either a few organisms or by a single, pregnant female. In these instances, genetic diversity would be relatively low, even without drift.
3. When individuals leave or join a population, they carry their alleles with them. *Migration,* or *gene flow,* is another way that populations evolve, but it is not adaptive. Recent migrations—from the mainland to the archipelago, or between islands—characterize Galápagos.
4. Lastly, populations can evolve through *adaptive* change, whereby individuals best suited to prevailing environmental conditions survive and reproduce, but those not as well adapted to the environment do not reproduce as much (or at all). This differential rate of reproduction associated with *natural selection* is the only scientifically supported explanation for adaptive evolutionary change.

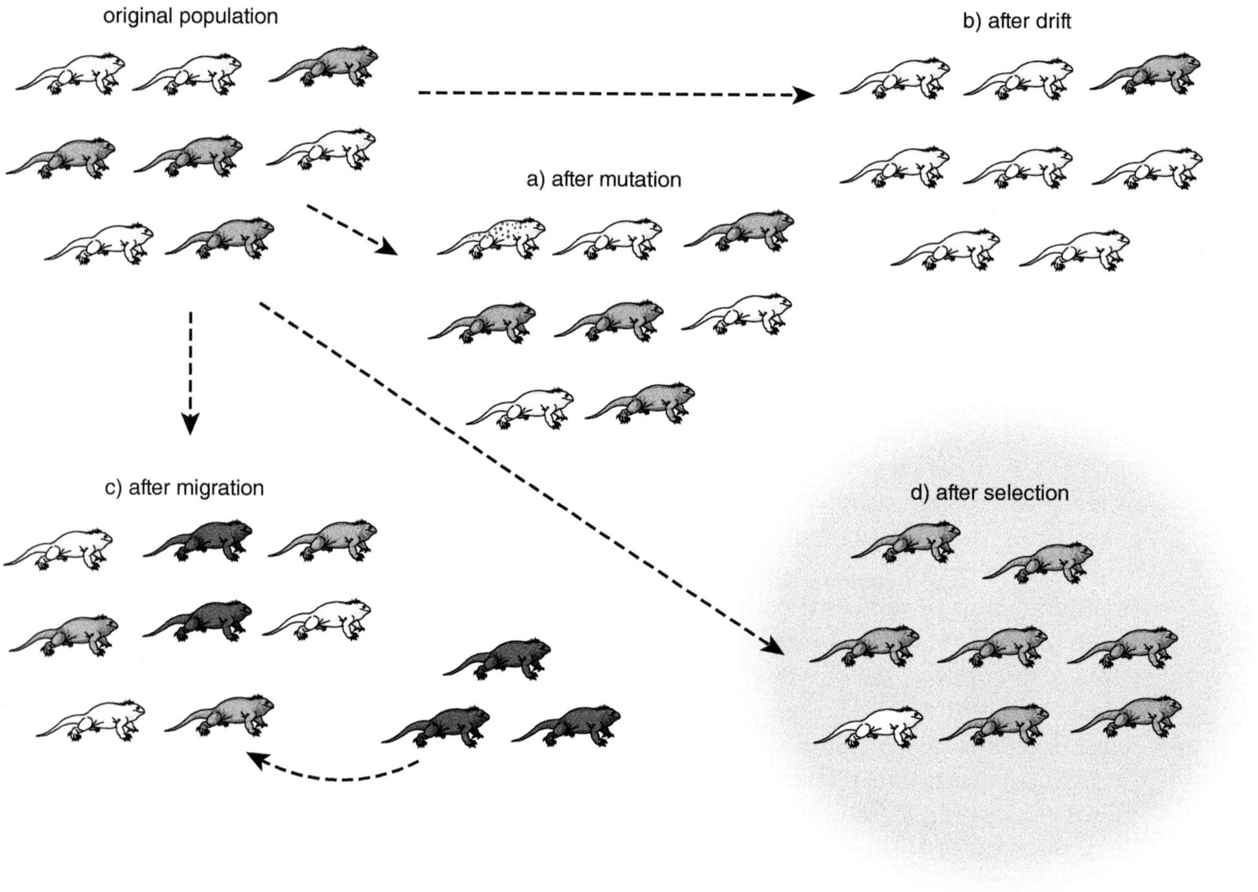

FIGURE 1.1 Populations can evolve through (a) mutation, (b) drift, (c) gene flow, and (d) natural selection. Although all of these mechanisms produce evolutionary change, only natural selection increases the frequency of adaptations that increase survival and reproduction.

EVOLUTION BY NATURAL SELECTION

Mutation, gene flow, and sexual reproduction produce variability, and some of this variability—for example, the ability to withstand salt water, or to survive for long periods without fresh water—may help individuals in a population to survive and reproduce. Any feature these survivors possess that contributes to their success is an *adaptation*. That is, individuals with a given adaptation are more likely to survive and reproduce than are other members of the population. If all or part of this feature is heritable, then it can be passed to the offspring of the survivors.

For example, the size of beaks in Galápagos finches is a heritable trait; finches with large beaks are likely to produce baby finches with large beaks.

Within a population, finches exhibit variation in the size of their beaks. Under some environmental conditions, a certain beak-size is advantageous. For example, a severe drought in Galápagos in 1977 killed most of the medium ground finches on Daphne Major. But survival was not random; medium ground finches having large, strong beaks—that is, birds that could crack open the only seeds available during the drought—were far more likely to survive than were birds having smaller, weaker beaks. Prior to the drought, the medium ground finches on Daphne Major had an average beak-depth of 9.42 millimeters; after the drought, surviving birds' beak-depths averaged 9.96 millimeters. In a world where seeds are often less than 1 millimeter in diameter, one-half of a millimeter can make a big difference: Birds with bigger beaks had a much better chance of surviving than did birds with smaller beaks. In this environment, larger, stronger beaks were an *adaptation* that helped the birds survive and reproduce.

Natural selection refers to adaptive evolutionary change that occurs when heritable variations confer a benefit to the bearer. The advantage could be survival itself, or increased reproductive success. Either way, individuals with the beneficial variant (i.e., *adaptation*) can pass on their genes to more offspring than can individuals without the adaptation. In the preceding example, ground finches with larger beaks survived to reproduce, passing on the genetic instructions for large beaks to their offspring. The environment, changing in response to a drought, selected *against* birds with small beaks, and selected *for* birds with larger beaks. We speak of these large-beaked individuals as having greater *fitness* because they produced more offspring—that is, they had more genetic representation in the next generation.

The Inevitability of Natural Selection

Scientists often speak of evolution by natural selection as an inevitable consequence of life. That is, if certain conditions exist, populations will evolve by undergoing natural selection. These specific conditions are:

1. *Organisms produce many more offspring than the environment can sustain.* Organisms have a tremendous capacity to reproduce, but resources are limited. As a result, any population of organisms, reproducing at full capacity, will outstrip the available resources within a few generations. This leads to a competition for resources. This competition may be overt (e.g., a physical battle for resources or mates) or not (e.g., some individuals may be better at hunting or fishing or building nests). Among competing organisms, there will be winners and losers; the winners survive and reproduce, while the losers die and are eliminated from the population before they reproduce.
2. *Competing organisms exhibit individual variability.* Variability between competing organisms may contribute to an individual's success or failure.

For example, some finches may have adaptations (e.g., larger beaks) that enable them to obtain food more efficiently than other finches.

3. *Some of the variability that organisms exhibit is heritable.* If some individuals survive and reproduce because of inherited traits that the nonsurvivors lack, then the winners will pass these features on to the next generation, thereby altering the gene frequencies of the population. This change in gene frequency, based on differential survival and differential reproductive success, is evolution by natural selection.

In summary, competition for resources, heritability of traits, and differential reproductive success inevitably lead to evolution by natural selection.

EVOLUTION IN GALÁPAGOS

Several themes emerge during any discussion of evolution, and the following will be on display in Galápagos:

1. *Selection acts on continuous traits.* While it is easy to think of variation in an either/or context, selection typically acts on continuous traits, which are the result of several genes, rather than on discrete traits from a single gene. In the finch-beak example discussed above, there are a range of beak sizes, not just small and large. Selection shifted the distribution of beak sizes from 9.42 millimeters to 9.96 millimeters—that is, average beak-sizes increased. This change in size was a result of *directional selection* on a continuous trait (beak size). Selection on multiple-gene traits can also be *disruptive,* whereby extremes are favored, and *stabilizing,* whereby the mean is favored and extremes are selected against.
2. *Nature and nurture combine to make phenotype.* As important as an individual's genes are, they are not everything. The environment can play a large role in shaping an organism's phenotype. In blue-footed boobies (discussed above), foot coloration was influenced by a bird's environment—specifically, the foods that the boobies ate. Consuming foods rich in beta-carotene gave male boobies the bright, greenish feet preferred by female boobies. Of course, a male's ability to secure the best foods is, to some extent at least, inherited, so distinguishing between what is genetic and what is not can be difficult. Studies of parent-offspring similarities can help determine the heritability of some traits.
3. *Selective forces can be abiotic.* Selection for or against a trait can come from *abiotic* (i.e., nonliving) forces such as wind, rain, and rock. In Galápagos, a dramatic abiotic selective force is a periodic—and

unpredictable—warming event (El Niño; Chapter 3). In subsequent chapters of this book, we will discuss how many of the island's fauna and flora are affected by these, and other, abiotic events.

4. *Selective forces are often biotic.* Selection is often the result of interactions with other organisms, either of one's own species (*intraspecific* interactions) or of another species (*interspecific* interactions). For example, male sea lions fight for dominance and access to females. As a result, there is competition among male sea lions for strength, size, and fighting ability. Plants compete for water, nutrients, and light. We will discuss many examples of natural selection via intraspecific interactions.

   We will also discuss various types of *symbioses,* which are the relationships between two different species (i.e., *interspecific* interactions). Symbioses can affect participants positively, negatively, or not at all.

   - In *competition,* both participants are harmed. Competition can be *direct* (e.g., physical aggression between symbionts) or *indirect* (e.g., when individuals of both species consume the same resources). For example, tortoises and land iguanas both eat a cactus common in Galápagos called *Opuntia* (prickly pear cactus). On small islands with a finite number of cacti, this *resource competition* may have a negative effect on both participants. Tortoise-iguana-cactus dynamics are discussed further in Chapter 6.
   - In *mutualism,* both participants benefit. Many mutualistic relationships are on display in Galápagos. For example, there are fish ("cleaners") that remove the parasites from other fish ("clients"). The cleaners get an easy, reliable meal, while the clients are rid of pesky parasites. Both the cleaner and the client benefit in the short term and, as a result, both may enjoy increased fitness in the long term. Cleaner-fish mutualisms are discussed further in Chapter 8.
   - During *predation* or *parasitism,* one participant benefits and the other is harmed. In Galápagos, shark predation is an important selective force of sea lion and fur sea-lion communities. You will probably see evidence of this in shark-bite wounds and scars on fur sea lions. The threat of predation has clearly been a selective force; for example, fur sea lions avoid hunting during a full moon when they would be more visible to sharks. Fur sea lions' adaptations to predation are discussed further in Chapter 7.
   - In *commensalism,* one symbiont benefits and the other is unaffected. When small fish surround a marine iguana scraping algae from underwater rocks, they are looking for food dislodged from the algae. The iguana is not significantly affected by this association, and the fish get an easy meal. However, with all commensalisms, there

may be a cryptic benefit or cost, to be disclosed by future research. Foraging iguanas are discussed in Chapter 6.

All of these symbioses are dynamic and can be shifted by environmental or genetic changes. For example, cleaner-fish mutualisms often become parasitic when the cleaner exploits the client by eating the client's scales or mucus. Without client vigilance, the mutualism is likely to fail.

5. *Evolution can help us understand speciation.* Mutation, drift, and migration all play a role in generating diversity, and selection eliminates some of this diversity by removing unsuccessful variants from the population. Over time, populations of the same species can diverge so much that they are considered separate species. For example, an ancestral population of finches from mainland South America colonized Galápagos millions of years ago. After many generations of mutation, drift, and selection, the colonizing population was sufficiently different from the parent population to warrant classification as a different species. This pattern was then repeated throughout the islands, with migration, drift, mutation, and selection all acting differently on the populations of different islands. Thus, from one colonizing species, the archipelago now has 13 different species of finches. All of these finches are *endemic* to the archipelago—that is, they are found nowhere else in the world. *Adaptive radiation* occurs when one ancestral species leads to several descendent species as new habitats are colonized. There are many examples of adaptive radiation and endemism in Galápagos, including finches, mockingbirds, lava lizards, bulimulid snails, *Scalesia,* and *Opuntia.* (The definition of *species* is discussed in Chapter 5.)
6. *Reproduction can be sexual or asexual.* For many organisms, sexual reproduction is worth the obvious costs, such as the energetic demands of acquiring a mate, the risk of transmitting disease, and the uncertainty associated with genetically novel offspring. Sex, with the production of variable gametes (sperm and egg) and the creation of unique combinations, produces genetic variation and vast diversity. There are different hypotheses for why sex is adaptive, including the idea that sexual recombination generates novelty, which makes it difficult for an organism's competitors (e.g., parasites and predators) to keep up. Regardless of the reason, some organisms reproduce sexually, whereas others forego sex and reproduce asexually. In Galápagos, many of the colonizing plants make gametes and have sex, but they are *self-fertile,* which eliminates the problem of having to find a mate in a new, barren landscape. Some organisms can reproduce sexually *and* asexually, while others reproduce only sexually. You will observe a diversity of reproductive strategies in the plants and animals of Galápagos.

7. *Sexual selection is a powerful evolutionary force.* Selection for traits that increase an individual's reproductive success is termed *sexual selection.* With *intrasexual selection,* individuals of the same sex (typically males) compete for dominance and access to the most, or the best, females. You may see male marine iguanas butting heads and biting each other to establish dominance. This aggression comes at a cost, because males often walk away with bloody wounds. The benefit—that is, access to females—must be worth the cost of the aggression.

   In *intersexual selection,* individuals of one sex choose partners of the other sex. In most cases, female choice is the selective force, and successful males (i.e., the males chosen by females) are those with, for example, the brightest feet (in blue-footed boobies) or the best nests (in small tree-finches). We will discuss many examples of sexual selection throughout this book.

8. *A diversity of mating systems occurs in Galápagos.* In Galápagos you will see examples of monogamy (in which individuals in a mating pair copulate only with each other), polygyny (whereby one male mates with two or more females), and polyandry (one female mates with two or more males). There is a difference between *genetic* monogamy (whereby the partners in a mating pair do not cheat) and *social* monogamy (in which cheating occurs, and paternity can be questionable). For example, genetic tests confirm that frigatebirds in Galápagos sometimes have sex outside of the pair bond, that sea lions and marine iguanas are polygynous, and that Galápagos hawks are polyandrous. These and many other examples of mating systems are discussed throughout the text.

9. *There are many different types of parental care.* Most of the fish and arthropods you will encounter in Galápagos provide no parental care. After eggs are fertilized, the parents move on. However, selection has favored degrees of parental care in many of the archipelago's inhabitants. With most mammals, the female provides all of the caregiving. This makes evolutionary sense, for gestating and lactating females have invested more resources in their offspring than males. Uniparental care (i.e., care by one parent) also occurs in Galápagos hawks and some fish. For example, male sergeant major fish hover over their egg masses to protect them, and male seahorses carry fertilized eggs in an abdominal pouch until the offspring hatch and disperse. With some organisms, biparental care (i.e., care by two parents) is necessary to ensure the survival of the offspring. In most birds, both the male and female parents incubate the eggs and feed the hatchlings. Although caregiving imposes limits on the number of offspring an individual can have, young that are protected and fed by their parents are more likely

FIGURE 1.2
A Galápagos hawk eating a dead sea-lion pup, in full view of other sea lions and tourists. You will see predation, death, and scavenging in Galápagos, all of which attest to the ruthlessness of natural selection.

to survive. Like so many other life-history features we will explore, parental-care regimes are determined by selection and a complex mix of environmental and biological constraints.

10. *Natural selection can be gruesome.* During natural selection, many individuals die—from direct competition, resource competition, disease, starvation, and environmental changes. In many cases, the individuals that survive these challenges have adaptations that give them an edge over others of their species. In Galápagos, death is everywhere. You may see dead marine iguanas decomposing on hot lava rock, a frigatebird carcass lodged in a *Parkinsonia* bush, or a dead sea-lion pup being eaten by scavengers such as mockingbirds and Galápagos hawks (Figure 1.2). Although these events are not unique to Galápagos, they are often removed or hidden in neighborhoods, parks, and tourist-frequented natural areas. Death is not hidden in Galápagos. This raw aspect to Galápagos is part of what makes the place special, and the evolutionary "losers" can help us fully understand the mechanics of natural selection.

This overview of evolution should equip you to appreciate the biological phenomena you will encounter in the islands. Fighting iguanas, dancing boobies, screaming fur sea lions, *Opuntia* in various sizes and shapes—natural selection can help explain it all. Understanding evolution by natural selection will help you to understand what you see in Galápagos.

CHAPTER 2

History of the Galápagos Islands

There is much more to the bleak Encantadas than meets the eye.
—Lillian Otterman, *Clinker Islands: The Mysterious Galápagos,* 1993

DESPITE ITS POPULARITY AS A TOURIST SITE AND SCIENCE SHRINE, many people harbor misconceptions about Galápagos and do not appreciate the importance of the islands' history for our modern understanding of evolution. Granted, many people link Galápagos with Charles Darwin and evolution, but even Darwin's famous visit to the islands is often misunderstood. Contrary to what many people believe, Darwin did not methodically survey—or even visit—all of the major islands, was not the first naturalist to visit the archipelago, was not the first naturalist to write a book about the islands, was not especially impressed with the islands' now-iconic finches during his visit, and did not conceive of evolution by natural selection while he was in Galápagos.[1] Nevertheless, these and other stories persist, thus obscuring the true—and far more interesting—history and importance of Galápagos to modern biology and society.

Charles Darwin *did* help make Galápagos famous; this is why he is memorialized throughout the archipelago (Appendix 3). However, there is much more to Galápagos than Charles Darwin. Other people who are memorialized in Galápagos, such as Margret Wittmer, José de Villamil, and Manuel J. Cobos, have also shaped the islands, and understanding what these people did will help you understand what you see today in Galápagos. Similarly, although the islands' many curious plants and animals have fascinated visitors for generations, there is more to Galápagos than its wildlife. Today's Galápagos has been shaped by myriad factors, ranging from its isolation and sparse supply of fresh water to the waves of plundering pirates, whalers, fishermen, colonists, scientists, and, most recently, tourists, each with social and/or economic agendas that have

[1] Although Darwin kept notes during his years aboard the *Beagle,* he did not notice (or consider notable) several iconic species while he was in Galápagos, including blue-footed boobies, flightless cormorants, or Galápagos penguins.

influenced the islands either directly or via the more than 1,400 introduced species that they have left behind.[2] Consider the linkages between tourism, agriculture, the Galápagos economy, and introduced species. When tourism surged in the 1980s, many Ecuadorians migrated to Galápagos to enjoy the higher standard of living afforded by the better-paying jobs there. This simultaneous increase in the number of tourists and residents forced Galápagos to import more food, fuel, water, and other supplies from the mainland, as well as to open more entry points to the archipelago (e.g., more flights, more airports, and more tour boats). The resulting increases in pollution, fuel consumption, waste production, and other problems associated with increased populations (e.g., the *Jessica* oil spill in 2001) would have been problematic enough, but an even bigger problem is the fact that 80% of the introduced species in Galápagos enter the islands on organic products imported from the mainland to support the archipelago's growing demands.[3] You will see introduced species at many sites in Galápagos, and your understanding of their historical context will help you appreciate not only the names of several streets in the towns of Galápagos, but also the islands' landscapes today.

In this chapter you will find details about the many major events that have shaped Galápagos, including its discovery, its annexation by Ecuador, the visit by Charles Darwin, the use of the islands as military bases in World War II, the development of the Galápagos National Park (GNP) and Charles Darwin Research Station (CDRS), and the political tensions associated with conserving the islands. We have also included entries showing how the floods of tourists have prompted residents to develop services and infrastructure that have "modernized" Galápagos. There are many examples of such developments, including the arrival of the first tour boats (1934), the availability of the Internet (2000), and how the numbers of hotels, restaurants, and other services have increased to accommodate the residents and visitors to Galápagos.

We also describe the many ways in which Galápagos has been modernized, including how it became a popular tourist destination, when it began appearing in mainstream publications, when dignitaries (e.g., the president of the United States) began visiting the archipelago, and how a variety of social ills such as drugs and prostitution have accompanied the development and ongoing flood of tourists to the archipelago. There have also been subtle influences, including how discoveries in Japan, Pennsylvania, and London helped diminish humans' impact on the archipelago, how the removal of land iguanas from Baltra later saved these iguanas from extinction, how tourism has been both a blessing and a curse to the islands, and how humans' activities have repeatedly altered the islands' biological landscape. The cumulative impact of

[2] A mural depicting the history of Galápagos is at the south end of Charles Darwin Avenue in Puerto Ayora.
[3] Only about 5% of the introduced species arrive on packaging, and another 5% arrives on wood and lumber.

these diverse events has been significant, and understanding them will help you understand what you see in Galápagos. GPS coordinates for many of the sites mentioned in this and other chapters are listed in Appendix 1.

HISTORICAL TIMELINE

1493 Pope Alexander VI issues the papal bull *Inter Caetera,* which gives Spain all of the land in the Pacific, including the islands that later became known as Galápagos. Spain, however, never claimed the archipelago, probably because the earliest visitors considered it to be worthless. In 1840, Spain formally transferred sovereignty of Galápagos to Ecuador.

1535 Dominican Fray Tomás de Berlanga,[4] fourth Bishop of Panama, discovers Galápagos while sailing from Panama to Peru to mediate a dispute between Francisco Pizarro and his officers after their conquest of the Incas (Berlanga's diocese included Peru).[5] When winds died and the boat was "becalmed for 6 days," currents took Berlanga's ship west to Galápagos. In his report to Spanish Monarch Charles V, Berlanga described the archipelago's giant tortoises (so large "that each could carry a man on top"), "iguanas that are like serpents," and other tame animals "so silly they do not know how to flee," and lamented the islands' harsh, desert-like environment as "worthless" and "filled with very great stones," noting that it "seems as if God had showered rocks." Berlanga desperately needed water, and two of his crew (and 10 horses) died of thirst before they found fresh water in the archipelago (many ate *Opuntia* pads, which Berlanga described as "somewhat juicy, though not very tasty"). Berlanga, who neither named nor claimed the islands for Charles V, later sailed to the mainland (it took him 24 days to get there) and never returned. Berlanga's report to Charles V was the first written record about Galápagos.[6] The archipelago's isolation, scarcity of

---

[4] Berlanga (1487–1551) was born Tomás Enríque; when he entered the Dominican Order, he changed his surname to de Berlanga. As the fourth Bishop of Panama, Berlanga urged King Charles I to build a road along what is now the Panama Canal. Berlanga's visit may have temporarily calmed tensions in Peru, but in 1538 a civil war erupted between Pizarro's troops and those of his rival, Diego de Almagro. Berlanga is memorialized with a statue and plaque ("Descubridor de las Islas Galápagos") in his native city of Berlanga de Duero, Spain.

[5] Incan stories from approximately 60 years earlier describe a voyage to the west by Inca king Túpac Yupanqui and the discovery of two "Islands of Fire." Some people later argued that these islands were Galápagos. However, these stories are vague, contain inconsistencies, and are documented by no credible physical evidence. There is no evidence of prehistoric settlements, but there are remains of earlier unknown visitors.

[6] Berlanga's letter to Charles V was not published until 1883, when it appeared in a collection of unedited documents in Madrid. Berlanga introduced many useful plants (e.g., bananas) to South America. Unlike for most Pacific islands, missionaries played negligible roles in the history of Galápagos, for the islands had no indigenous people to convert or subvert.

fresh water, and harsh environment would delay the establishment of the first permanent settlement in Galápagos for another 297 years.

1546 Don Diego de Rivadeneira and 22 sailors flee the Peruvian civil war and arrive in Galápagos aboard a ship that has lost its way. Rivadeneira, who later named the islands "Enchanted Islands," urged Peru to colonize Galápagos and name him governor, but his efforts failed.

1562 Galápagos first appears on a map in an anonymous Spanish manuscript. Eight years later, a map by cartographer Gerardus Mercator showed two groups of islands labeled "Ysolas de los Galopegos" separated by several hundred miles.

1570 Dutch cartographer Abraham Ortelius includes "Ins. de los Gallopegos" in his *Theatrum Orbis Terrarum (Theatre of the World),* the first modern atlas. By 1574, Ortelius listed the islands as "Isolas de Galapagas," and by 1589 the islands were referred to as "Las Islas Encantadas."[7]

1594 While fleeing the Spanish Armada, Sir Richard Hawkins becomes the first English captain to visit Galápagos. Hawkins described the islands as "all desert [that] beare no fruite." The scarcity of fresh water became a defining feature of the archipelago, not just for the plants and animals, but also for the islands' visitors and would-be settlers. As William Beebe noted in *Galápagos: World's End* (1924), "Everyone who ever came to the Galápagos arrived thirsty; most voyagers left the islands in the same condition."

1600s Galápagos becomes a sanctuary for pirates raiding Spanish galleons carrying treasures from South America to Spain. The pirates, like subsequent visitors, soon realized that tortoises were an excellent source of fresh meat during their voyages. The tortoises were also valuable on the islands, for visitors could often follow the trails of tortoises to water.

1684 The ships *Bachelor's Delight* (a renamed 40-gun Danish ship) and *Nicholas* arrive in Galápagos carrying, among others, William Dampier, Ambrose Cowley, and John Edward Cook. These voyages, which were the first intentional and extended visits to Galápagos, discovered and named many sites in the archipelago. That same year, Cowley produced one of the first navigational charts of Galápagos,[8] naming several islands after English noblemen and kings. Cowley remarked that tortoises were tasty and abundant, adding "I believe there is not any place in the world that have such plenty of these Creatures." These and

[7] Ortelius was also the first person to suggest that once-joined continents had drifted apart to their current positions (i.e., continental drift).

[8] Cowley's map was crude and inaccurate, but it would be more than a century before a more accurate map was produced. The islands appeared on an undated vellum chart suspected to be from the 1530s, but an artist probably added the islands after the chart's production.

future visitors killed tens of thousands of tortoises in the archipelago. Many of the names assigned by Cowley are today those most closely associated with the islands.

1697 British naturalist William Dampier, who had sailed west "to try if we could find those islands the Spaniards call Gallappagos or Enchanted Islands," publishes his *A New Voyage Round the World*. This book was the first English description of Galápagos, and it included discussions of the islands' plants, animals, geology, and climate. Dampier, who coined the words *sea lion, subspecies,* and hundreds more, was especially impressed by the calmness of the islands' animals and the taste of their meat ("they are so tame that a man may knock down twenty [of the animals] in an hour's time with a club").[9] Dampier's book, which led to his election into the Royal Society of London (England's premier scientific organization), established a new style and standard for travel books. When Charles Darwin left England in 1831 aboard the *Beagle,* he took Dampier's books with him.[10] Today, Dampier's picture in London's National Portrait Gallery describes Dampier as a "Pirate and Hydrographer."

1708 Scotsman Alexander Selkirk visits Galápagos after being picked up in the Juan Fernández Islands. Selkirk inspired Daniel Defoe to write *Robinson Crusoe,* but later died in poverty.

1727 English buccaneer Woodes Rogers—who disliked Galápagos and distrusted the accounts provided by previous visitors—recognizes the volcanic origins of Galápagos. Rogers described Galápagos as "nothing but loose Rocks, like Cynders, very rotten and heavy, and the Earth so parch'd, that it will not bear a man."

1789 Soon after oil from sperm whales is shown to be much better than oil from baleen whales for candles, soaps, lubricants, and lamps, London's *Amelia* becomes one of the first ships to harvest sperm whales along the western coast of South America. Several years later, the *Beaver* became the first American whaling ship to reach the Pacific. By 1846, there were 735 whaling boats in the Pacific, many of which took hundreds of tortoises from Galápagos (see below). Tortoises in the archipelago became a well-known source of fresh meat for the ships' hungry crews.

---

[9] Dampier's travels and work as a naturalist were widely known. For example, Jonathan Swift mentioned Dampier in *Gulliver's Travels* (1726), Samuel Coleridge praised Dampier's work, Captain James Cook relied on Dampier's books to sail the world, and Alexander von Humboldt labeled Dampier "the finest of all travel writers." Dampier, who sailed more than 200,000 miles in his lifetime, was the first person to map winds of the Pacific and the first to circumnavigate the world three times. Dampier also noted the islands' often-poor anchorage. This results from lava cooling faster in the sea than in air, thus giving volcanoes steeper sides below sea level than above sea level.

[10] Dampier's now-famous books made relatively little money when they were published.

1789 Peru resident Alonso de Torres leads the first expedition to Galápagos for naturalists. In 1949, Torres's visit became the basis for a claim that Peru should own Galápagos.

1790 Spain sends Sicilian naturalist Alessandro Malaspina and colleagues on the first scientific expedition to Galápagos. Malaspina was later imprisoned and his records were lost.

1790s Whalers begin replacing pirates in Galápagos. The whalers, who were more numerous than the pirates, removed hundreds of thousands of tortoises from the islands during upcoming decades as they harvested sperm-whale oil, which was of higher quality than other lamp oils available at the time. This oil helped fuel the Industrial Revolution in Europe.

1791 There are six American whaling ships in the Pacific. By 1820, there were hundreds, many of which visited Galápagos to harvest tortoises for their meat.

1792 Spurred by the success of earlier missions, Englishman James Colnett and the HMS *Rattler* are sent by the British Navy and Samuel Enderby & Sons to Galápagos to investigate the feasibility of large-scale commercial whaling in the archipelago. Colnett's descriptions of some of the islands' plants and animals were later cited by Charles Darwin.

1795 Captain George Vancouver and naturalist Archibald Menzies of HMS *Discovery* (and later of HMS *Chatham*) collect the first plants from Galápagos, including the iconic *Scalesia* (Chapter 10). Vancouver described Galápagos as "the most dreary barren and desolate country I ever beheld," but also noted that the adjacent ocean "swarmed with large Lizards . . ."

1797 Isabela's Volcán Wolf erupts. This is the earliest observed volcanic eruption in Galápagos.[11]

1798 In *A Voyage to the South Atlantic,* Captain James Colnett of HMS *Rattler* publishes the first accurate map of Galápagos. This map, which was prepared by Aaron Arrowsmith, was the first major improvement of a map of Galápagos since Cowley's map from 1684.[12] Colnett's book reported abundant whales and, in doing so, helped begin the modern era of Pacific whaling, which had a dramatic impact on Galápagos.

[11] More recent eruptions of Isabela's volcanoes were in 1813 (Volcán Darwin), 1979 (Volcán Chico), 1982 (Volcán Wolf), 1993 (Volcán Alcedo), 2005 (Volcán Sierra Negra [Figure 3.1]), and 2008 (Volcán Cerro Azul). Volcán Ecuador is extinct. Nearby Volcán La Cumbre on Fernandina last erupted in 2009.

[12] Colnett's map was later corrected and supplemented by the captains of several ships, including those of *Tagus* (1814) and *Indefatigable* (1815). Although Colnett is often credited with creating the Floreana post office, the chart in Colnett's book does not include the post office. Moreover, Colnett did not mention "post office" in his book, and *Rattler* did not visit Floreana. The first person to mention a post office on Floreana was David Porter in 1815 (see below).

1800 After driving fur sea lions to extinction in the Falkland Islands in only 16 years, a 60-ship sealing fleet begins hunting in Galápagos. During the next century, more than 17,400 fur sea lions were harvested in the archipelago.

1801 Captain and fur sea lion hunter Amasa Delano (a maternal ancestor of future U.S. president Franklin Delano Roosevelt) of *Perseverance,* who circled the world several times during his life, watches volcanic eruptions on Santiago Island from James Bay. He later described Galápagos's lava lizards and a freshwater ravine south of Tagus Cove for the first time. Delano, who noted the islands' "diver" birds (i.e., blue-footed boobies), recognized the importance of the islands for whalers: "Were it not for the advantages which these islands afford to the American and English whaling ships, in getting a supply of fresh provisions when in these seas, I know not what they would do."

1804 Cameron Hathawson of the ship *Halard* leaves the oldest known legible graffiti in Galápagos in a cave on the west side of Santiago's Cerro Cowan.

1805 Irish sailor Patrick "Irish Pat" Watkins is marooned on Floreana, thus becoming the first permanent human inhabitant of Galápagos. Watkins, who sold vegetables to visiting ships, lived on Floreana until he stole *Black Prince* in 1809 and moved to the mainland. Watkins, the inspiration for "Oberlus" in Herman Melville's *Las Encantadas,* was last seen in a jail at Paita, Peru.

1809 Charles Darwin is born into a life of wealth and privilege in Shrewsbury, Shropshire, England. As a young man, Darwin was often unfocused and, when he dropped out of medical school at the University of Edinburgh, his father predicted "you will be a disgrace to yourself and all your family." Darwin later returned to college and became the most famous biologist in history.

1810 During the next three decades, American whalers extract more than four million gallons of whale oil from whales around Galápagos.

1813 The 860-ton, 44-gun USS *Essex* is sent to Galápagos to protect American ships and harass British whaling ships. Thanks to information he obtained from letters left at Floreana's post office, Captain David Porter and his crew of 250 sailors captured 12 British ships (i.e., more than half of the British whaling fleet in the Pacific) during his first five months in Galápagos; this shifted control of the whaling industry from the British to Americans.[13] Porter, who described an eruption

[13] Aboard *Essex* was 12-year-old midshipman David Farragut, Porter's adopted son. Farragut later became the U.S. Navy's first admiral. While in Galápagos, *Essex* collected 14 tons of tortoises, which sailors referred to as "Galápagos mutton." Porter's success—largely due to information he gathered at Floreana's Post Office Bay—prompted the British to leave the Pacific. Porter described marine iguanas as "the most timid of animals, and in a few moments we knocked down hundreds of them with our clubs."

on Floreana (the only known historic eruption of this volcano), dismissed Galápagos as "dreary, desolate, and inhospitable . . . unsuited for the residence of man," but later urged the United States to annex any unclaimed islands. Porter accidently released four goats and some sheep at James Bay.[14] When *Essex* was in Galápagos, Lt. John S. Cowan was killed in a duel with Lt. John M. Gamble and was buried "among some green bushes near the beach" of James Bay. Numerous subsequent visitors in Galápagos—including President Franklin Roosevelt—sought, but did not find, Cowan's grave.

1815 David Porter's *Journal of a Cruise Made to the Pacific Ocean* notes differences between populations of tortoises on different islands of Galápagos,[15] and includes the first mention of a post office on Floreana ("a box nailed to a post, over which was a black sign on which was painted *Hathaway's Postoffice*"). Porter had visited Floreana two years earlier, where he found a letter in the box dated June 14, 1812. Thus, someone had created the now-famous Floreana Post Office by 1812. Contrary to many claims, Porter did not create Floreana's post office—he merely used it to determine the locations of British ships that he was chasing. Hathaway's identity remains unknown.

1817 More than 7,500 fur sea lions are harvested in Galápagos by a single boat. In 1825, another ship killed 5,000 fur sea lions. Sea lions escaped a similar fate because their fur—unlike that of fur sea lions—often falls apart when it is dried.

1819 HMS *Siren* discovers profitable whaling grounds northwest of Japan. Within a few years, many American and British whalers began harvesting whales there, thus reducing their visits to (and impact on) Galápagos.

1820 Aaron Arrowsmith reprints a chart by James Colnett and adds the words "Post Office" to describe the now-famous Visitor Site on Floreana. This was the first appearance of these words on a Galápagos chart, but the site was not described in the book's text. When the *Beagle* visited Floreana in 1835, Captain Robert FitzRoy—like Porter had done two decades earlier—described the post office as "a box . . . on a post," not a barrel.

1820 Harvests of tortoises, fur sea lions, and other animals continue unabated in Galápagos. For example, in 1830, *Essex* harvested more

[14] Although these introductions probably failed to reproduce, subsequent ones did not; between 1968 and 1982, GNPS rangers killed more than 19,167 goats on the island.

[15] Darwin, who saw only dome-shaped tortoises while in Galápagos, used Porter's descriptions of saddleback tortoises to document interisland variation among tortoises in the second edition of his *Journal of Researches* (1845). Porter's *Journal of a Cruise Made to the Pacific Ocean* was surpassed only by Darwin's *Journal of Researches* for informing readers about the uniqueness of Galápagos.

than 300 tortoises from Española, and an additional 60 from Floreana. Two years later, *Wasp* harvested 100 tortoises from Floreana and 5,000 skins from fur sea lions, and in 1831 *Uncas* harvested 416 tortoises from Santiago. In 1834, *Moss* took 350 tortoises from Floreana, and in 1837 *Omega* took 240 tortoises from San Cristóbal. By this time, whalers and pirates had killed at least 100,000 tortoises in Galápagos.

1820 HMS *Beagle*—a 90-foot-long, double-mast, 235-ton "Cherokee Class" warship built at a cost of £7,803—is launched from the Royal Naval Dockyard in Woolwich, England. In 1831, the *Beagle*—one of several British ships to have had that name—began a voyage that later helped Charles Darwin formulate his ideas about evolution by natural selection.

1825 British naturalists David Douglas and John Scouler aboard the brig *William and Ann* collect plants from Santiago Island. In 1847, Joseph Hooker included 18 of these plants with his descriptions of plants collected by Charles Darwin while aboard the *Beagle.* Today, Scouler is honored by *Cordia scouleri* (Scouler's cordia) and *Croton scouleri* (Galápagos croton; Chapter 10).

1825 Captain Benjamin Morrell and the crew of the New-York-based schooner *Tartar* watch as Fernandina's Volcán La Cumbre erupts like "ten thousand thunders . . . with accumulated vengeance." Five hours later, water near the boat was 150°F, and air temperatures reached 147°F as the boat's rigging began to melt. Two years earlier, Morrell and his crew had harvested 5,000 fur sea lions in Galápagos.

1826 HMS *Beagle,* with Captain Pringle Stokes at its helm, begins its first survey of the coastline of South America. Two years later, Stokes killed himself off the coast of Tierra del Fuego, and the young, aristocratic Robert FitzRoy became captain of the ship. FitzRoy, too, would later commit suicide, but not while captain of the *Beagle.*

1830 During the next three decades, the 500 whaling ships near Galápagos harvest vast numbers of tortoises each year for their hungry crews. For example, *Isabela* took 335 tortoises from Hood in 1831, *Hector* took 237 tortoises from Charles in 1832, *Abigail* took 132 tortoises from Santa Cruz in 1834, and *Bengal* took 200 tortoises from Charles in 1834. The following year (1835), Charles Darwin noted, "formerly single vessels have taken away as many as 700 [tortoises], and that the ship's company of a frigate some years since brought down in one day 200 tortoises to the beach."

1830 Wealthy Louisiana native José de Villamil begins planning to colonize Galápagos.[16] Two years later, he claimed the land by organizing the Galápagos Settlement Company, which was approved by General Juan José Flores (Ecuador's first president).

[16] Villamil was born in 1789 in Louisiana, which was a Spanish colony until the Louisiana Purchase in 1803.

1831 Charles Darwin earns a bachelor's degree from Cambridge, but is drawn to the prospect of travel. Earlier in the year, Cambridge botanist John Henslow had received a letter from Cambridge astronomer George Peacock (1791–1858) describing an opening for a naturalist on the upcoming voyage of the *Beagle.* After considering the offer himself, Henslow approached his brother-in-law, Leonard Jenyns (1800–1893; like Henslow, a vicar with a background in natural history) about his interest in the position. Jenyns declined, after which Henslow contacted his former student, 22-year-old Charles Darwin, noting that although Darwin was not "a finished naturalist," he was certainly "qualified for collecting, observing, and noting anything worthy to be noted in Natural History. . . . I think you are the very man they are in search of." Darwin wanted "to add a few facts" to natural science and was "ambitious to take a fair place among scientific men." On December 27, the *Beagle,* manned by Captain Robert FitzRoy and a crew of 73 sailors, sailed from Devonport to survey the Cape Verde Islands, the South American Coast, the Strait of Magellan, Galápagos, Tahiti, New Zealand, Australia, the Maldives, and Mauritius. The *Beagle*'s voyage was among the most ambitious and best-financed survey voyages sponsored by the Admiralty Board in more than a decade. The *Beagle*'s official naturalist (and surgeon) was Robert McCormick (1800–1890). However, FitzRoy could have little social contact with the ship's crew and needed a companion for his voyage, and FitzRoy wanted this person to be a naturalist. Although FitzRoy initially had reservations about Darwin serving in this role (he believed the shape of a person's skull indicated the person's character, and concluded that Darwin's nose meant that Darwin was lazy and hesitant), Darwin was eventually accepted. While at sea, Darwin—who subsequently referred to December 27 as "my real birthday"[17]—was often sick; he later noted, "I loathe, I abhor the sea." Darwin's 40,000-mile voyage aboard the *Beagle*—at a cost of about £1,000, paid by his father—lasted 58 months, 43 of which were spent in South America. Although the *Beagle*'s trip is often described as a sea voyage, Darwin spent over three of the years on land. Darwin's book about his voyage, which became famous as *The Voyage of the Beagle,* remains one of the world's great travel books. Darwin later supported FitzRoy's election to the Royal Society (FitzRoy invented weather forecasting) and named a species of dolphin after him, despite the fact that the two differed about Darwin's eventual conclusions based on the observations and collections made during the voyage. Darwin later noted in his autobiography, "the voyage of the *Beagle* has been by far the most important event in my life and has determined my whole career."

[17] Darwin knew that joining the crew of the *Beagle* was a turning point in his life, noting in a letter to his sister Susan that "there is indeed a tide in the affairs of men, and I have experienced it."

FIGURE 2.1 Although many people believe that this intersection in Puerto Ayora (and a similar one in Puerto Baquerizo Moreno on San Cristóbal) honors Charles Darwin's birthday (February 12, 1809), it does not. Instead, Avenida 12 de Febrero commemorates February 12, 1832, the date that Ecuador annexed Galápagos. February 12 is "Galápagos Day" in Ecuador.

1832 José de Villamil convinces the Ecuadorian government to claim Galápagos, and then departs (with 12 craftsmen, several convicts and pardoned political prisoners, and numerous donkeys, pigs, goats, and cattle) for Floreana to create the first settlement in Galápagos; that settlement, which was near a spring of fresh water, was named Asilo de la Paz (Haven of Peace).[18] There, at a ceremony on February 12, Ecuador annexed Galápagos and renamed it Archipiélago de Ecuador. (On that day, Charles Darwin celebrated his 23rd-birthday aboard HMS *Beagle*, sick in the North Atlantic; Darwin would not reach Galápagos for another three years.) The ceremony on Floreana was officiated by Colonel Ignacio Hernández (the first justice of the peace in Galápagos), whose first official act was to name the island "Floreana" in honor of Ecuadorian president Juan José Flores. (Their ship, the "war schooner" *Mercedes*,[19] was named in honor of Flores' wife.) Attendees included crews of *Levant* and *Richmond*, two U.S. whaling ships that were at Floreana. Villamil, who became the first general governor of Galápagos (and a founder of the Ecuadorian navy), later established the first penal colony in Galápagos on Floreana, and is a national hero in Ecuador. Coincidentally, February 12 is also Charles Darwin's birthday (Figure 2.1).

1835 Well into the fourth year of its voyage, HMS *Beagle*, sailing from Callao, Peru, arrives at the northwest end of San Cristóbal (Chatham) Island

[18] Galápagos officially became a penal colony in 1833, and pardoned prisoners on Floreana grew crops, raised cattle, and harvested orchils (Chapter 10). The banishment of prisoners to Galápagos set a precedent that later led to the creation of several penal colonies in the archipelago. Floreana was later abandoned in the 1850s and again in the 1870s. When Heinz and Margret Wittmer came to Floreana in the 1930s, they used the name "Asilo de la Paz" for their farm. By the time Villamil died in 1866, he had spent much of his personal wealth trying to settle Galápagos.
[19] Until 1880, *Mercedes* was the only operational ship in Ecuador's Navy.

at "black, dismal-looking heaps of broken lava, forming a shore fit for Pandemonium." Although the 26-year-old Darwin was anxious to get home, he had become enthused about Galápagos after reading books by Dampier, Colnett, and Byron while aboard the *Beagle;* as he noted in a letter to his sister Catherine, "I am very anxious for the Galápagos Islands,—I think both the Geology & Zoology cannot fail to be very interesting." Nevertheless, when Darwin saw San Cristóbal, he noted, "nothing could be less inviting than the first appearance." Darwin and the *Beagle* spent five weeks in Galápagos, during which Darwin spent 19 days (some only in part, such as at Tagus Cove) at four islands: Chatham (San Cristóbal, from September 16–22; Color Image 1), Charles (Floreana, from September 24–27), Albemarle (Isabela, from September 29 to October 2), and James (Santiago, from which he saw Halley's Comet, from October 8–17), each of which had been visited previously by humans.[20] He described Galápagos as "a little world within itself; the greater number of its inhabitants, both vegetable and animal, being found nowhere else. . . . I never dreamed that islands 50 or 60 miles apart, and most of them in sight of each other, formed of precisely the same rocks, placed under a quite similar climate, rising to nearly equal height, would have been differently tenanted." Although virtually all biology textbooks later cited Galápagos as central to Darwin's theory, Darwin's own *On the Origin of Species* (1859) mentioned Galápagos only six times. Navigational charts of Galápagos prepared by the *Beagle*'s captain Robert FitzRoy were the best available until World War II (Figure 2.2); the English names of the islands have remained unchanged since FitzRoy's visit.

1836 Crewmembers of the *Phoenix* etch "Phoenix 1836" under an overhang at Tagus Cove. This graffiti is the oldest legible graffiti at Tagus Cove.

1836 Nine months after departing Galápagos as the *Beagle* sailed from the Cape of Good Hope to Saint Helena and Ascension islands,[21] Darwin copies his "Ornithological Notes" (for colleagues in England) and inserts his first tentative reference to the possibility of evolutionary change. After noting that "the Spaniards" at Santiago could identify which island a tortoise came from, Darwin confesses that "if there is the slightest foundation for these remarks the zoology of Archipelagos—will be well worth examining; for such facts [would] undermine the stability of species." In his *Journal of Researches* (1839; see below),

[20] Although Darwin landed on only four islands, he would have had good views of at least eight other islands in the archipelago.

[21] These volcanic islands, like those of Galápagos, were created by a "hotspot" (Chapter 3). At Ascension Island, Darwin received a letter from his mentor John Henslow promising that Darwin would be famous upon his return to England.

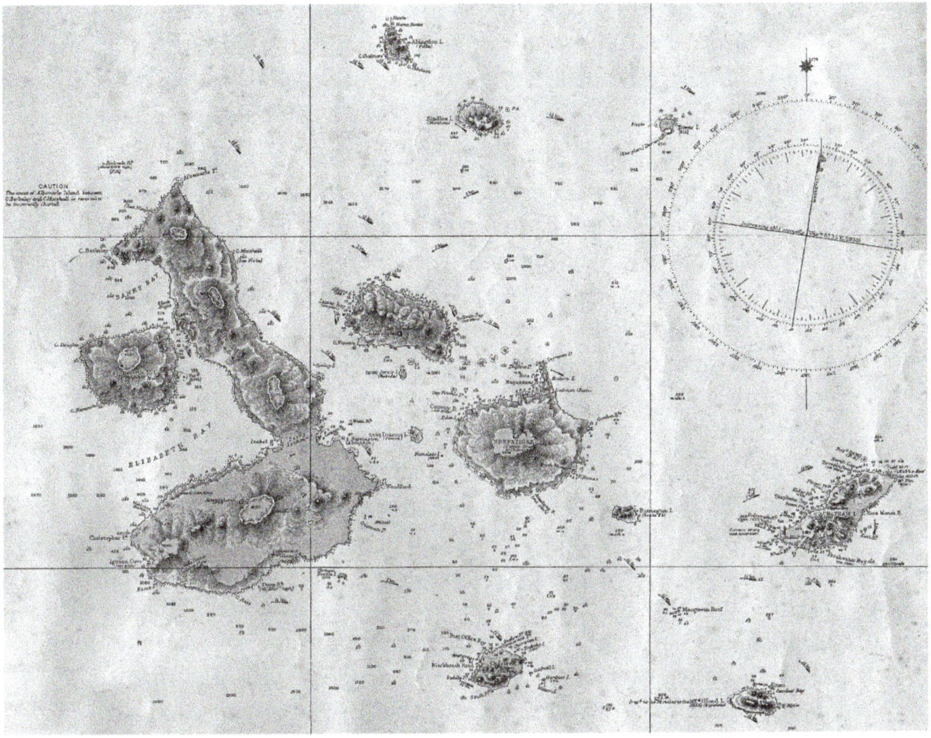

FIGURE 2.2 Robert FitzRoy's map of Galápagos, which was published in 1836, was the most accurate map of the archipelago for more than a century.

Darwin credited Floreana's Nicholas Lawson (an Englishman) with having told him of island-to-island differences among tortoises, but Darwin apparently had more than one such discussion while in Galápagos.

1836 HMS *Beagle* returns to England; its projected two-year voyage had lasted four years, nine months, and five days. Although the *Beagle* had left England with 74 people aboard, only 62 returned (five had died, and others had abandoned the trip; 65 were aboard when the *Beagle* was in Galápagos). After disembarking from the *Beagle,* Darwin never again left England. The *Beagle* spent only about 2% of its voyage in Galápagos, but as American explorer and writer Victor Wolfgang von Hagen noted, "the HMS *Beagle* [gave] the Galápagos scientific immortality." The *Beagle*'s time in Galápagos would also have a disproportionate impact on Darwin. In 1870, the *Beagle* was sold to scrap merchants for £525. A keepsake box made of the *Beagle's* timber is displayed at the National Maritime Museum.

1837 Nearly a year and a half after leaving Galápagos (and five months after returning to England), Charles Darwin begins his "Transmutation of Species" notebook, in which he describes the "species on Galápagos Archipelago" as the primary source "of all my views." That same year, Darwin finished writing *Journal and Remarks,* which described his adventures aboard HMS *Beagle*. Darwin's book, which would not appear until 1839 as part of a three-volume series edited by the *Beagle*'s Captain Robert FitzRoy, became the most popular book about Galápagos, and today is known as *The Voyage of the Beagle.*

1837 Charles Darwin gives renowned ornithologist John Gould (1804–1881) the birds he and others aboard the *Beagle* had collected in Galápagos.[22] One week later Gould reported that the collection included three distinct species of mockingbirds ("mocking-thrushes")—one species from Floreana, one species from Isabela, and one species that lived on both Santiago and San Cristóbal. Gould also reported that the collection included "a series of ground Finches which are so peculiar" because they formed "an entirely new group, containing 12 species." Two months later, Darwin met with Gould, who reported that the Galápagos "wren" was another species of finch. That same year, Gould described Darwin's birds in the *Proceedings of the Zoological Society of London.* Herpetologist Thomas Bell, who had been identifying the reptiles collected by Darwin in Galápagos, gave Darwin a parallel conclusion—namely, that at least two species of tortoises were native to Galápagos (and not brought there by buccaneers, pirates, or whalers, as Darwin and others had assumed). Darwin later noted in his *Journal of Researches into the Geology and Natural History of the Various Countries Visited by the H.M.S.* Beagle *under the Command of Captain FitzRoy, R.N., from 1832 to 1836* that finches of Galápagos appeared as if "one species had been taken and modified for different ends," but he was not yet ready to say how this might have happened. During the next six years, the *Beagle*'s fossils were described by Richard Owen, its mammals by George Waterhouse, and its fish by Leonard Jenyns.

1837 José de Villamil resigns as governor of Floreana and, the following year, is replaced by English military officer James Williams, who soon becomes a cruel ruler of the island. Williams was ousted in a bloody revolt in 1841. By 1852, the settlement had disappeared.

1838 French herpetologist Gabriel Bibron suggests to Charles Darwin that there might be at least two species of tortoises endemic to Galápagos. That same year, when pondering the similarities of species on islands with those of neighboring continents, Darwin realized "This fact speaks volumes. . . . My theory explains this but no other will."

1839 Robert FitzRoy edits the three-volume (plus an appendix to Volume II) *Narrative of the Surveying Voyages of H.M.S. Ships* Adventure *and* Beagle.

[22] Although Darwin carefully labeled his geological specimens, he did not label his bird specimens carefully, so he gathered specimens from three of his shipmates—Captain Robert FitzRoy, Harry Fuller (a sailor), and Syms Covington (Darwin's servant)—who *had* labeled their collections carefully. Darwin's incomplete collections of birds may seem amateurish by today's standards, but they may have been reasonable for a naturalist managing several different priorities during a short visit to the archipelago.

FitzRoy, like William Dampier had earlier, believed that Galápagos tortoises had been introduced from the Indian Ocean, and would eventually revert to the "specific type" once the environmental factors causing the morphological changes were gone. FitzRoy's contributions (especially those titled "Remarks with Reference to the Deluge") were ridiculed by some naturalists for adhering to biblical literalism; as geologist Charles Lyell noted, "[i]t beats all the other nonsense I have ever read on the subject." Charles Darwin contributed the third volume, which was titled *Journal and Remarks;* this was later published as a separate book titled *Journal of Researches into the Geology and Natural History of the Various Countries Visited by H.M.S.* Beagle *from 1832 to 1836.* This book, which is largely a narrative controlled by chronology and place, established Darwin as a serious scientist. Since then, Darwin's book has had four different titles, and today is known as *The Voyage of the* Beagle.[23] FitzRoy later dissociated himself from Darwin's published account of the *Beagle*'s voyage because he believed Darwin's work was undermining the Bible.

1841 An executive decree places the Galápagos colonies under the jurisdiction of Guayaquil.

1841 Twenty-two-year-old Herman Melville visits Galápagos for a month while aboard the whaler *Acushnet.* Whereas Darwin used straightforward, objective prose to describe the archipelago's natural history, Melville lamented the islands' "emphatic uninhabitableness. . . . Man and wolf alike disown them," equated the islands with "evilly enchanted ground," and described the islands as being more desolate than "abandoned cemeteries of long ago" and which "ruin itself can work little more upon them." After describing the archipelago's plants as "ungrateful," Melville claimed, "In no world but a fallen one could such lands exist." A decade later, Melville—who referred to the archipelago by its Spanish name, "The Encantadas, or Enchanted Isles"—used his experiences in Galápagos as inspiration for his greatest work, *Moby-Dick* (which quoted Darwin's *Journal of Researches*).

1844 Charles Darwin writes a 250-page essay in which he describes "a natural means of selection" that occurs on oceanic islands and continents.

1844 Joseph Hooker tells Darwin that different islands of Galápagos house different species of plants. For example, 30 of 38 species on San Salvador (Santiago) were found nowhere else; the ratio was 22/26 for Isabela, 12/16 for San Cristóbal, and 21/29 for Santa María (Floreana). The next

[23] This book showcased Darwin as a nature writer instead of a proponent of a new theory, as he was in his subsequent *On the Origin of Species* (1859). Darwin was exceedingly proud of his first book; as he later noted, "The success of my first literary child always tickles my vanity more than that of my other works."

year, Darwin wrote to Hooker that "geographical [distribution] . . . will be the key which will unlock the mystery of species."

1845 The first immigrants from mainland Ecuador settle in San Cristóbal. Many of these (and subsequent) settlers were disappointed with their new home.

1845 When the second (and only other) edition of Darwin's book about his *Beagle* adventure is published, Darwin—emboldened by the findings of Joseph Hooker—expands his discussions of Galápagos by almost 50% and includes this prophetic evolutionary conjecture: "the natural history of these islands is eminently curious and well deserves attention. Most of the organic productions are aboriginal creations, found nowhere else; there is even a difference between the inhabitants of the different islands; yet all show a marked relationship with those of America. . . . Hence, both in space and time, we seem to be brought somewhat near to that great fact, the mystery of mysteries, the first appearance of beings on this earth . . . one might really fancy that from an original paucity of birds in this archipelago, one species had been taken and modified for different ends." Darwin added, at his own expense, a plate showing the heads of four finches from Galápagos, and used David Porter's *Journal of a Cruise* (1815) to describe dome- and saddle-shaped tortoises. The title of this edition transposed the words "geology" and "natural history," producing the title *Journal of Researches into the Natural History and Geology of the Various Countries Visited by the H.M.S.* Beagle *under the Command of Captain FitzRoy, R.N., from 1832 to 1836.* The spine was labeled *Naturalist's Voyage,* and by 1860 it was labeled *Naturalist's Voyage Round the World.* Darwin sold the copyright for his book for £150 (by 1860, sales exceeded 10,000 copies, and by 1879 they exceeded 14,000). The title *The Voyage of the Beagle* did not appear until 1905.

1846 Although the *Beagle*'s Robert FitzRoy had reported in 1835 that tortoises were common on Floreana and that their shells were used as flowerpots, Berthold Seeman, a naturalist aboard HMS *Pandora,* reports that there are no tortoises on Floreana (but there are dogs, goats, pigs, and 2,000 cattle). When cattle and other introduced species had eaten almost all of the cacti on the island, Floreana mockingbirds disappeared from the island. Near this time, tortoises also disappeared from Jervis (Rábida) and Barrington (Santa Fé) Islands.

1846 The first American edition of Darwin's *Journal of Researches* is published.

1847 Joseph Hooker publishes *An Enumeration of the Plants of the Galápagos Archipelago,* the first scientific guide to Galápagos plants and a pioneering work in biogeography. Hooker described 236 species of plants from Galápagos, many of which were collected by Darwin while aboard HMS *Beagle*. Hooker's earlier lobbying had helped convince British authorities to add Galápagos to the *Beagle*'s itinerary, "if the season permits."

1848 The whaler *Good Return* reports no tortoises on Santiago. In 1813, 13 tons of tortoises had been collected from Santiago by one ship.

1851 Ecuador considers giving Galápagos to Great Britain to settle a loan incurred decades earlier by Simón Bolivar during the War of Independence, but takes no action.

1851 Herman Melville publishes *Moby-Dick,* which was partly inspired by his experiences in Galápagos and an account of the sinking of *Essex,* an American whaling ship sunk by a sperm whale in 1820 in the South Pacific. Melville, who rejected several of Darwin's scientific observations of the islands, parodied Darwin's claims and grouped Darwin's work with other "barren, bootless allusions from some few passing voyagers or compilers."

1852 During a 10-day visit to Galápagos, Swedish biologist N. J. Andersson discovers 50 new species of plants.

1852 Convicts led by famed bandit Manuel Briones capture the United States whaler *George Howland,* after which the United States demands $40,000 from Ecuador to compensate the boat's owner (Matthew Howland) and crew. This was the first international political exposure for Galápagos.

1854 In his fanciful *The Encantadas, or Enchanted Islands,* Herman Melville describes Galápagos as "five-and-twenty heaps of cinders dumped here and there in an outside lot," and Galápagos tortoises as "newly crawled forth from beneath the foundations of the world." Melville, after noting that buccaneers had carved seats into the rocks of islands so they could look out at the archipelago's beautiful landscapes, found it "hard to impute the construction of these romantic seats to any other motive than one of pure peacefulness and kindly fellowship with nature."

1856 While working in his apartment, London teenager William Perkin serendipitously synthesizes mauveine, the first synthetic organic dye. The subsequent large-scale production of mauveine helped launch the synthetic chemical industry and later doomed several of the earliest settlements in Galápagos, which were based on gathering lichens called orchils that contained a similar dye.

1858 A visitor to Galápagos finds orchils, a group of lichens (e.g., *Roccella;* Figure 2.3) from which can be extracted a valuable purple dye. In some areas, these lichens were so abundant that they broke the branches of trees. Several businesses—most notably Manuel Julián Cobos's

FIGURE 2.3 Some of the earliest settlements in Galápagos were based on harvesting dye-containing lichens called orchils ("dyer's moss"). Shown here is one of the most abundant of these orchils, *Roccella*. The synthesis of artificial dyes, along with the discovery of dye-rich lichens in California, later led to the collapse of the orchil industry in Galápagos.

Industrial Orchil and Fishing Company—soon brought workers to Galápagos to harvest the lichens (which were called "dyer's moss"). *Roccella tinctoria* is an orchil that is a source of litmus, a water-soluble mixture of dyes whose color is sensitive to pH.

1858 Charles Darwin receives a letter from often-impoverished Alfred Russel Wallace describing Wallace's idea about evolution by natural selection. Wallace's idea was almost identical to that conceived by Darwin; as Darwin noted, "I never saw a more striking coincidence; if Wallace had my M.S. sketch written out in 1842 he could not have made a better short abstract." Darwin immediately began writing what became *On the Origin of Species,* which was published late the following year. Darwin and Wallace's papers were jointly presented on July 1, 1858, at a meeting of the Linnean Society that was chaired by Thomas Bell, who had earlier examined the reptiles that Darwin had collected in Galápagos.

1858 The United States asks for Galápagos as collateral for a $3-million loan to Ecuador. Public outcry in Ecuador ended the negotiations.

1859 As sperm whales become increasingly scarce in the Pacific, Edwin Drake devises a new way to drill for oil in Pennsylvania. Investors soon shifted their money from whaling to petroleum, thereby reducing the whalers' impact on populations of whales and Galápagos tortoises. Between 1860 and 1900, about 13,000 tortoises were killed by whalers (down from more than 100,000 tortoises in previous decades).

1859 Charles Darwin publishes *On the Origin of Species,* his masterful "one long argument" for evolution by natural selection. As he predicted in 1845, Darwin's primary evidence was biogeography. Darwin devoted only about 1.1% of the text to Galápagos; this was little more than he devoted to New Zealand, and three-times less than he devoted to pigeons. Darwin's six comments about Galápagos, many of which were inspired by Joseph Hooker's studies of Galápagos plants, discussed the islands' high rates of endemism, the absence of some groups of organisms, and the similarity of the archipelago's plants and animals with those of South America. Darwin did not mention Galápagos finches.

1860 Darwin's *Journal of Researches* is again reprinted with an added postscript (dated February 1, 1860). This is the version that is most common today.

1860 Tortoises in Galápagos are increasingly hard to find; five-day harvests have dropped from 200–300 tortoises in the 1830s to fewer than 100 in 1860.

1861 Ecuador's president Gabriel Garcia Moreno is denied permission to sell Santa Cruz to American Carlos Stewart. That same year, Moreno declared Galápagos a province, with Floreana as its capital. Moreno was assassinated in 1875 during his second term as Ecuador's president.

1865 Former captain of the *Beagle*, Robert FitzRoy, like his predecessor Pringle Stokes, commits suicide and is buried with little fanfare in London. (FitzRoy had been dealing with failing health, financial problems, and a history of depression.) FitzRoy's friend Bartholomew Sulivan (for whom Bartolomé and Sulivan Bay are named, and who invented minesweeping and iron cladding for ships) set up a charity to ensure that FitzRoy's wife and daughter did not live in poverty. Charles Darwin contributed £100 (equivalent to more than $10,000 today) to that charity.

1868 During the past 37 years, 79 American boats have removed 13,000 tortoises from Galápagos. Boats from other countries took even more.

1869 Spanish businessman José de Valdizán pays $4,000 for the exclusive rights to harvest orchil lichens in Galápagos for making dyes and inks. Valdizán used prisoners as workers, but the operation failed when Europeans developed cheaper methods for making similar dyes (e.g., mauveine). Nine years later, Valdizán was murdered by one of his workers (Lucas Alvarado) and buried near his house on Floreana, but his body was later moved to a family plot in Guayaquil.

1869 Ecuadorian entrepreneurs Manuel Julián Cobos, José Monroy (Cobos's brother-in-law), José Valdizán, and 10 workers build a series of canals to divert some of San Cristóbal's freshwater streams to a sugar-refining operation called Centro del Progreso (known today as El Progreso; Figure 2.4). By 1889, the plantation's 350 workers were producing 15 tons of refined sugar per month "from January to January." Cobos later began growing other crops on San Cristóbal, including the coffee trees that are harvested today at El Cafetal (see Chapter 4, "Coffee at El Cafetal"). Cobos, who tightly controlled the island and issued his own money (a common practice among Central American plantation owners

FIGURE 2.4
This machinery is a remnant of Manuel Cobos's agricultural empire on San Cristóbal. Cobos's settlement, called El Progreso, was the first permanent colony in Galápagos, and today includes several GNPS Visitor Sites.

since there was no national currency),[24] often beat, imprisoned, exiled, and executed disobedient workers.[25] Like Valdizán, Cobos was later killed by his workers, but his settlement (El Progreso) continued producing sugar for more than two decades. Today, El Progreso is a quiet village.

1870 Albert Günther of the British Museum (Natural History) uses museum specimens for the first systematic study of Galápagos tortoises. Günther later concluded that there are six species of the tortoise. Today, biologists recognize 14 distinct groups of *Geochelone nigra,* four of which are extinct.

1871 Charles Darwin includes the word *evolution* in his book *The Descent of Man.* The following year, he also included *evolution* in the sixth (and final) edition of *On the Origin of Species.*

1872 Famed biologist and Darwin adversary Louis Agassiz visits Galápagos for 10 days aboard *Hassler.* Although Agassiz and his wife Elizabeth's visit occurred during a drought, they reported that the archipelago's marine organisms "delight in living" and were "having a great jubilee." This description sharply contrasted Darwin's Malthusian "war of nature, from famine and death." Agassiz hoped to show "that the transmutation theory is wholly without foundation in facts," and in the journal *Nature* claimed "Darwin's hypothesis of gradual variation of species, and the natural selection for preservation of those whose variations were favorable to them in the struggle for life, seems to me to have few facts to sustain it, and very many to oppose it." Agassiz,

---

[24] The sucre did not appear as Ecuador's currency until 1884, several years after Cobos had established himself in Galápagos.

[25] For example, in 1900, Cobos exiled Camilo Casanova to Santa Cruz, where he was rescued in 1904 after Cobos was killed. Cobos had placed a sign on the island telling visitors "Please do not take this man off the island for he is twenty-times a criminal."

who died a year after visiting Galápagos, was the last famous American biologist to proclaim the special creation of species.

1875 Geologist Theodor Wolf, who first came to Galápagos in 1870 to help establish Escuela Politécnica Nacional, again visits Galápagos aboard the schooner *Venecia* and finds 14 to 18 people living on Floreana. Wolf also reported feral animals (donkeys, cattle, cats, chickens, dogs, goats, hogs, and horses) on Floreana, Isabela, Santa Cruz, San Cristóbal, and Santa Fé. One group of more than 60 orchil-collectors had lived several months on Española by eating eggs taken from albatrosses. Wolf verified the volcanic origin of the islands and deduced that cold currents from Antarctica produce the archipelago's unique climate and, as a result, the rainfall patterns that influence the distribution of plants and animals he noted during his visits. Wolf, who later published *Map of Galápagos,* is honored by the names of Wolf Island and Volcán Wolf.

1876 Darwin tells German naturalist Moritz Wagner that Galápagos "chiefly led me to study the origin of species."

1876 English naturalist Osbert Salvin's monograph *On the Avifauna of the Galápagos Archipelago* is the first published evolutionary account of Galápagos finches. Salvin described Galápagos as "classic ground" and praised Darwin's insights in Galápagos, "the importance of which in their bearing upon the study of natural history has never been equaled."

1877 When German zoologist Otto Zacharias asks Darwin whether Darwin had been a convinced evolutionist while aboard the *Beagle,* Darwin notes, "I believed in the permanence of species, but, as far as I can remember, vague doubts occasionally flitted across my mind. On my return home in the autumn of 1836 I immediately began to prepare my journal for publication, and then saw how many facts indicated the common descent of species . . ."

1879 Englishman Thomas Levick, who ruled the colony on Floreana after José Valdizán was murdered, moves 94 women and 5 men to San Cristóbal to work for Manual Cobos.

1881 In *Island Life,* Alfred Russel Wallace—the co-discoverer with Charles Darwin of evolution by natural selection—uses Galápagos to show how natural selection produces new species when small populations of organisms arrive on isolated oceanic islands.

1881 George Earl Church, sent to Ecuador by Secretary of State James G. Blaine, concludes "Ecuador's occupation of the Galápagos does not exist to a sufficient degree to give them rights over other nations." Two years later, a report to the U.S. Senate claimed that Galápagos is *res nullius* (i.e., is not owned by anyone). Ecuador protested, and the United States did nothing to claim the archipelago.

1882 On April 19, after telling his wife Emma to "remember what a good wife you have been" and that he "was not the least afraid to die," Charles Darwin dies in his upstairs bedroom at Down House. Although Darwin had wanted to be buried in St. Mary's churchyard in Downe beside his children, he was buried in a white-oak coffin in the northeast corner of the nave of London's Westminster Abbey beside astronomer John Herschel and near his friend Sir Charles Lyell. Darwin's tombstone, which forms part of the Abbey's floor, bears the simple inscription: "Charles Robert Darwin, Born 12 February 1809, Died 19 April 1882." Although Darwin was never honored by the government during his lifetime, three of his sons were knighted. Darwin was the first and only biologist to be buried in Westminster Abbey.

1885 Ecuador offers 20 hectares of land to all prospective settlers of Galápagos, regardless of their nationality.

1885 Ecuador's president José María Plácido Caamaño repeals García Moreno's decree that had made Galápagos a province, and again places the archipelago under the jurisdiction of Guayaquil, with governmental offices in what is present-day Puerto Baquerizo Moreno on San Cristóbal. The following year, the 200 residents of San Cristóbal built the first school in Galápagos. Today, Puerto Baquerizo Moreno is the capital of Galápagos.

1888 Charles Haskins Townsend of the New York Aquarium makes the first of two visits to Galápagos aboard the U.S. Fisheries' state-of-the-art steamer *Albatross*. The *Albatross* expedition, led by Alexander Agassiz (Louis Agassiz's son), took 18 tortoises for the National Museum in Washington, DC, but all of the tortoises soon died. Townsend, who was the first person to try to map the depths of the Pacific, later publicized the extinction of the islands' tortoises. His most famous book, *The Galápagos Tortoises in their Relation to the Whaling Industry,* appeared in 1925.

1890 Volcanic activity near this time produces the lava flow into Sulivan Bay, a popular Visitor Site at Santiago Island (Figure 3.4).

1891 France and the United States request permission to do scientific research in Galápagos, but an ulterior motive was likely to determine the geopolitical value of the archipelago.

1892 As part of the celebrations associated with the 400th anniversary of the discovery of the New World by Christopher Columbus (Cristóbal Colón), Ecuador renames Galápagos the Archipiélago de Colon; this remains the official name of the islands. Ecuador also formally renamed 13 of the islands to honor various aspects of Columbus' life (e.g., Floreana was renamed Santa María; Chapter 4).

1893 When rumors again start circulating about the sale of Galápagos to the United States, Ecuador's minister of external affairs (Vicente Lucio

Salazar) tells the U.S. ambassador in Quito (Towland Mahany) to deny the rumors. Two years later, a consortium of European nations offered to buy Galápagos for $25,000,000, France offered to buy all of the islands for 100 million francs, and England offered to buy the archipelago for £3 million.

1895 Lord Byron, son of the famous poet, visits Tagus Cove aboard HMS *Blonde* and notes that Galápagos "is like a new creation; the birds and beasts do not get out of our way; the pelicans and sea lions look in our faces as if we had no right to intrude on their solitude; the small birds are so tame that they hop upon our feet; and all this amidst volcanoes which are burning round us on either island. Altogether it is as wild and desolate a scene as imagination can picture."

1895 The plant *Blutaparon rigidum,* which is endemic to Santiago, is collected for the first time. Eleven years later, it was collected for the last time, and is now one of three endemic plants in Galápagos believed to be extinct.

1897 After failing to resettle Floreana, Guayaquil businessman don Antonio Gil founds Puerto Villamil on the southeastern coast of Isabela. Gil grew crops and harvested coral, which he burned to produce lime. Five years later, 20 kilometers inland on the slopes of Volcán Sierra Negra, Gil began raising cattle and mining sulfur at Santo Tomás. Gil's Pacific Sulfur Trading Company later failed. Local residents ate tortoise meat and used tortoise shells as tables, flowerpots, cradles, and other objects. For decades, the trail from Puerto Villamil to Santo Tomás was littered with shells of hundreds of tortoises killed for meat and oil to light the island's streets and homes (Figure 2.5). A 40 pound (18 kg) tortoise provides about 2 quarts (2 L) of oil.

FIGURE 2.5 Shells of tortoises killed for food and oil in 1901 on Villamil Mountain (Volcán Sierra Negra) on southern Isabela.

1897 During the 8-month Webster-Harris Expedition aboard *Lila and Mattie,* explorers witness an eruption on Santiago and report that dogs are eating Galápagos tortoises "as fast as they [are] hatched." This expedition later found the remains of Raimundo Guardado, whom Manuel Cobos had ordered be marooned on the island. Scientists on the expedition collected more than 3,000 bird skins and 65 live tortoises. The following year, scientists from Harvard University and Stanford captured 1,200 reptiles (including 26 tortoises), thousands of invertebrates, and 23 new species of fish in Galápagos.

1898 Eccentric and wealthy collector Walter Rothschild tells herpetologist Albert Günther that "in 3 years time there will not be a living giant land tortoise of any kind in the Galápagos Islands, 'What a damnable shame,' is it not?" This prophecy helped motivate the famed expedition in 1905–1906 to Galápagos by the California Academy of Sciences, which produced much of our modern knowledge of the archipelago.

1899 Archibald Sampson, the newly appointed U.S. minister in Ecuador, tells officials in Washington, DC, that Galápagos could be strategically important (e.g., for refueling transoceanic voyages) when the proposed Panama Canal is completed. Soon thereafter, the United States negotiated (unsuccessfully) to lease San Cristóbal for 99 years for $5,000 per year. When Ecuador priced the major islands at $50 million, negotiations stalled.

1900 *Julia E. Whalen* harvests 249 fur sea lions, but finds only four tortoises on Pinzón. *Julia E. Whalen* was the last recorded fur sea lion boat to visit Galápagos, where previous harvests had driven fur sea lions to the brink of extinction. Indeed, in 1906, the *Academy* expedition (see below) reported finding only one fur sea lion; in 1923, William Beebe saw only two; and in 1930, Charles Townsend noted that "although [fur sea lions were] formerly abundant, [they are] probably near extinction as none have been seen during recent years."

1901 *Mary Sachs* harvests 27 living (and 23 dead) tortoises (including the only known female from Pinta) for Baron Walter Rothschild's private museum. Today, these tortoise shells are part of the British Museum (Natural History). After being blackmailed for an affair with a married woman, Rothschild sold more than 250,000 of his bird specimens (including all of those from Galápagos) to the American Museum of Natural History for $225,000.[26] Later, David Lack used these and other specimens to study speciation of Galápagos finches; Lack's work produced the book *Darwin's Finches* (1947).

[26] Ernst Mayr of AMNH used Rothschild's specimens to write *Systematics and the Origin of Species from the Viewpoint of a Zoologist* (1942). This book became a foundational document of the "Modern Synthesis" that integrated disciplines such as evolution, ecology, and systematics. *Systematics and the Origin of Species* pioneered Mayr's new concept of species based on factors such as interbreeding and reproductive isolation.

1903 Government officials encourage Galápagos's residents to raise cattle.

1904 After refusing to pardon the 400 lashes that were to be given to worker José Prieto, San Cristóbal businessman Manual J. Cobos is hacked to death with machetes by his workers at his house and buried in the San Cristóbal highlands where he had beaten and executed several of his workers. Cobos's body was later moved to a family burial plot in Guayaquil's General Cemetery.[27]

1905 A delegation of eight scientists and three crew led by Rollo Beck[28] begin a 17-month expedition by the California Academy of Sciences to Galápagos aboard the 27-meter-long schooner *Academy.* This expedition, which was the longest and most thorough in Galápagos expedition history, came ashore near the present-day CDRS and collected plants and animals from throughout Galápagos for a year and a day. The collectors did much for science but little for conservation; they collected more than 70,000 specimens (more than any Galápagos expedition before or since) from 23 islands, including 8,691 birds, 4,000 reptiles (including 264 living tortoises from 10 islands), 800 clutches of eggs, 4,000 reptiles, 10,000 plants, and other organisms. Only 15 tortoises survived *Academy*'s return to California (but later died in their pens in Golden Gate Park), and the tortoises' shells are stored in the California Academy of Sciences' museum. The California Academy of Sciences expedition, which reported that tortoises were rare on Rábida and Fernandina, publicized the remarkable nature of the archipelago, and was reported by the San Francisco *Chronicle* as "the most important ever sent out from the Pacific Coast." Scientists on this expedition, like those on countless others, justified their killing of the archipelago's wildlife by claiming it was destined for extinction and, therefore, was better dead and preserved in a museum collection than left to the vagaries of life in the then-unprotected Galápagos. (Ironically, this policy ensured that species would disappear even faster.) The California Academy of Sciences expedition first documented the use of tools by woodpecker finches, collected the first (and only) tortoise from Fernandina,[29] enhanced our understanding of the systematics of giant tortoises, and documented the ecological zonation of plants from the

[27] Cobos's plot (02°10.971S, 79°53.183W) in Guayaquil is near Gate 3 of the cemetery and is not far from the monument to Vicente Rocafuerte, the first native Ecuadorian to serve as Ecuador's President. Rocafuerte, who is memorialized at Santa Cruz's Punta Rocafuerte, was preceded and succeeded as president by Juan José Flores, for whom Floreana Island is named. Cobos's tomb on San Cristóbal is a GNPS Visitor Site.
[28] Each of the eight scientists had a specialty: birds, reptiles, insects, fossils, rocks, mollusks, mammals, and plants. Beck, who had visited Galápagos earlier aboard *Lila and Mattie* (1897) and *Mary Sachs* (1901), became the premier bird-collector of his time, despite the fact that he had dropped out of school at age 12 (just shy of finishing the eighth grade).
[29] This tortoise was collected by Beck on April 3, 1906, and today its shell is housed in the California Academy of Sciences (specimen #8101).

coasts to the highlands (Chapter 10). Galápagos became famous largely due to the huge collections of this expedition.

1905 Darwin's travelogue of his adventures aboard HMS *Beagle,* originally published as *Journal of Researches* in 1839, first appears with its now-famous title, *The Voyage of the Beagle.* This book, which is almost entirely about Darwin's observations while on land rather than aboard the *Beagle,* is one of the world's best-known travel books.

1905 Geologic activity raises the anchorage at Puerto Villamil so much that it becomes too shallow for ships to enter safely. Today, boats at Puerto Villamil anchor farther from shore (Appendix 2).

1906 Ecuador's president General Eloy Alfaro tries to mortgage the "useless" Galápagos Islands for $10 million.

1908 Elihu Root, the U.S. secretary of state, notes that "all the North-West Coast of South America could be dominated" from Galápagos. Two years later, the U.S. Navy identified Galápagos as a potential site for a military base.

1909 To celebrate the 100th anniversary of Darwin's birth and the 50th anniversary of his *On the Origin of Species,* the AMNH in New York City opens its Darwin Hall of Invertebrate Zoology. A bronze bust of Darwin, commissioned for $1,000 from American sculptor William Couper (1853–1942), was given to AMNH by the New York Academy of Sciences to celebrate the Hall's opening. A replica of this bust, with a legend written by Darwin's son Leonard, was dedicated on San Cristóbal in 1935;[30] this bust of an elderly Darwin is located outside the entrance of the Second Zone Naval Base, just west of Puerto Baquerizo Moreno at the end of Avenida Armada Nacional (Figure 2.9). (Not far away, in a small plaza along Charles Darwin Malecón, is another bust of Darwin; Appendix 3) Other replicas of the bust are displayed at the University of Guayaquil,[31] Down House, Christ's College of Cambridge University, and in the Great Hall of the National Academy of Sciences in Washington, DC.

1911 Ecuador grants tax exemptions to settlers of Galápagos. That same year, the United States offered to lease Galápagos for 99 years for $15 million. In 1913, Ecuador's president Leonidas Plaza Gutiérrez (for whom Plazas Islands are named) rejected the offer.

[30] Darwin Hall was dismantled in 1940, and AMNH returned the original bust to the New York Academy of Sciences in 1960. Today, you can see that bust in the lobby of the New York Academy of Sciences in New York City.

[31] This bust is at the corner of Calle Chile and Chiriboga (02°11.860S, 79°53.018W) in downtown Guayaquil. A sketch of this monument has appeared on Ecuadorian postage stamps that feature Galápagos wildlife.

1912 German geophysicist Alfred Wegener challenges ideas about Earth's stasis by suggesting what later was called *continental drift*. Wegener's ideas about "continental displacement," which were popularized in 1922 when his *The Origin of Continents and Oceans* (1915) was translated into English, are today known as *plate tectonics*. Wegener's claim that continents had once been massed together in a single supercontinent named Pangaea, and then drifted apart, was initially rejected and ridiculed, and was not accepted until the late 1960s, when scientists began using submarines and sonar to map the oceans' floors. Today, Wegener's claim is a foundation of modern geology. Galápagos is formed by the tectonic movement of the Nazca Plate over the Galápagos Hotspot (Chapter 3).

1913 The British Navy ends regular visits to Galápagos to search for shipwrecked sailors.

1914 After a decade of construction, the Panama Canal opens, thereby increasing ships' access to Galápagos.

1917 Alfredo Baquerizo Moreno visits San Cristóbal's Puerto Chico and, in doing so, becomes the first president of Ecuador to visit Galápagos. Moreno's visit, which lasted only a few hours, prompted the town's surprised citizens to rename their town Puerto Baquerizo Moreno.

1917 Disgruntled Ecuadorians, hoping to incite a political revolution, circulate pamphlets claiming Galápagos has been sold to the United States.

1920s Tortoise oil is no longer exported from Galápagos, but continues to be used by residents of the islands. Small tortoises and their dried hind feet were sold as souvenirs in Galápagos until the early 1960s.

1923 An expedition to Galápagos organized by the New York Zoological Society (and funded by patron Harrison Williams) is led by ornithologist William Beebe. Although their 250-foot-long steam yacht *Noma* visited Galápagos for only four days (during which it left graffiti at Darwin Bay), AMNH president Henry Fairfield Osborn claimed that the visit "was like rubbing the Aladdin's lamp of science!" Beebe, an advocate of Lamarckism, used the experience to write 10 articles and the embellished *Galápagos: World's End* (1924), whose frontispiece featured a land iguana from Baltra. Beebe's best-selling book publicized Galápagos, prompted several people to move to the archipelago, and later helped inspire the beginning of ecotourism in Galápagos. (Today, tourism dominates the economy of Galápagos.) Among his many observations, Beebe noted, "The most astonishing thing about various islands of the Galápagos is their superficial similarity and their actual diversity." This would later be a key observation in Peter and Rosemary Grant's understanding of the evolution of Darwin's finches.

1924 The British Scientific Expedition erects a new barrel at Post Office Bay on Floreana.

1925 After studying captains' logs from numerous whaling ships, Charles Townsend concludes that whalers took over 200,000 tortoises from Galápagos between 1830 and 1860. Townsend, who understood the impact of invasive species, proclaimed "if [Galápagos] could be cleared of the pests introduced by civilized man and the original conditions restored! This is now impossible on the Galápagos. The only remaining hope for the race is the establishment of survivors elsewhere." Three years later, the New York Zoological Society removed 180 tortoises from Galápagos and distributed them throughout the United States, Bermuda, Panama, and Australia. Almost 50 years later, one of those tortoises—which had been given to the San Diego Zoo and who today is known as "Super Diego" (Figure 2.12)—helped save the Española race of tortoises (see below).

1925 Author Zane Grey visits Galápagos and describes the archipelago as "one of the wildest and lonesomest places in all the Seven Seas."

1925 Inspired by William Beebe's *Galápagos: World's End* (1924) and encouraged by promoters Harry Randall and August Christensen (the Ecuadorian ambassador in Oslo), 22 Norwegians aboard the schooner *Floreana* settle at what is now Puerto Ayora on Santa Cruz. A second group of 83 people arrived in Galápagos the following year. After a year of processing fish, they moved to the highlands and became farmers, but most left the island by 1930. Today, Puerto Ayora is the largest town in Galápagos.

1925 William Beebe again visits Galápagos, and the following year publishes *The Arcturus Adventure* describing his experiences. Near this time, newspapers along the west coast of the United States began publishing advertisements tempting people to come to the "tropical paradise" of Galápagos. Beebe described Galápagos as "the one place in the world that remains unchanged."

1926 The United Maritime Society of Guayas is created in Guayaquil to encourage commerce in Galápagos, and its first operation (headed by Guayaquil businessman Davío Egas Sánchez) is a salt mine at Santiago's James Bay. The mine failed three years later when salt prices dropped on the mainland.

1927 The United States makes the first of several offers to install a military base in Galápagos that "would not affect [Ecuador's] sovereignty" (e.g., such as the United States's perpetual lease at Guantánamo Bay, Cuba).

1927 A group of 60 Norwegians, after landing at Floreana with more than 200 tons of prefabricated houses, a few generators, and a tractor, set up "Casa Matriz," a fish-processing business at Post Office Bay (Figure 2.6).

FIGURE 2.6
Just a few meters from the Post Office Barrel (background) are the remains of a fish-processing company founded on Floreana in 1927. The concrete blocks that litter this area supported a large house called "Casa Matriz" that included the company's head-quarters. The company failed two years after it was founded. For another view of the Post Office Barrel, see Figure 4.2.

When the venture failed, many of the settlers returned to Norway, some moved to San Cristóbal, and others joined the Norwegian settlement at Academy Bay on Santa Cruz.

1929 After reading William Beebe's *Galápagos: World's End,* several more groups of Europeans begin migrating to Floreana. The first of these settlers were the spouse-trading Berlin doctor Friedrich Ritter and his patient Dore Strauch Koerwin, who named their home "Friedo" (a combination of the first syllables of their names, and similar to the German word for "peace") in Floreana's highlands. Ritter and Strauch practiced Nietzsche's doctrine of the superman and lived in one of Patrick Watkins's caves. Ritter and Strauch's occasional nudity prompted the international press to label them "The Adam and Eve of Galápagos." Ritter's articles about Galápagos, which were published in Germany and elsewhere (including *Atlantic Monthly*), lured other settlers to the archipelago. Ritter and Strauch used the buildings, supplies, crops, and animals left behind by the Norwegians a few years earlier.[32]

1930 The population of Galápagos is 571; 2 people live on Floreana, 8 on Santa Cruz, 155 on Isabela, 389 on San Cristóbal, and 17 on Santiago.

1931 After determining that water between Galápagos and Cocos Island is up to 3,700 meters deep, Norwegian geologist Jacob Horneman dismisses as "nonsense" William Beebe's claim that a land bridge had once connected Galápagos to Cocos Island.

1932 Ed Palmer of the *Velero III* films shipmate John Garth holding a land iguana on Baltra. This is the first known film made in Galápagos.

[32] Floreana was inhabited when Ritter and Strauch arrived; they were met on Floreana by Hugo (a 14-year-old Indian boy) and an "old Norwegian" who sold fish and cow hides.

1932 Lured by stories written by earlier settlers such as Friedrich Ritter, Heinz and Margret Wittmer—with Heinz's vision-impaired teenaged son Harry from a previous marriage—move from Germany to Floreana Island to "lead our own lives in peace and quiet." They settled in the highlands, about a mile south of Friedo. As Margret later noted, "I felt as if I were no longer on this earth but had landed on some distant and deserted planet." Harry later drowned when his fishing boat capsized. When the Wittmers arrived, Ritter and Strauch were the only people on Floreana. However, soon after the Wittmers's arrival, Austrian self-appointed "baroness" Eloisa Wehrborn von Wagner-Bousquet (Figure 2.7) settled at "Hotel Paradise" in the highlands with three male companions (Robert Philippson and Alfred Rudolph Lorenz from Germany, and Felipe Valdivieso Borja, whom they picked up in Guayaquil). Newspapers throughout the world featured the baroness and her plans to build an exclusive resort on Floreana for American millionaires; these articles often depicted the baroness as a sexual vixen—clad in silk panties and wielding a pearl-handled revolver—who had imprisoned fishermen as her slaves. However, Valdivieso left Floreana a month after arriving, and Philippson and the baroness mysteriously disappeared in 1934, prompting speculation that continues to this day about their fates. After the baroness disappeared, Lorenz left Floreana. Four months later, the desiccated, mummified corpses of Lorenz and Norwegian fisherman Trygve Nuggeröd[33] were found on a Marchena Island beach (more than 165 km from Floreana) near a pile of French money, several letters written by Margret Wittmer, Lorenz's passport, and their last meal, the butchered bodies of an iguana and a sea lion (Figure 2.8).[34] Ritter, a vegetarian and dentist who had pulled all of his teeth to avoid dental problems, was allegedly killed by food poisoning after eating chicken.[35] Strauch—who blamed the deaths of Lorenz and Nuggeröd on the Baroness and Philippson—returned to Germany. The so-called "Galápagos Affair" deaths, and their associated mysteries, generated sensational press coverage throughout the world. Today, Floreana is the quietest of the five inhabited islands in Galápagos; most of

[33] Nuggeröd regularly visited Floreana aboard his boat *Dinamita*. His final visit brought coffee plants to the baroness.

[34] These deaths produced worldwide headlines; for example, the *Los Angeles Times* produced a series of stories with headlines such as "Baronesa Believed to Hold Key to Marchena Island Mystery," and *Time* magazine ran a full-page story titled "Death in Galápagos."

[35] Ritter, who boasted that he would live for more than 150 years, later claimed that it was his Christian duty to eat meat so nothing would be wasted. Others charged that Ritter was a vegetarian "on account of being a poor hunter." Some accounts attributed Ritter's death to a stroke or pneumonia. Ritter was buried in an oasis on Floreana above Black Beach that is now owned by the Claudio Cruz family, who often treats visitors to lunches that include "Ritter's Chicken." The original cross on Ritter's grave has been replaced by one from the cemetery of Floreana's penal colony.

FIGURE 2.7 The arrival on Floreana of self-proclaimed "Baronesa" Eloisa Wehrborn von Wagner-Bousquet (right) and her two lovers—Alfred Lorenz (middle) and Robert Philippson (left)—brought worldwide publicity to Galápagos. The Baroness, who hoped to build a resort on Floreana, soon began dispatching stories with titles such as "Empress of Floreana," "The Baroness Takes Control of Galápagos," and "Pirate Queen of the Galápagos." However, the Baroness and Philippson disappeared in 1934, the same year that Lorenz was found dead on Marchena Island (Figure 2.8). The Baroness remains a mythic figure in Galápagos, and a scenic GNPS Visitor Site ("Mirador de la Baronesa"; Color Image 4) is named for her.

its 160 residents are subsistence farmers. Relatively few tourists visit the settlement on Floreana.

1933 On New Year's Day just four months after arriving in the archipelago, Margret Wittmer (with the help of neighbor Friedrich Ritter) gives birth in a pirate cave on Floreana to the first known native-born citizen of Galápagos, Rolf Wittmer. In 1982, Rolf Wittmer helped launch an eco-tourism company in Galápagos (Tip Top Fleet), which he later turned over to his children before his death in 2011. Today, Margret's daughter Ingeborg "Floreanita" Wittmer (b. 1937) continues the family tradition at Floreana's Wittmer Pension with her daughter Erika.

1933 As harvests of tuna plummet in southern California after more than a decade of overfishing, companies begin moving their boats to Galápagos. For the next several years, most tuna processed in California was from Galápagos, and by 1996, more than one-third of Ecuador's catch of yellowfin tuna was from Galápagos. Industrial tuna-fishing remained popular in Galápagos until passage in 1998 of the Special Law for the Preservation and Sustainable Development of Galápagos.

FIGURE 2.8 The discovery of the desiccated corpses of Rudolph Lorenz (right foreground) and Trygve Nuggeröd (left background, face-down by the boat) on a Marchena Island beach became an international story that contributed to the mystique of Galápagos. This photo was taken by members of Allan Hancock's *Velero III* expedition on December 3, 1934.

1933 The *Los Angeles Times* describes a movie by Hollywood producer Emory Johnson depicting Floreana's Baroness and Robert Philippson as pirates who, after being jailed, escape to freedom in a raft. Photos from the film, showing the Baroness and Philippson with guns, were soon published in newspapers and magazines throughout the world. The next year, the Baroness and Philippson disappeared from Floreana.

1934 *Stella Polaris,* a TransPacific cruise ship, becomes the first tour-boat to stop in Galápagos.

1934 The crew of the tuna boat *Santa Amaro* discovers the desiccated bodies of Rudolph Lorenz and Trygve Nuggeröd on a beach on Marchena (Figure 2.8). Soon thereafter, the bodies were photographed by members of a Hancock expedition aboard *Velero III.* Their deaths were announced in *Time* magazine and the *New York Times,* soon after which headlines throughout the world implicated the baroness for the deaths.[36] After Margret Wittmer claimed that "conditions are peaceful, now that our overbearing neighbor has gone," other writers blamed Wittmer for the baroness's death. These mysterious deaths remain part of the mystique of Floreana.

1934 The U.S. Navy sends the destroyer *Hale* and two hydroplanes to Tagus Cove to help William Robinson, who is having an acute attack of appendicitis. These were the first airplanes to visit Galápagos. The following year witnessed the first delivery of mail to the archipelago by plane.

[36] For example, a headline in the December 6, 1934, issue of the *Milwaukee Sentinel* announced, "Galápagos Empress Exiled Aid to Death—Baroness Drove Lorenz from Island without Food or Water."

1935 To celebrate the 100th anniversary of Charles Darwin's visit to Galápagos, Ecuador passes the first legislation (Part Three of the *Regulations for Fisheries and Marine Game*) to protect wildlife on Hood, Barrington, Seymour, Jervis, Daphne, Duncan, Santiago, Marchena, Pinta, Tower, Culpepper, and Wenman Islands, along with parts of northern Isabela (Santa Cruz was added to the list the following year). By this time, however, three of the archipelago's 14 races of tortoises were extinct, and others were threatened with extinction (including the Pinta species, of which the late Lonesome George was the last living member until his death in 2012). Little was done to enforce the new laws. Other endemic species (e.g., the native rice rat) were also extinct on several islands. As part of Ecuador's celebration of Darwin's visit to Galápagos, anthropologist and travel writer Victor Wolfgang von Hagen dedicated the monument to Charles Darwin that stands just outside the naval station at Puerto Baquerizo Moreno (Figure 2.9). To commemorate the dedication, Ecuador issued a set of six commemorative stamps featuring Darwin and the plants, animals, and landscapes of Galápagos.

1935 When Ecuador's president Pererico Páez asks for a loan from the United States, President Roosevelt suggests making Galápagos an international wildlife refuge.

1936 Pan American-Grace Airways (Panagra) begins offering occasional flights from Quito to Galápagos.

1936 Ecuador's Executive Decree 31 proposes a national park in Galápagos, but the park isn't officially created until 1959.

1936 In *Satan Came to Eden,* which was marketed as "The Story of the Only Survivor," Dore Strauch describes her years (and the many deaths and other mysteries) in Galápagos. Strauch died in Germany during a bombing raid in 1942. In 2012, Geller/Goldfine Productions announced plans to produce a full-length movie titled *Satan Came to Eden: The Galápagos Affair.*

1937 Two years after their parents sold their house to buy a boat, the Angermeyer brothers—Karl, Gus, Hans, and Fritz—use the boat to flee the political upheaval in Nazi Germany and settle on Santa Cruz.[37] After farming in the highlands for a year, they moved to the coast at what is now Puerto Ayora. The "cave" in which boisterous Gus lived is now part of Angermeyer Waterfront Inn, and the bar at Angermeyer Point is

[37] Karl, who painted with his fingers and promoted himself as "Duke of the Galápagos" (brother Gus was the self-described "King of Galápagos"), lived in a stone house near present-day Angermeyer Point. The nearby beach (Playa de los Alemanos) was labeled "Private Beach—For Germans Only," and photos of Angermeyer often appeared on covers of adventure magazines in America and Germany. In Puerto Ayora, Hotel Silberstein (on Charles Darwin Avenue) is prominently marked "Residencial Angermeyer," and originally consisted of six huts made of Baltra pine that rented for $1 per night. The property was managed by Lucrecia Angermeyer, the wife of Augusto "Gus" Angermeyer.

FIGURE 2.9 This monument on San Cristóbal was dedicated in 1935 as part of the celebration of the 100th anniversary of Charles Darwin's visit to Galápagos. At the ceremony, Victor Wolfgang von Hagen urged officials "to help save the fauna of the most fascinating theater of living evolution in the world" and suggested that the U.S. Navy help Ecuador establish a research station in Galápagos, but his suggestion was ignored. Near the base of the monument is a plaque written by Darwin's son, Leonard, that includes a relief of a younger Darwin with this text: *CHARLES DARWIN landed on the Galápagos Islands in 1835 and his studies of the distribution of animals and plants thereon led him for the first time to consider the problem of organic evolution. Thus was started the revolution in thought on this subject which has since taken place. Erected September 17th 1935, by the members of the Darwin Memorial Expedition: Victor Wolfgang von Hagen, Alexander R. Brown III, Christine Inez Brooks, Christine Inez von Hagen, Dard Hunter.* Of the people whose names are listed on the plaque, only Dr. and Mrs. von Hagen were at the dedication. Replicas of this monument are found in Guayaquil, New York, England, and elsewhere. Not far from this monument, along Malecón Charles Darwin, is another monument of a younger Darwin (Appendix 3).

named "La Cueva de Gus" ("Gus's Cave"). Descendants of the Angermeyers still operate hotels and boats in Puerto Ayora, and Johanna Angermeyer (Hans' daughter)[38] later described their life in her book *My Father's Island* (1990).

1938 The United States begins secret negotiations to build military bases in Ecuador. That same year, U.S. president Franklin D. Roosevelt, touring Latin America on the third presidential voyage aboard the warship USS *Houston,* was among the first tourists to visit Galápagos. Roosevelt never left the boat, but he caught several fish (including a 45-pound tuna) that are now part of the Smithsonian's collection. When he was unable to contact the Wittmers, Roosevelt left whiskey, chocolate, medicines, and other supplies at Post Office Bay for Floreana's residents.[39] Roosevelt hoped that his expedition could find and return the

[38] Johanna Angermeyer (b. 1948) now lives in England.

[39] Roosevelt's charity produced headlines such as "Roosevelt Feeds Starving Galápagans" and "President Brings Relief to Hungry Islanders." Thanks to the personal intercession by biologist Waldo La Salle Schmitt (a curator of the U.S. National Museum in Washington) with Roosevelt, the Wittmers were not deported from the archipelago. Later, Margaret Wittmer admited that her family had "our friendship with Professor Schmitt to thank for our survival." The United States also began building an airstrip in the northeastern corner of Floreana, but this work was terminated in 1942.

remains of James Cowan to the U.S. Naval Academy for interment, but a three-hour search by 50 officers for Cowan's grave failed.

1940 An "inspection flight" by the Ecuadorian army finds "30 or 40 American Marines" living on North Seymour Island.

1941 The United States announces that it will lease Baltra, which *Time* magazine calls "the fantastically storied haven of pirates and more modern escapists." On December 11—just four days after Japan attacked Pearl Harbor—Ecuador gave the United States permission to use Galápagos and the Ecuadorian coast for "the establishment of such military bases as may be necessary." The following day, 36 American soldiers and technicians disembarked on Baltra. Construction of the air base (Figure 2.10) began two months later, and the first B-24 planes landed on Baltra's two runways three months later (when they were built, these runways were the longest airstrips in South America). More than 200 wooden houses and other buildings (e.g., a munitions dump, church, theater, radio station, and a clubhouse named "World's End") were constructed on the island.[40] The agreement, which was completed the following year, also allowed the United States to build installations on other islands (e.g., Isabela).

1942 Ecuador and the United States formally agree on the *Treaty of Occupation of South Seymour* (Baltra), which grants the United States permission to establish an air base on Baltra and build several radar stations (three on Baltra, three on Isabela, and one on Española), all of which were completed in seven months at a cost of $8 million. *Time* magazine described the Baltra base as "a beachhead on the Moon," and troops stationed at Baltra referred to the base as *Beta, Base Beta,* and *The Rock.*[41] That same year, the U.S. Navy reported that land iguanas were abundant on Baltra, but these iguanas were gone from the island four years later. Troops at Baltra published a mimeographed newspaper called *Goat's Whiskers,* which was the first newspaper published in Galápagos.[42] After World War II, the radar stations were abandoned and the air base was given to Ecuador. Today, Ecuador maintains a small military base on Baltra, along with its commercial airport. Airports were later added on San Cristóbal and Isabela.

[40] Workers at the base lived in rent-free houses, enjoyed free meals, health care, and laundry, and were paid up to 30 sucres per day (1 sucre could buy a pack of cigarettes). Each resident was given a carton of cigarettes per week and two bottles of beer per night. After three months, workers got two weeks of vacation, along with free transportation to Panama. For the first two months of 1942, water for the base came from Panama. Thereafter, water for the base came from San Cristóbal, where workers extended the waterworks built earlier by Manual J. Cobos (from the highlands to El Progreso) to Puerto Baquerizo Moreno, where the water was loaded onto two large barges (each having a capacity of 215,000 gallons).

[41] *Alpha Base* was Hawaii, *Beta Base* was Baltra, and *Charlie Base* was Panama. Today, *The Rock* is the name of a popular restaurant on Avenida Charles Darwin (at its intersection with Calle Bolivar Naveda) in Puerto Ayora. The Baltra airbase (Figure 2.10) launched daily maritime patrols, each of which was 5-9 hours long.

[42] Today, the Galápagos newspaper, which is published monthly, is *El Colono.*

FIGURE 2.10 In the 1940s, the United States operated a military base on Baltra, from which it patrolled the Pacific to protect the Panama Canal. In the air are two Consolidated B-24 Liberators, and near the runway is the only B-15 bomber ever built. When it was designed, the B-15 was the most massive and most voluminous plane ever built in the United States. This photo was taken on April 3, 1945.

1942 Paul Foster, commander of the U.S. Navy, urges U.S. president Franklin Roosevelt to "seize the opportunity" to buy Galápagos because "under no circumstances should our government relinquish our air base [there] upon the termination of the war."

1942 The improved charts of Galápagos prepared by USS *Bowditch* replace those made by Captain Robert FitzRoy of HMS *Beagle* more than a century earlier.

1942 The Smithsonian Institution, noting that the Galápagos animals "are fearless to an extraordinary degree," urges troops at Baltra to implement "measures for the protection of the peculiar fauna of these islands. Many species found here are encountered nowhere else in the World, and if they are destroyed here there will be definite loss to science and the human interest as well as much resentment on the part of Ecuadorian and other authorities. . . . It is important that action be prompt . . . unless [the wildlife is] protected they will be exterminated in a very short time." That same year, the U.S. secretary of war instructed commanders of the Baltra air base to "take appropriate action to prevent any unnecessary molestation of the wildlife in the Galápagos Archipelago and prohibit the introduction of domestic animals that may prey on the native fauna."

1943 The United States offers $10 million to lease Baltra and Santa Cruz for 99 years, but a political revolution in Ecuador ends the negotiations. The next year, the U.S. Congress—citing the Monroe Doctrine—discussed "obtaining the Galápagos Islands as permanent possessions of the United States." The more than 2,400 U.S. servicemen and 750 civilian workers at the U.S. base on Baltra were the largest colonization of the islands. (The other islands had a combined population of approximately 810.)

1943 Isabela's Volcán Azul erupts, prompting soldiers at the Punta Cristóbal radar station to request an emergency evacuation.

1944 When a Joint Commission of the U.S. Army and Marines argues that "immediate steps should be taken to acquire permanent rights" to use Galápagos, a Nebraska senator notes that Ecuador "is not capable of protecting [Galápagos] and any well-armed country could take them away." That same year, Ecuador's president José María Velasco Ibarra visited Galápagos and ordered American troops to begin evacuating the islands. After U.S. president Franklin Roosevelt visited Galápagos, he suggested that Galápagos "be preserved for all time as a kind of international park" and asked Ecuador "to abandon any thought of agriculture or cattle raising in the Islands." Roosevelt, who wanted to continue to use Galápagos to protect South America and the Panama Canal, said he "would die happy" if the U.S. State Department could make Galápagos a protected wildlife preserve. (Roosevelt's wife, Eleanor, described Galápagos as "one of the most discouraging spots in the world.") On August 16, 1944, the U.S. Department of State declared that the U.S. military facilities in Galápagos were "auxiliary," thus officially ending their strategic importance.

1945 U.S. planes drop pamphlets on Floreana informing the Wittmers and other residents that Germany has surrendered. When a Panama City newspaper claimed that Hitler had escaped in a submarine and was hiding near the Wittmers's home, U.S. soldiers searched Floreana. Hitler was not there.

1946 Two years after assuming control of Galápagos from the Ecuadorian Army, the Ecuadorian Navy becomes the first institution to take responsibility for public health in the archipelago.

1946 Canadian scientist Abraham Gesner discovers how to distill kerosene from petroleum. Because of the abundance of petroleum and the fact that kerosene burned much cleaner than whale oil, Gesner's discovery later produced a rapid decline in the whaling industry in Galápagos and elsewhere.

1946 When U.S. troops leave Isabela, Ecuador's president José María Velasco Ibarra establishes a penal colony at an abandoned base there, after which *Calderón* eventually brings 300 prisoners and 30 guards to the island. (This was the archipelago's third penal colony; the first had been at Floreana, and the second at San Cristóbal.) The penal colony on Isabela included four compounds: one along the coast for fishing, two in the highlands for farming, and one in the arid scrublands to "re-educate" prisoners being punished for breaking laws and disobeying orders. Prisoners at Isabela's notorious "punishment colony," under the supervision of Warden Durán, spent three years (1946–1949) building a wall of volcanic rocks today known as El Muro de las Lágrimas (The Wall of Tears; Figure 2.11); this monument to the prisoners' suffering is now a popular Visitor Site.[43] The convicts there, who failed in their attempt to hijack the yacht *Valinda* to escape to the mainland, were treated harshly, and the colony was described as a place "where the strong cry and the weak die."[44] Prisoners who tried to escape were killed and buried where they fell.

1946 Eight officers, four enlisted men, and several political prisoners provided by the Ecuadorian government begin preparing the Baltra air base for its transfer to the Ecuadorian government.[45] Most domesticated animals were rounded up and killed, and soldiers who had died in Galápagos were disinterred and taken to Panama for reburial.[46] Near this time, the Pacific Science Board recommended using part of the air base as a research station, but the idea was soon forgotten. When the Baltra air base closed, two Ecuadorian officers and ten servicemen who remained on the island became the first Ecuadorian military garrison on Baltra. Soon thereafter, Jorge Páez, the governor of Galápagos, declared that homeowners in Galápagos would each be given one of the base's buildings, which were relocated to the homeowners' islands. (Most went to San Cristóbal, which was the most populous island at the time.)[47] This decision changed the landscapes of Galápagos; as one official noted, "from [1946] onwards, the beach was filled with new houses made with wood from Baltra." Heinz and Margret Wittmer chose the base's post office, which was dismantled into 120 bundles and taken on *Inca* to Floreana, where it was reconstructed (and still resides) on Black Beach. During the 1950s, "Baltra pine" was an important building material in Galápagos; many of these relocated buildings remain in Galápagos.[48]

---

[43] The guards lived in wooden buildings abandoned by the U.S. military from World War II; prisoners, who were from several countries, lived in tents.

[44] A later attempt by some prisoners to escape aboard the fishing boat *Teresita* was successful. In 2012, Victor Lopez, the captain of that boat, was living in Puerto Ayora.

[45] After the base was inactivated in 1946, several of the base's more than 200 buildings were bulldozed, and the remnants dumped offshore (where they continue to be found by divers). The base's Stone House (La Casa de Piedra) built by Americans as a beer garden and officers' club with the longest bar in the world, which was sometimes stocked with prostitutes from Panama, became part of the Ecuadorian military base. However, fire destroyed the building in 2006.

[46] According to the U.S. National Archives *Index,* there were 42 deaths in Galápagos during World War II.

[47] For example, in 1950, 811 people lived on San Cristóbal, 198 on Santa Cruz, 216 (plus 92 convicts) on Isabela, and 21 on Floreana. The population of Santa Cruz did not surpass that of San Cristóbal until the mid-1960s.

[48] One such home in Puerto Ayora is on Calle Enrique Fuentes (local residents call this street "Saigon Street") between Bolivar Naveda and Avenida Baltra. "Baltra pine" is part of many buildings along Charles Darwin Avenue in Puerto Ayora.

FIGURE 2.11
A popular Visitor Site on Isabela is El Muro de las Lágrimas (Wall of Tears), a monument to a harsh prison that operated there after U.S. troops abandoned their base on the island. This prison was one of several penal colonies in Galápagos, all of which were closed by 1959.

1947 David Lack's *Darwin's Finches* popularizes the Galápagos finches to which Darwin paid little attention more than 100 years earlier. Lack, whose work persists in modern biology textbooks, accurately predicted that "many of the conclusions apply, I believe, much more widely than to the Galápagos finches." *Darwin's Finches,* which helped bring competition theory into animal ecology by showing that interspecific competition for food is a major diversifying force in the radiation of finches in Galápagos, remains one of the most influential books in evolutionary ecology.

1949 An article titled "The Enchanted Islands of Peru" in *Continente* argues that Galápagos belongs to Peru because of a visit there in 1789 by Peruvian resident Alonso de Torres. However, Peru did not try to claim the islands.

1953 Thor Heyerdahl's Norwegian Archaeological Expedition aboard *Don Lucho* uncovers 131 shards of pre-Columbian pottery on Santiago, Santa Cruz, and Floreana Islands, and the following year the Walt Disney Expedition finds similar artifacts at Santa Cruz's Cerro Colorado.[49] Heyerdahl concluded that Galápagos was visited as early as 300 AD during Peru's Gallinazo period. However, no corresponding graves or constructions have ever been found, and Heyerdahl's conclusions have since lost favor.

1954 Land iguanas on Baltra are declared extinct (Chapter 6, "Did American Troops Kill All of Baltra's Land Iguanas?").

1955 Irenäus Eibl-Eibesfeldt urges the International Union for Conservation of Nature (IUCN) to save Galápagos by "the setting aside of nearly all

---

[49] The expedition also hoped to learn more about the stone face that had been photographed near the Wittmers's farm on Floreana. Although some experts debated whether it had been carved by Incas or Polynesians, Heinz Wittmer told the expedition that he had carved the face.

undeveloped land as an inviolate nature sanctuary . . . and establishing an international science station to study the wildlife."

1955 The fishing company La Predial (Sociedad Pesquera de Galápagos), the first that could freeze fish on San Cristóbal, collapses. Remains of the company's offices are still visible near the pier used to access the Interpretation Center on San Cristóbal.

1956 Ecuador's president José María Velasco Ibarra visits the Wittmers on Floreana Island. Floreana's Puerto Velasco Ibarra is named in Ibarra's honor.

1957 Biologist Robert Bowman, after spending more than four months in Galápagos, reports "out of the 89 species and subspecies of birds that nest on the islands, 77 exist nowhere else in the world." This visit was instrumental in establishing the Charles Darwin Foundation (CDF). *Life* magazine later opened its popular *Life Nature Library* series of books with an essay about Galápagos titled "A Showcase of Evolution."

1958 Julian Huxley proposes the creation of a research station in Galápagos "as a memorial to Darwin's great achievement." This ultimately produced CDRS in Puerto Ayora.

1958 Although informal tourism began in Galápagos years earlier when adventurous backpackers came to the islands on fishing boats, organized tourism begins in Galápagos with the establishment of CETUGA (Compañia Ecuatoriana de Turismo in Galápagos), bringing visitors to the islands from Guayaquil. That same year, *Life* magazine promoted "The Fantastic Galápagos: Darwin's Treasure of Wildlife."

1958 The Astral Fruit Trading Corporation begins shipping bananas and feral cattle between Isabela and Guayaquil. Among the employees was Norwegian Jacob Lumb, who later became the first naturalist guide in Galápagos.

1959 A utopian organization called Filiate Science Antrorse Island Development Company ("Together with Science We Move Forward")—hoping "to demonstrate to all nations that it is possible to live together without selfishness and greed"—lures 106 Americans (each paying $2,500) to Galápagos with this advertisement in several West Coast newspapers: *Wanted: Swiss Family Robinson. Is your family one of the 50 adventurous families with the spirit of America's early pioneers needed to establish a model community on a beautiful Pacific island?* The company, led by high school dropout Donald Harrsch, wanted to make "everyone a millionaire . . . [and] create and leave for their children a better world to live in." Another brochure declared "we hope to show that a pioneer colonizing group dedicated to furthering scientific research can succeed with this motivation replacing the religious or political motivations of many such new colonies in the past." The settlers were disappointed when they arrived at their new homes, which were the ruins of Manual J. Cobos's empire at El Progreso. On the mainland, many Ecuadorians considered the American settlers to be Yankee imperialists. Headlines such as "Galápagos Colonists Suffer Togetherness" soon followed, and within two years all but one of the settlers had returned to the United States.

1959 As part of the 100th anniversary of the publication of Charles Darwin's *On the Origin of Species,* the Charles Darwin Foundation (CDF) for the Galápagos Islands is incorporated in Belgium, with Belgian scientist Victor Van Straelen, who had organized the Congo National Park, as its first president. CDF is an international, nonprofit, independent research organization dedicated to conserving Galápagos. The Charles Darwin Research Station (CDRS), which is headquartered in Puerto Ayora, is the operational branch of CDF. CDF works closely with the Galápagos National Park Service (GNPS), which is in charge of conservation and natural-resources concerns in Galápagos. The foundation began operating in the 1960s and, four years later, the CDRS was founded. Today, CDF continues to be based in Brussels, and the king of Belgium signs all of the CDF's statutes.

1959 Prisoners overpower guards at penal colonies at Alemania and Santo Tomás (both on Isabela) and capture the *Viking* near Punta Moreno. One of the hostages taken by the revolting prisoners was Arnoldo Tupiza, for whom the tortoise breeding station at Isabela is now named. (Tupiza was also the first GNPS warden on Isabela.) The prisoners' uprising made international news and led to the closure of the last penal colonies in the archipelago (the remaining prisoners were transferred to Guayaquil), thereby ending 127 years of Galápagos being used as a prison. One of those penal colonies, Isabela's infamous "Wall of Tears" (Figure 2.11) remains a popular GNPS Visitor Site.

1959 Ecuador's president Camilo Ponce Enríquez issues Emergency Law-Decree No. 17 ratifying GNP as Ecuador's first national park. All land in Galápagos not currently inhabited or used for agriculture—about 97% of the 7,800 square kilometers of land in the archipelago—was made part of the park.

1959 Photographers and filmmakers from throughout the world—including those from Walt Disney Productions making its *True-Life Adventures* series—begin converging on Galápagos to make documentaries.

1959 The first giant tortoise hatches in captivity at the San Diego Zoo.

1960 Julian Huxley reports to UNESCO[50] that a primary purpose of the proposed GNP could be "the enjoyment and interest of visitors," later adding that the park should conserve "resources for present and future use and enjoyment." That same year, the Galápagos Tourism Company

[50] UNESCO, established in 1945, is the United Nations Educational, Scientific and Cultural Organization. Huxley, a founder and the first director-general of UNESCO, expanded the organization's mission to include conservation. Huxley was often described as "the man who put the 'S' in UNESCO."

in Guayaquil began taking visitors to Galápagos aboard the *Cristobal Carrier* (the renamed *Manual J. Cobos*). This boat became an early icon of tourism in the archipelago. Ecotourism later became the foundation of the Galápagos economy.

1960 Thanks to funding from UNESCO, construction begins on the CDRS near Puerto Ayora on Santa Cruz Island. Swiss biologist Raymond Lévêque was appointed the first director of CDRS. When CDRS officially opened early in 1964, the director was British ornithologist David Snow (Snow had worked at the Edward Grey Institute, which was directed by David Lack).

1960 The first Galápagos hotel and restaurant, with several rooms made of "Baltra pine" scavenged from the abandoned U.S. air base on Baltra, opens at Pelican Bay at Puerto Ayora. That hotel, named Hotel Galápagos, was operated by American sailor and expatriate Forrest Nelson. Hotel Galápagos, which is southeast of the cemetery along the way to CDRS, closed in 2005.[51] On the grounds, you can see Nelson's tiny tractor, which was one of the first motorized vehicles on Santa Cruz.

1960 Margret Wittmer publishes *Floreana* (original title: *Postlagernd Floreana*). The popular book, which was later translated into 14 languages, provided no answers to the public's lingering questions about the deaths and other mysterious events associated with "The Galápagos Affair" (Figures 2.7, 2.8). In subsequent years, the growing popularity of Galápagos—and especially the mystique of Floreana—made Wittmer increasingly famous, and she began charging for interviews. Ecuador later gave Wittmer 7,315 square meters of land on Floreana.

1961 Heinz Wittmer becomes the first known resident of Floreana to die of natural causes. Wittmer died of a cerebral hemorrhage; the failure of the island's radio delayed a doctor from reaching Wittmer for three days.

1961 David Lack's *Darwin's Finches* (1947) is reissued, on the condition that the text not be changed. The famous book was again reissued in 1983.

1962 CDRS begins collecting vertebrates in Galápagos. Today, this collection (consisting of more than 1,500 specimens of more than 135 species) ranges from the tiny warbler finch (*Certhidea olivacea,* whose live weight is 8 g) to Bryde's whale (*Balaenoptera edeni,* whose live weight often exceeds 20 metric tons).

1963 Isabela's Volcán Sierra Negra erupts for more than a month.

1963 Canadian geophysicist J. Tuzo Wilson proposes the "hotspot theory" to account for volcanic activity that occurs away from boundaries of tectonic plates. The Galápagos Islands have been produced by the Galápagos Hotspot (Chapter 3).

[51] Hotel Galápagos on Santa Cruz was the first licensed hotel in Galápagos. The older Angermeyer Hostal, which was not licensed, was at the present-day site of the Silberstein Hotel in Puerto Ayora (Appendix 1).

1964 Great Britain's Prince Philip, Duke of Edinburgh, visits Galápagos aboard the royal yacht *Britannia* and films a documentary about conservation. Philip's visit was later commemorated by naming a Visitor Site at Genovesa in his honor (i.e., Prince Philip's Steps; Figure 4.4). Philip visited Galápagos again in 1981.

1964 The California Academy of Sciences sends another large expedition to Galápagos. This expedition began the modern era of scientific research and conservation in the archipelago.

1964 Biologist Ernst Mayr describes Darwin's *On the Origin of Species* as "the book that shook the world."

1964 The CDRS opens in Puerto Ayora on Santa Cruz. The following year, the CDRS began collecting and hatching tortoise eggs, and later returned the tortoises to their respective islands. A similar program was subsequently started to protect land iguanas; these and other conservation programs slowed the damage done to native species by feral pigs, dogs, cats, donkeys, and other invasive species that previous visitors had introduced in the archipelago. At a symposium dedicating the CDRS, Julian Huxley proclaimed "the general evolutionary battle has been won. . . . Man is the latest and highest type produced by the evolutionary process, and that his destiny is to guide its future course on this planet." Huxley hoped that Galápagos would be a centerpiece of future evolutionary studies. As part of the inauguration of CDRS, 60 scientists spread throughout the archipelago to survey the biodiversity of the islands.

1964 While hiking with six others to the Wittmers's farm on Floreana, 70-year-old tourist Saydee Reiser disappears. Reiser's disappearance made international news and reminded many people of Floreana's past mysteries, and some ships refused to anchor there overnight. Reiser, whose body was not found until 1980, is buried on Floreana.

1966 Ecuador begins promoting international tourism in Galápagos. That same year, CDRS offered its first courses for teachers and citizens, and the first nature-based tour boat arrived in the archipelago as CDRS began its first conservation education programs.

1967 Reconstruction of the airport begins at Baltra. TAME, the national airline of Ecuador, soon began charter flights to Galápagos; these flights, which occurred twice per month, each carried 62 passengers. Weekly flights to Galápagos began in 1975.

1968 The GNPS near Puerto Ayora on Santa Cruz Island begins administering the conservation work started earlier by the CDRS. GNPS had one boat to patrol the park (an area about the size of Pennsylvania).[52] Today, GNPS has more than 290 staff, four technical offices, and three community

[52] That boat, which was donated by an Asian businessman with connections to organized crime, was usually docked in Academy Bay because of insufficient funds to buy fuel.

centers. As noted by Eliécer Cruz, former director of GNPS, "the goal of the GNPS is to balance the needs of the local community with the protection of the fragile ecosystems of the islands on which we all depend."

1968 When Fernandina erupts, the 7-square-kilometer floor of its caldera drops almost 300 meters and earthquakes continue for a week (the largest being magnitude 5). Land iguanas began repopulating the caldera within two years.

1968 Alan and Joan Root's film *The Enchanted Islands* becomes one of the first British wildlife films to be shown in the United States. Three years later, famed oceanographer Jacques Cousteau's movie *The Dragons of Galapagos* further promoted the wonders of Galápagos.

1968 The 10-passenger, three-masted schooner *Golden Cachalote* begins offering the first-ever regularly scheduled tourist cruises in Galápagos. Two years later, the 60-passenger *Lina-A*—the first major tourism venture in Galápagos—began regularly scheduled three- and four-day cruises around the archipelago. *Lina-A,* which could carry 58 passengers, was owned by Quito-based Metropolitan Touring, which built several of the landing sites in the archipelago.

1969 Ecuador's Ministerial Accord 690A formally confirms boundaries of GNP, enables GNPS to control and conserve the park, and adds a two-mile marine zone (later expanded to 40 nautical miles) to GNP. The accord was renewed in 1989 for five years, and again in 1991 for 25 years.

1969 Ecuador's military resurfaces one of the two airstrips at Baltra. Charter flights continued to bring small groups of tourists to Galápagos as several companies (most notably Metropolitan Touring and Lindblad Expeditions) began offering boat tours of the archipelago. Lars-Eric Lindblad later funded an early study of the impact of tourism in Galápagos and paid the salaries of Juan Black and José Villa, the first two officials of what became GNPS.

1970 Ecuador's Law for the Protection of Wild Animals and Fisheries Resources bans the capture and selling of rare species, the use of chemicals that may hurt plants and animals, and all activities that damage the national park and its natural resources.

1970 In its first complete year of regularly scheduled commercial flights to Galápagos, TAME brings 1,500 visitors to the archipelago; this is approximately one-third of the total visitors to the islands. Baltra's airport consisted of little more than a recently refurbished airstrip and a few meager buildings; goats wandering near the tarmac were scared away by airport personnel. The archipelago was served by only six small tourboats and one cruise-ship. During the upcoming decade, the population of Galápagos doubled to 5,800, and the number of visitors to Galápagos tripled to more than 17,000.

1970 *The New York Times* tells readers that Galápagos is where they "can now walk in the company of Darwin."

1970 The first census of albatrosses in Galápagos reports 14,346 of the birds in the archipelago. That same year, the first census of penguins in Galápagos reported more than 3,000 of the flightless birds in the archipelago. By 2008, the population of penguins had dropped to only 1,539 birds.

1970 Tortoise eggs from Santiago are incubated at CDRS. By 1999, 487 tortoises had been repatriated to the island. Also in 1970, 20 tortoises raised at the CDRS were released on Pinzón, and the last 14 tortoises (12 males and 2 females) from Española were moved to CDRS for breeding. Since then, more than 1,700 tortoises bred and hatched at CDRS have been repatriated to Española, 500 to Pinzón, and 600 to Santiago.

1970 The first Master Plan for Galápagos recommends limiting tourism to 12,000 tourists per year.

1971 CDRS begins a scholarship program that involves Galápageños in the station's work.

1971 In "Galápagos—The Way the World Was" published in *Holiday* magazine, Richard Atcheson labels Galápagos a "Dream Place of the '70s" while lamenting that humans might destroy the archipelago ("Man has already destroyed countless species; now I'm afraid he'll do a job on the Galápagos. . . . On islands elsewhere in the world, more people mean fewer native species"). Articles making similar claims soon began appearing in other popular magazines and newspapers.

1971 While looking for snails on Pinta, biologist József Vágvölgyl and his wife Maria find an emaciated tortoise that later became famous as "Lonesome George." The following year, Lonesome George was located by GNPS rangers (while they were hunting goats) and taken to the CDRS on Santa Cruz.[53] After convening a symposium to discuss Lonesome George's fate, authorities decided to keep George at the CDRS as a fund-raising tool. George died in 2012 and produced no offspring, yet was the most popular reptile in the archipelago. According to many residents, George was named after Jorge Ballesteros, a ranger who found George on Pinta. Other stories attribute George's name to American actor and comedian George Goebel, who had often referred to himself as "Lonesome George" in television interviews.

1972 David Lack receives the Royal Society's Darwin Medal for his studies of evolution in Galápagos.

1972 The World Wildlife Fund gives CDRS its first boat, *Beagle III*. That same year, Jaime Torres became the first superintendent of GNPS. Torres was soon assisted by two conservation officers and six rangers stationed on

[53] By the end of the 1970s, GNPS had killed 41,390 goats on Pinta.

Santa Cruz. Within a year, GNPS had 18 staff. In 1975, GNPS approved its first management plan, which included 40 rangers to help implement the plan; this plan recommended extending the park boundaries to 12 miles outside the islands and suggested a maximum of 8,800 tourists per year. By 1981, this "limit" had been raised to 25,000 as the maximum sustainable number of tourists per year. Today, more than 170,000 tourists visit Galápagos every year (Figure 1).

1972 A census estimates there are 800 pairs of flightless cormorants in Galápagos.

1973 After three years of construction, the Van Straelen Interpretation Center opens at CDRS. The center was renovated in 1997.

1973 There are 13 tour-boats in Galápagos. That same year, commercial diving cruises began in the islands. By 1991, there were 5,000 divers visiting Galápagos annually. In upcoming years, the numbers steadily increased—to 8,000 divers in 1993, to 10,000 in 1995, to 15,000 in 1996, to 20,000 in 1998, and to 25,000 in 2005.

1973 Ecuador's president Guillermo Rodriguez makes Galápagos a province (Provincia de Galápagos), despite it having a population of only 4,000 people. This soon produced excessive governmental regulation and made it difficult to limit the activities of Ecuadorian citizens in the archipelago. This was remedied in 1998 by passage of the Special Law for the Preservation and Sustainable Development of Galápagos, which limited migration to Galápagos, consolidated the National Park and Marine Reserve, and provided a framework for conserving the archipelago.

1973 Juan Black's book *Galápagos, Archipiélago del Ecuador* is the first natural history of Galápagos written in Spanish. One thousand copies were given to Ecuador's schools.

1974 There are approximately 3,000 wild tortoises in Galápagos. Today, thanks to protections, repatriation, and breeding programs (Chapter 11), there are more than 20,000.

1975 A road opens linking Puerto Ayora with Itabaca Channel on the north side of Santa Cruz.

1975 GNPS formally requires all visitors to Galápagos to be accompanied by a certified naturalist guide trained by CDRS. Today, each guide can supervise no more than 16 visitors on a single tour.

1975 Park officials build fences on Santiago to protect endemic plants from feral animals. That same year, park officials found two land iguanas in southern Santiago, where they were believed to be extinct. Also, the first of several species of geckos was discovered at Puerto Ayora, and the oldest of the tortoises from Santiago and Española that had been raised at CDRS were repatriated to their original islands.

1976 A report by former U.S. Peace Corps volunteer Gerard Wellington documents the high percentages (20–65%) of endemism of marine life in Galápagos. The report recommended extending the protective zone around each island by two miles and protecting migratory species such as whales.

1976 In *Galápagos: A Creationist Visits Darwin's Islands,* Lester Harris describes "how a loving God has enabled [life in Galápagos] to adapt in a sin-damaged world" while urging readers to "praise God for His ingenious creative power in giving us the Sally Lightfoot crabs" and "pleasant variations on God's original lizard theme." Harris, who pleaded with readers to learn "what lessons God is trying to teach us" in Galápagos, claimed "Darwin jumped to the wrong conclusion. His ideas caused thousands upon thousands of persons to reject God and His Word."

1977 The San Diego Zoo sends an adult male Española tortoise to CDRS to diversify the breeding of tortoises for repatriation to the island. (In the late 1920s, the zoo had been given several Española tortoises, but this tortoise was the only survivor.) Although Lonesome George received much more attention, the saddleback tortoise from San Diego has fathered more than one-third of the 1,781 tortoises (as of 2012) repatriated to Española, thereby earning him the nickname "Super Diego" (Figure 2.12).

1978 Galápagos is designated one of the first four UNESCO World Heritage Sites.[54] The following year, a plaque commemorating this designation was unveiled near the offices of the GNPS along Charles Darwin Avenue in Puerto Ayora.

1978 GNPS meets with local residents to discuss recommendations for expanding the Galápagos Marine Reserve. Many residents resisted the recommendations.

1979 There are now 35 tour-boats in Galápagos. After 18,000 tourists visit Galápagos, GNPS increases its visitor limit from 12,000 to 25,000 tourists per year. During the 1970s, an average of 17,500 visitors came annually to Galápagos. In the upcoming decade, the population of Galápagos approached 10,000 people as tourism doubled to more than 40,000 per year.

1980 Ecuador creates INGALA (Instituto Nacional Galápagos) to plan, implement, develop, regulate, and finance projects in Galápagos.

1980s Ecuador's national currency (sucre) begins a precipitous decline that eventually takes more than 90% of its value. The resulting inflation and loss of real per-capita income prompted many residents to flee to Galápagos, where jobs and pay were relatively plentiful.

[54] A World Heritage Site is a natural site having physical or cultural significance, such as rivers, forests, or human-made sites such as monuments. As of 2012, there were 936 World Heritage Sites.

FIGURE 2.12
Super Diego, the last survivor of a group of Galápagos tortoises given to the San Diego (California) Zoo in the 1920s, was returned to Galápagos in 1977 to help repatriate tortoises to Española Island. Super Diego, who in 2012 was living quietly in a pen at CDRS, has fathered hundreds of offspring now living on Española.

1982 El Niño in Galápagos produces 1.36 meters of rain, an amount 2.5-times greater than the previous record from 1952–1953. Almost 80% of the seeds eaten by finches at Daphne Major were soft, and the population of finches who ate these seeds increased by 400%. However, the population of flightless cormorants was reduced by 45%, the population of marine iguanas by 50%, and the population of penguins by 78%. Rains flooded roads on Baltra.

1982 There are 152 registered fishermen in Galápagos. By 1993, there were 392, and by 2002 there were 956.

1982 Isabela's Volcán Wolf erupts.

1983 As GNPS regulations begin to restrict (and sometimes criminalize) their harvests, fishermen on San Cristóbal form the first fishermen's cooperative in Galápagos. Other cooperatives later formed on Santa Cruz (1993) and Isabela (1995), and a second San Cristóbal cooperative formed in 1996. These organizations, which became strong advocates for fishermen, often opposed conservation measures implemented by GNPS.

1984 A second management plan sets a maximum carrying capacity of 41,767 tourists per year in Galápagos.

1984 Galápagos joins over 100 countries in the International Man and Biosphere Program. A plaque commemorating this program is in the roundabout at the public plaza on Santa Cruz.

1985 In his novel *Galápagos*, Kurt Vonnegut describes how survivors of "The Nature Cruise of the Century" evolved into a peaceful race of people after all other people were killed by disease, famine, and war.

As Vonnegut noted, "Darwin did not change the islands, only people's opinion of them."

1985 The largest brush fire on record in Galápagos erupts on the south side of Isabela's Volcán Sierra Negra. The fire lasted for five months, during which GNPS officials evacuated several tortoises from the area.

1986 Ecuador establishes the Galápagos Marine Resources Reserve, one of the largest marine reserves in the world. The reserve, which covers approximately 70,000 square kilometers, extended 24 (and later 64) kilometers beyond the baseline connecting the outermost islands. The announcement of the reserve coincided with the 25th anniversary of the World Wildlife Fund. The Ecuadorian Navy (and not GNPS) patrols the area. Cargo boats must stay at least 100 kilometers away from the archipelago.

1986 INGALA completes construction of an airport at San Cristóbal.

1987 Early in the year, Ecuador limits the number of tourists in Galápagos to 25,000 per year. Later in the year, the number of tourists exceeded that limit.

1988 An eruption of Volcán La Cumbre (Fernandina) widens its caldera by 300 meters and forms a 250-meter-high pile of debris on the caldera's floor.

1988 *Iguana* sinks near Academy Bay, spilling 50,000 gallons of diesel fuel.

1989 45,800 visitors come to Galápagos.

1990s Increased demand for specialty coffee revives the Galápagos coffee industry (e.g., near Bellavista on Santa Cruz; at El Cafetal on San Cristóbal). All coffee grown in Galápagos is organic "Arabica" coffee (Chapter 4, "Coffee at El Cafetal").

1990 On its way from Genovesa to Baltra, the 74-foot-long *Bartolomé* catches fire and sinks, killing five passengers and one crew member. The accident made news worldwide.

1990 CDRS builds a new trail to show Lonesome George and other tortoises to visitors. That same year, GNPS rangers went on strike for higher wages.

1990 Galápagos is declared a Whale Sanctuary. That same year, Ecuadorian officials banned the capture, processing, and trade of all species of marine turtles in Ecuador.

1991 Puerto Ayora gets reliable telephone service and its first bank (Banco del Pacifico), but it still lacks paved streets and reliable indoor plumbing. That same year, Galápagos received its first satellite television programs from foreign countries (mostly from Peru), but did not get Ecuadorian television programming until 2002. In 1991, 39,000 tourists spent $32.3 million on their visits to Galápagos. In future years, these numbers would skyrocket.

1992 Oceanographer D. M. Christie and colleagues report ancient "drowned islands" of Galápagos between Española and mainland South America that are five to nine million years old. The authors concluded "the time available for speciation of Galápagos organisms is much longer than the age range of the existing islands" (Chapter 3).

1992 Ecuador approves an administrative plan for the Galápagos Marine Reserve and establishes the Galápagos Ministry of Tourism to promote tourism in the archipelago and mainland. That same year, 80 tour boats were cruising Galápagos.

1992 The first dive center opens in Puerto Ayora. The center offered dive courses and day trips to several Marine Visitor Sites.

1992 The first Master Plan for the Marine Resources Reserve highlights the many conflicts of interest facing the reserve, especially illegal fishing. The plan was revised in 1997–1998, and an expanded Marine Resources Reserve followed the next year. This new reserve extended the former boundaries from 24 kilometers to 64 kilometers from the extreme points of the islands. The two most persistently poached organisms in Galápagos are sharks and sea cucumbers.

1992 The notorious brothel "Quatro y Media" (also known as the Bada Bing Club) opens 4 kilometers north of Puerto Ayora. Soon thereafter, and despite protests by residents, other brothels (all legal) opened on Santa Cruz and San Cristóbal.

1992 The Seventh Day Adventist Church opens a kindergarten, high school, and Loma Linda Adventist College "fundamentada en principios" in Puerto Ayora.[55] Campus included a wooden billboard proclaiming Genesis 1:1 ("In the beginning . . .") that today remains prominent on Charles Darwin Avenue. Not far away (where Charles Bindford Avenue intercepts Bolivar Naveda Avenue) is a mural of the Christian creation story at Escuela Fiscal Mixta. Today in Galápagos, few schools teach students about evolution.

1993 GNPS increases the entrance fees for foreigners from $40 to $100, and for Ecuadorian nationals from $0.60 to $6.00. Despite this increase, the number of visitors increased 15%.

1994 GNPS allows, for the first time, entry into Galápagos of two cruise ships carrying more than 1,200 passengers. In 2007, such ships were banned from the archipelago.

[55] In 2012, the 18 public elementary schools and nine high schools in Galápagos enrolled 7,321 students (2,324 on San Cristóbal; 4,362 on Santa Cruz; and 635 on Isabela; classes began on Floreana in 2012). There are relatively few private schools in the islands (e.g., Tomás Berlanga School and Loma Linda Adventist School, both on Santa Cruz), and Galápagos students must visit the mainland for college. The University of San Francisco in Quito has a campus on San Cristóbal, but this campus is for preparatory classes and thesis work. More than 70% of teachers in Galápagos have college degrees.

1996 Resolution R-DE 058 officially makes the Galápagos Marine Reserve part of GNP and bans all fishing in the park except artisanal fishing.

1996 61,000 visitors come to Galápagos, which is now served by 88 tour boats. That same year, *Business Week*, noting the many threats facing Galápagos, urged its readers to visit "evolution's lab while it lasts. . . . Go soon."

1997 "El Grupo Nucleo" publicizes how illegal fishing, increasing tourism, and neglected laws are damaging Galápagos.

1997 Americans Julia Hawke and Susannah Stephen are lost at sea while diving in Darwin Bay. Today, a plaque on Genovesa commemorates their deaths.

1997 Full-time electricity becomes available in most inhabited areas of Galápagos, thereby enabling residents and businesses to store perishable foods. Two years later, several hotels added direct-dial telephones and the Internet.

1997 In an article titled "Last Look at Paradise?" *International Wildlife* laments how illegal fishing has promoted societal problems such as drugs and prostitution in the archipelago. The next year, Columbian Jorge Castrillón Henao was arrested and charged with exporting drugs to Galápagos. Similar problems have continued, with several additional arrests and the dissolution of a child pornography ring in Puerto Ayora in 2003.

1998 CDF opens Environmental Education Centers on Isabela, San Cristóbal, and Santa Cruz.

1998 Ecuador's "Special Law for the Conservation and Sustainable Development of the Galápagos Province" protects the wildlife of Galápagos so that "the evolutionary process" can proceed "under a minimal human interference." The law, which was influenced by data and analyses provided by CDRS, did not set a ceiling on the number of tourists, but it prohibited further settlement on the islands, controlled fishing, restricted migration to the islands,[56] decentralized governmental regulation, returned all visitors' entrance-fees to the archipelago, discouraged the introduction of new species, and expanded the Galápagos Marine Reserve (GMR) to 40 miles outside the current boundary. Today, the reserve encompasses 138,000 square kilometers and is administered by GNPS.

1998 Four Americans die when *Moby Dick* capsizes near Santa Fé. That same year, two filmmakers—including the inventor of the IMAX camera—were killed when their helicopter crashed at Isabela's Cerro Azul.

[56] People born in Galápagos, and those who lived there for at least five years before March 18, 1998, were made permanent residents; these are the only people automatically allowed to live and work in Galápagos. People who moved to Galápagos after March 18, 1998, were made temporary residents who could obtain renewable work permits. The rest could be deported (INGALA deported an average of 94 people per month in 2000). The long-term enforcement of this provision has been problematic. When the Special Law was passed, protesters dressed in black carried coffins through Puerto Ayora's streets on a "march to mourn the archipelago."

1998 In *Atlantic Monthly,* American writer Kenneth Brower describes Galápagos as "the most photographed archipelago on the Earth."

1998 *Sports Illustrated* visits Galápagos to produce its annual swimsuit issue. On the cover, model Heidi Klum and her cleavage are "straddling the equator."

1998 When the eruption of Volcán Cerro Azul threatens the population of "aplastada"[57] tortoises, GNPS officials and the Ecuadorian Army move the tortoises to the GNPS breeding center near Puerto Villamil.

1998 The Interpretation Center opens at Puerto Baquerizo Moreno. That same year, water taxis became common at Puerto Ayora.

1998 The population of Galápagos is 14,661. Of this number, 8,512 live on Santa Cruz.

1999 Galápagos authorities again allow fishing for sea cucumbers; three years later, the harvest would top eight million. Today, fishermen legally harvest fewer than one million sea cucumbers per year from the archipelago, but illegal fishing probably exceeds that total.

1999 In an article titled "Galápagos, Paradise in Peril," *National Geographic* describes the many threats to the archipelago, including overfishing, introduced species, and rampant tourism.

1999 The population of Galápagos reaches 16,000. Tourism accounts for 40% of the jobs for Galápagos residents.

2000 During the past three years, the number of fishing boats in Galápagos has increased from 270 to 341, and the number of registered fisherman from 457 to 682.

2000 Ecuador switches its currency from the sucre to the U.S. dollar.

2000 In the same year that it produces a list of allowed and restricted products, GNPS confiscates 804 products (mostly at the Baltra airport). Residents were responsible for 74% of the introductions, foreign tourists for 17%, and domestic tourists for 9%.

2000 Margret Wittmer dies after living on Floreana for 68 years. Today, many tourists—especially those who have read Wittmer's *Floreana*—visit Wittmer's gravesite on Floreana.

2000 On average, residents of Isabela's highlands are eating approximately one tortoise per week.

2000 *Scuba Diving* magazine ranks Galápagos as the world's top dive site.

2001 Captain Tarquino Arevalo of the Ecuadorian oil tanker *Jessica*—coming from Guayaquil carrying 160,000 gallons of diesel fuel for Baltra and 80,000 gallons of bunker fuel for the tour-boat *Galápagos*

[57] Aplastada tortoises are distinguished by the flattened shapes of their shells.

Explorer II—mistakes a signal buoy for a lighthouse and runs aground in Wreck Bay, 800 meters offshore from Puerto Baquerizo Moreno (Figure 2.13). Although currents minimized the impact of the spill on land (and bright sunlight accelerated evaporation of the fuel's volatile compounds), leaking fuel covered many of the local animals, and within a year more than 60% of the marine iguanas near the spill died. Headlines lamenting the disaster (e.g., "For Humanity, Save the Galápagos from Humanity" in the *Chicago Tribune*) were common throughout the world, and Captain Arevalo and his crew were arrested. Later, the GNPS sued the insurers of *Jessica* (British company Terra Nova) for $10 million. The *Jessica* disaster—for which Arevalo admitted, "I am completely to blame"—produced the first international lawsuit to be settled in favor of environmental damage reparation. Despite concerns raised by the spill, the following year Galápagos imported six million gallons of fuel.

Where Is Jessica?

The grounding and sinking of the oil tanker *Jessica* in 2001 in Wreck Bay (San Cristóbal) is a landmark event in the history of Galápagos. The sinking, which made headlines worldwide, highlighted the dangers associated with the demands created by the flood of tourists who visit the islands. Given this notoriety, many visitors to San Cristóbal ask where *Jessica* sank. What they're told is often wrong.

Wreck Bay is marked by several lighted buoys that steer ships eastward around a shallow reef to a deeper channel that captains use to enter the bay. (*Jessica* crashed against this reef.) Most water-taxi drivers and tour-boat captains tell visitors that the wreckage of *Jessica* is below the red-and-black, lighted buoy that is directly in front of the public pier at the end of Avenida Española. In fact, this buoy is attached to the reef, not to *Jessica*.

Jessica's remains, which can be seen from planes landing at the nearby airport (and barely visible from a water taxi if the ocean is calm), are about 50 meters northwest of the buoy where waves break in the bay. If you scuba dive, it's easy to find and examine *Jessica's* wreckage (it's about 8 m deep), but don't expect a memorable dive; the wreckage is flattened and scattered, and there is little to see.

There are relatively few sites for wreck diving in Galápagos. However, if you're looking for a better dive in Wreck Bay than the *Jessica* wreckage, go to the unlighted, balloon-like buoys just northwest (i.e., toward Frigatebird Hill) of *Jessica*. Here, you'll find the wreckage of *Karahua*, which sank more than 70 years ago in 14-meter-deep water. This scenic wreck, which is home to lots of wildlife (e.g., seahorses, sea slugs), is an ideal and easily accessible dive. For other scenic dives in Galápagos, see Appendix 5.

2001 During the next five years, the number of tourists in Galápagos increases 83%, the amount of cargo imported to Galápagos increases 94%, and the number of commercial flights to Galápagos increases 193%. During this same period, the number of inspectors dropped by 20%.

FIGURE 2.13 On January 16, 2001, the oil tanker *Jessica* ran aground just off the coast at Puerto Baquerizo Moreno, dumping more than 200,000 gallons of fuel into the ocean. This accident, which made news around the world, highlighted the perils of the archipelago's dependence on imported fuels to meet the demands of its rapidly growing population and tourist industry.

2001 CDF receives the J. Paul Getty Wildlife Conservation Prize for its work to conserve Galápagos.

2001 The UNESCO Galápagos World Heritage Site is expanded to include the Galápagos Marine Reserve.

2002 Peter and Rosemary Grant receive the Royal Society's Darwin Medal "for their fundamental work on the ecology, breeding and evolution of Darwin's finches on the Galápagos Islands. This work has become the classic example of Darwinian evolution in the wild."[58]

2002 CDF receives the Society for Conservation Biology Award for Distinguished Achievement. That same year, CDRS received the International Cosmos Prize from Japan.

2002 The Ecuadorian Navy tanker *Taurus* spills 5,000 gallons of diesel fuel near Puerto Villamil.

2004 More than 170 GNPS rangers strike for higher wages. That same year, Galápagos fishermen staged two strikes to relax restrictions on fishing harvests.

2004 On the 100th anniversary of Manuel J. Cobos's death, El Progreso is declared a National Cultural Heritage Property.

2004 108,436 tourists visit Galápagos. The following year, the average total expenditures of foreign tourists visiting Galápagos increased to $4,180 per person (up from $3,677 in 1999).

[58] Other winners of the Darwin Medal included in this book are Alfred R. Wallace (1890), Joseph D. Hooker (1892), Thomas H. Huxley (1894), Henry F. Osborn (1918), David Lack (1972), and Ernst Mayr (1984).

2005 CDRS commemorates the 1905–1906 expedition to Galápagos by the California Academy of Sciences by dedicating a plaque near the station's iguana pens.

2005 Galápagos Marine Reserve is named a Particularly Sensitive Sea Area by the International Maritime Organization because of its ecological and scientific importance and its vulnerability to damage by marine activities. (In 2012, there were 14 areas with this designation worldwide.) That same year, GNPS opened the Miguel Cifuentes Community Center for Environmental Education on Santa Cruz.

2005 Residents of Ecuador comprise 19% of the more than 123,000 visitors to Galápagos. The population of Galápagos was increasing by 6.4% per year, and visitation was increasing by 9% per year. Nonresidents owned 60% of the boats in the archipelago, and land-based tourism contributed only 10% of tourism's gross revenues. Authorities warned that if the current growth rate of tourism continued, the year 2031 would see 969,000 visitors coming to Galápagos; and if the current rate of population growth continued, the population of Galápagos in 2031 would be 118,000. During the past six years, the gross domestic product of Galápagos has increased by 78%, and the population has increased by one person for every $3,000 earned in the archipelago.

2006 A survey reports that most residents of Galápagos accept the bans on shark fishing, cutting native timber, year-round fishing, and removing sand from beaches. However, most residents did not accept quarantine measures, restrictions on sea cucumber fishing, immigration restrictions on family members, and restrictions on the number of tourists admitted to the islands.

2006 Raquel Molina becomes the first woman to direct GNPS. In 2008, after becoming the first director in five years to stay in office for more than a calendar year, she was fired. GNPS, with a budget of $11.1 million (almost half of which is derived from park entrance fees), was and is now the largest and most influential organization in the archipelago. There were 303 active guides in Galápagos in 2006.

2006 There are now 2,051 vehicles in Galápagos; more than half of them are in Santa Cruz.

2006 Regularly scheduled tours of Isabela begin. That same year, a small airport opened on Isabela for inter-island flights.

2006 Thirty flights per week serve Galápagos (up from 15 per week in 2001). To accommodate the burgeoning number of tourists, Puerto Baquerizo Moreno on San Cristóbal opened a tourist-friendly public dock.

2007 National Geographic Channel produces *Galápagos,* the first full-length movie in more than two decades about the archipelago.

2007 An international workshop on Santa Cruz produces the Regional Strategy for the Management of Rodents in the Galápagos Islands.

2007 Tourism now accounts for about 65%, and fishing only 4%, of the Galápagos economy. During the past decade, governmental expenditures in Galápagos have grown from $10 million to $40 million. Galápagos is regulated by more than 50 central and local governmental agencies, 40 of which have offices in the islands.

2007 The death of Fritz Angermeyer—the last surviving Angermeyer brother—closes the era in which Galápagos was settled and inhabited by hardy, adventurous foreigners. The Angermeyers are buried in a family plot in the Bellavista cemetery (Santa Cruz).

2007 GNPS officials increase their enforcement of park laws with several prominent confiscations and arrests. Among these was a ship carrying a submarine, for which passengers had paid $120,000 to cruise. Submarines are forbidden in GNP.

2008 A census reports 1,539 penguins in Galápagos, a drop of more than 50% since the 1970s. More than 90% of the penguins live on Fernandina and western Isabela.

2008 Graham Watkins, executive director of CDF, tells *The New York Times* "What we have [in Galápagos] is an unsustainable model of development." Later that year, *Frommer's* listed Galápagos as one of their "11 places to see before they disappear."

2008 Ecuador becomes the first country in the world to explicitly grant constitutional rights to nature.

2008 CDRS and GNPS set a goal to make Galápagos the first goat-free archipelago in the world by 2010.

2008 The U.S. National Science Foundation awards $21.7 million for research in Galápagos. That same year, Peter and Rosemary Grant received the Linnean Society's Darwin-Wallace Medal for their studies of evolution in Galápagos.[59]

2008 GNPS bans tours that combine scuba diving and land excursions.

2009 Fernandina's Volcán La Cumbre erupts for the first time since 2005. The eruption produced a 200-meter-long tunnel down the side of the volcano to the ocean.

2009 In *Lonely Planet* magazine, David Attenborough describes tourism in Galápagos as a "necessary evil," adding that "these islands are an example, a parable, for how we treat the natural world."

---

[59] The first recipient of the Darwin-Wallace Medal was Alfred R. Wallace (1908), who received a gold medal. Other recipients, who have received silver medals, include Joseph D. Hooker (1908), Victor van Straelen (1958), Julian Huxley (1958), Ernst Mayr (1958), and Stephen J. Gould (2008). The award was given every 50 years until 2008; in 2010, the Linnean Society began awarding the medal every year.

2009 GNPS becomes the first marine reserve in the world to use satellite technology to monitor all ships in the archipelago.

2009 *The New York Times* ranks Galápagos number 3 in its "Places to Go" list.

2010 Fishermen find dozens of bombs from World War II submerged between Baltra and Bartolomé, apparently dumped there (with tons of other equipment) when the U.S. military left Baltra in 1946. Two years later, 11 of the bombs were detonated on Baltra.

2010 Galápagos continues to be recognized as a top destination for tourists. For example, *Travel and Leisure* magazine ranked Galápagos as the "Best Island" for the third consecutive year. *Virtuoso Life* magazine ranked Galápagos number 3 as "Best Family Destination," number 2 as "Best Outdoor Adventure," and number 1 as "Most Alluring Island Escape." *Scuba Diving* magazine again ranked Galápagos as the best dive site in the Pacific.

2010 GNPS and other organizations convene the First Sustainable Tourism Summit in Galápagos.

2010 The drowning of a scuba diver who had been touring dive sites aboard *Aggressor II* renews complaints about lax safety standards for diving in Galápagos.

2010 Although the Floreana giant tortoise has been assumed extinct for more than 150 years, geneticists discover nine giant tortoises that are hybrids having a Floreana ancestry. In 2012, another genetic study indicated that a Floreana giant tortoise may remain alive at Isabela's Volcán Wolf (Chapters 4 and 6).

2010 A total of 173,296 visitors come to Galápagos, an increase of 6% over the previous year. Of these visitors, 36% (i.e., 61,574) were Ecuadorians, 27% (i.e., 46,093) were from the United States, and 5% (i.e., 9,045) were from the United Kingdom. Baltra was the entry point for 78%, and San Cristóbal for 21%, of the visitors. The official census of Galápagos reported 22,770 permanent residents (an increase of 21% since 2001), but many residents in remote areas claimed they were not recorded.

2011 Approximately 7,000 people live on San Cristóbal, 21,000 on Santa Cruz, 2,400 on Isabela, and 110 on Floreana.

2011 As part of its stepped-up efforts to combat illegal fishing, the GNPS and Ecuadorian Navy confiscate the *P/P Fer Mary I* near Genovesa and find 379 sharks in its hold (it is illegal to capture, sell, or transfer sharks of any kind in Galápagos Marine Reserve). The captain and 30 members of the crew were arrested. The dorsal fins of the sharks had been severed to sell in shark-fin soup, a high-demand Asian delicacy.

2011 CDRS, which greets more than 7,700 tourists per month, closes access to the walk-through tortoise corrals in Puerto Ayora after numerous violations by visitors and park guides (e.g., littering, touching and riding the tortoises, walking on the tortoises' feeding platform). This was the only walk-through corral at any of the archipelago's breeding centers. In 2012, CDRS housed 925 juvenile and 70 adult tortoises.

2011 Another monument to Charles Darwin is unveiled at the "Darwin arch" in Puerto Ayora on the path to CDRS.

2011 Ecuador declares a state of emergency after a tsunami produced by the Töhoku (Japan) earthquake[60] raises water 1.8 meters above high-tide levels in the archipelago, thereby threatening coastlines in Galápagos and on the mainland. In Santa Cruz, water flooded some offices at CDRS and shops along Charles Darwin Avenue.

2011 After noticing that there are few juveniles of blue-footed boobies at key areas, CDRS begins the first comprehensive study of the distribution of these birds in Galápagos. The following year, CDRS reported that there are 6,000 to 8,000 of the birds in the archipelago (down from 25,000 in the 1970s), and that they found fewer than 100 juveniles.

2011 *Wanderlust* magazine names Galápagos the top travel experience in the world.

2012 Scientists at the British Geological Survey find more than 300 fossils collected by Charles Darwin, some of which were collected during the expedition of the *Beagle*. The specimens had been lost because Joseph Hooker had not catalogued them properly.

2012 Construction begins on a new $24 million airport on Baltra that will be powered by solar energy.

2012 Officials confirm 17 cases of dengue fever on Santa Cruz and begin removing more than 30,000 discarded tires (a breeding ground for mosquitoes that spread the disease) from the island.

2012 *Smithsonian* magazine features a marine iguana from Galápagos ("Darwin's Laboratory") on its cover promoting "must-see stops" of its "evolution world tour."

2012 Lonesome George, the last of the Pinta tortoises and an international icon of conservation, dies of natural causes at CDRS. George, who had lived at CDRS since 1972, was visited by thousands of tourists per year. George's body was discovered by Fausto Llerena, his longtime caretaker and the namesake for the CDRS tortoise breeding station at Puerto Ayora GNPS later implemented its Giant Tortoise Recovery Plan, a 10-year program to restore Galápagos ecosystems by restoring populations of giant tortoises in the archipelago. GNPS is constructing a new Visitor Center at the entrance to the Tortoise Center that will house the body of Lonesome George, just a few meters from where he lived for 40 years.

---

[60] This earthquake (magnitude 9.0), which was the most expensive natural disaster in world history, killed 15,854 people, collapsed 129,225 buildings, and produced 40-meter (133-foot) waves in Japan.

CHAPTER 3

Geology and Climate

Galápagos are no ordinary islands—a mysterious prehistoric world, a landscape that profoundly influences life. These islands are plumbed directly into the heart of the Earth.

—BBC/National Geographic, *Born of Fire,* 2007

THE GALÁPAGOS ISLANDS ARE THE PEAKS OF GIANT UNDERWATER volcanoes. Several of these volcanoes have merged to form the Galápagos Platform (depth = 180–920 m), on which the archipelago's central islands rest. (Wolf and Darwin are the only islands in Galápagos not resting on the Galápagos Platform.) Several of the islands—for example, Las Plazas, Baltra, and North Seymour—are uplifted submarine lava. Volcanic eruptions in Galápagos have often been spectacular; for example, an eruption of Fernandina's Volcán La Cumbre in 1825 raised the air temperature at Punta Tortuga to 147°F, raised the water temperature there to 150°F, and melted the rigging of the schooner *Tartar,* which was anchored nearby (Chapter 2). More recently, the eruption in 2005 of Isabela's Volcán Sierra Negra launched plumes of ash more than 7 kilometers into the atmosphere and produced rivers of lava more than 60 meters wide (Figure 3.1). During the early hours of the eruption, more than a million cubic meters of lava were released every hour.

Although you will probably not see a volcano erupt during your visit to Galápagos, you *will* see the products of past eruptions—that is, craters, lava fields, lava tubes, calderas, and black sand beaches—throughout the archipelago. You will even *hear* the lava; walking across the lava rocks on several trails produces a distinctive clinking sound, thus explaining why these rocks are often called "clinkers." In Galápagos, lava is everywhere.

When Charles Darwin visited Galápagos in 1835 (Chapter 2), he was especially interested in the geology of the archipelago. His notes are so extensive that you can retrace his footsteps on some islands. (His descriptions are also often vivid, such as when he described some landscapes as "cultivated parts of the infernal regions.") Darwin's interest in geology served him well, for he knew that life both conforms to, and is ultimately constrained by, geology.

FIGURE 3.1 The Galápagos Islands are the peaks of underwater volcanoes, and the western islands of the archipelago (e.g., Fernandina and Isabela) remain volcanically active. This eruption of Isabela's Volcán Sierra Negra occurred in 2005.

A Geographic Sketch of Galápagos

Location: Approximately 930 km (578 mi) west of Ecuador, straddling the equator

GPS coordinates of the archipelago: 1°40′N to 1°22′S, 89°16′ to 92°01′W

Center of the archipelago: 00°32.132S, 90°31.156W

Closest land to the southwest: Easter Island = 3,584 km (2,227 mi)

Closest land to north: Cocos Island = 720 km (447 mi)

Closest land to the west: Marquesas Islands = 5,667 km (3,513 mi)

Number of named islands and islets: 128

Number of islands covering >100 km^2: 19

Number of islands covering >500 km^2: 5

Largest island: Isabela (4,588 km^2), a composite island formed by the coalescence of six originally separated volcanoes, is larger than all of the other islands combined. The next six largest islands, in descending order, are Santa Cruz (986 km^2), Fernandina (642 km^2), Santiago (585 km^2), San Cristóbal (557 km^2), Floreana (174 km^2), and Marchena (130 km^2).

Length of archipelago: 431 km (from the southeastern corner of Española to the northwestern corner of Darwin)

Width of archipelago: 213 km (Punta Cristóbal [Isabela] to Genovesa)

Total land area: 7,882 km^2 (4,897 mi^2), 96.4% of which comprises Galápagos National Park

Total marine area: 45,000 km^2 (17,375 mi^2)

Length of coastline: 1,336 km (830 mi)

Origin of the islands: Volcanic cones and volcanic uplift

Average depth of ocean surrounding Galápagos: 3,048 m

Distance from Galápagos to Quito: 1,238 km (769 mi)

Distance from Galápagos to Guayaquil: 1,089 km (677 mi)

Consequently, he knew that an understanding of the geology of Galápagos would help him understand life in the archipelago. It will help you, too, to understand what you see in Galápagos.

HOW DID GALÁPAGOS FORM?

Earth's crust consists of 30 or so stiff *tectonic plates* that move around. In the Pacific, the Nazca Plate moves to the southeast, and the adjacent Cocos Plate moves to the northeast (Figure 3.2). The volcanic seam where these plates are pulled apart erupts lava that forms new seafloor at the Galápagos Spreading Center.[1] About 250 kilometers south of the Galápagos Spreading Center, on the Nazca Plate and beneath Fernandina and Isabela Islands, is the *Galápagos Hotspot,* which is an area where magma from Earth's interior pushes against the bottom of the overlying plate.[2] This hotspot, which is also referred to as the *Galápagos mantle plume,* is a column of hot rock approximately 100 kilometers in diameter rising from within the Earth; the rock rises because it is as much as 250°C hotter, and therefore less dense, than the surrounding rock. When the rising plume is about 150 kilometers from the surface, some (10–25%) of the rock melts because of reduced pressure and forms magma, which erupts into the sea and forms a volcano made of basalt. Initially, these volcanoes are low mounds on the seafloor. However, subsequent eruptions progressively build these mounds, and some of them rise above sea level and appear as oceanic islands (Figure 3.3).

There are two important things to note here:

1. Unlike continental islands such as Japan and the British Isles,[3] the Galápagos Islands were never attached to a continent. Instead, the Galápagos Islands were formed by volcanoes erupting from the ocean's floor (many from depths ranging from 2,000 to 3,500 m). As a result, the Galápagos Islands were originally barren and had to be colonized (see below). This origin profoundly influenced the plants and animals living in the archipelago today.

---

[1] In 1976, oceanographers exploring the seafloor near this site discovered crabs, mussels, and anemones growing around deep-sea hot springs. These chimney-like vents, which were isolated from ecosystems in shallower parts of the ocean, were later discovered in other volcanically active areas, and are now icons of oceanic ecosystems.

[2] Unlike many volcanically active areas, Galápagos is not at a junction of tectonic plates. Instead, it is on the northern part of the Nazca Plate (just south of the junction of the Nazca and Cocos Plates), which is one of Earth's most volcanically active areas (Figures 3.1, 3.2). (The collision of the eastward-moving Nazca Plate with South America has built the Andes.) At their current rate of movement, today's islands in Galápagos will be subducted into the Peru-Chile Trench and disappear in about 20 million years. Similarly, the Cocos Plate is sinking to the northeast beneath Central America.

[3] Other examples of continental islands include Madagascar and Newfoundland. These islands, which were separated from their nearby continents by rising seas or tectonic activity, share species with the continents from which they separated.

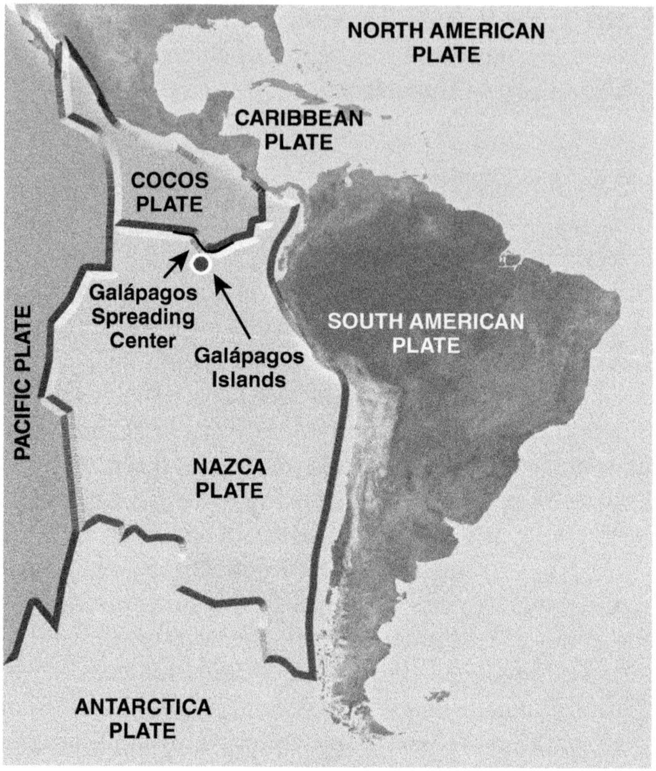

FIGURE 3.2
The Galápagos Islands are located near the northern edge of the Nazca Plate, which is moving eastward toward the South American Plate at a rate of approximately 7 centimeters per year. When Charles Darwin visited Galápagos in 1835, the islands were approximately 12 meters further west of South America than they are today. The collision of the Nazca Plate with South America has formed the Andes mountain range.

2. The Galápagos Hotspot is stationary. That is, the Galápagos Islands have not been formed by a moving hotspot, but instead by a stationary hotspot over which the Nazca Plate has moved (Figure 3.3). Because the Nazca Plate is moving eastward and because volcanoes get older in the direction of plate motion, the oldest islands of Galápagos are those in the eastern part of the archipelago; for example, Española is 3.2 million years old, and San Cristóbal is more than 2.4 million years old. Between Española and the coast of South America are older, submerged (some only 600 m down) islands that were once Galápagos islands. These so-called *seamounts,* which form the 9-million-year-old Carnegie Ridge, were, like the modern-day islands of Galápagos, produced by the Galápagos Hotspot.[4] Although these islands have now disappeared after millions of years of tectonic movement and erosion by the sea, they were probably stepping stones for colonization by the ancestors of many of

[4] Because the Cocos Plate and Carnegie Ridge disappear into subduction zones, the precise age of the Galápagos Hotspot is unknown. However, since the oldest (and now submerged) islands of Galápagos are at least 8 million years old, the Galápagos Hotspot must be at least 8 million years old. The ages of the islands, which have been determined by potassium-argon radiometric dating, correspond to when each island first appeared above sea level, not the first eruption of each volcano. Geologists have not yet determined how far down the mantle plume goes, but it could extend more than 3,000 kilometers.

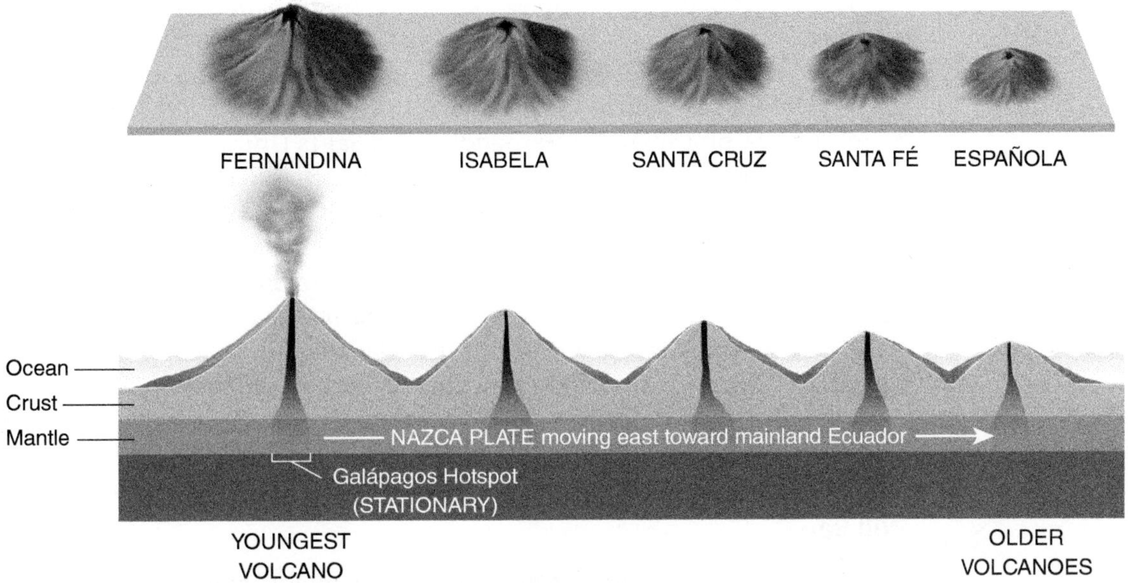

FIGURE 3.3 The Galápagos Islands have been formed by successive eruptions of the stationary Galápagos Hotspot. When the islands are moved eastward by the Nazca Plate, their volcanoes become inactive and the islands are eroded, finally disappearing below the ocean.

today's plants and animals of Galápagos.[5] So, too, will today's islands in Galápagos disappear in a few million years. The youngest islands in Galápagos, namely, Fernandina and Isabela (both less than a million years old), are in the western part of the archipelago. Although Fernandina is a single volcano, Isabela consists of six volcanoes that have merged, thus creating a single island from what had originally been six smaller islands. Five of these six volcanoes on Isabela remain active (Chapter 4).[6]

After an island forms in Galápagos, the tectonic movement of the Nazca Plate carries the island eastward and off the hotspot, as if it were on a conveyor belt. This extinguishes the island's volcano and explains why eastern islands such as Española are not volcanically active. The island is then eroded by waves and wind as the cooled Nazca Plate sinks beneath it. This gradually flattens the island, after which it disappears below sea level. As the oldest islands disappear, new islands form over the hotspot (Figure 3.3). Today, Fernandina, which sits atop the Galápagos Hotspot, is the most volcanically active island in the archipelago.

[5] We know that some of these seamounts were once Galápagos islands because their lava typifies that which is formed in air, not water.

[6] If you drilled into the center of Volcán Sierra Negra (Isabela) or Volcán La Cumbre (Fernandina), you would encounter about 2 kilometers of solid rock, followed by the magma chamber (a few hundred meters deep), several kilometers of partially crystallized magma, and finally the Galápagos plume in Earth's mantle.

The largest islands in Galápagos are *shield volcanoes,* which are massive accumulations of basaltic lava built from the ocean floor. These islands usually have gently sloping sides (slope $<10°$) that form a low-lying cone, much like a warrior's shield that is laid on the ground with its convex side upward (e.g., Sombrero Chino; Color Image 2). Fernandina and Santa Cruz are shield volcanoes.

Inside most shield volcanoes is a *caldera,* which is a large circular or oblong depression at the top of a volcano whose floor is made of lava. Calderas form when the roof of a subsurface magma chamber collapses during or after an eruption. Such an event occurred in 1968 on Fernandina, when the caldera's floor dropped 300 meters. The largest caldera in Galápagos is that of Volcán Sierra Negra, which is in southeastern Isabela between Volcán Alcedo and Volcán Cerro Azul. This caldera has a diameter of 3.2 kilometers. The caldera of Genovesa is below sea level; it opens to the sea to form Darwin Bay.

How the Formation of the Galápagos Islands Influences What You Will See

The volcanic nature of the Galápagos Islands explains many of the most striking features of the archipelago:

Because the Galápagos Islands are isolated volcanic islands that arose from beneath the ocean, they were never connected to a continent. That is, the islands of Galápagos were originally lifeless. The organisms that first chanced to colonize the islands had to somehow *get to* the islands. Organisms having the best chance of doing this were those that had the best chance of crossing an ocean—namely, plants, birds, and small insects. Lightweight seeds, spores, and insects were blown to Galápagos, and birds could fly there, often carrying seeds and insects in their digestive tracts or on their muddy feet.[7] Terrestrial mammals and freshwater organisms—that is, organisms that cannot withstand long exposures to seawater—are not, with few exceptions (e.g., rice rats), native to Galápagos because they could not get to Galápagos.

Galápagos lacks many types of organisms that are common on continents and continental islands (i.e., islands that were once attached to continents). Whereas most continents and continental islands house a wide variety of animals, Galápagos lacks some of the same types of animals, especially freshwater fish, land mammals, and amphibians. This unbalanced nature of wildlife in Galápagos has resulted from many animals simply not getting to Galápagos because, as Charles Darwin noted, "no terrestrial mammal can be transported across a wide space of sea." Indeed, although aquatic mammals such as sea lions are common in Galápagos, native terrestrial mammals are scarce. This also

[7] Approximately 60% of the endemic plant species arrived in Galápagos via birds, 31% arrived via winds, and 9% arrived by rafting across the ocean.

explains why Galápagos and other oceanic islands have smaller numbers of species than do comparable areas of the mainland; for example, there are approximately 600 species of native plants in Galápagos, but more than 20,000 in mainland Ecuador. As Charles Darwin noted in *On the Origin of Species* (1859), "we can see clearly why those animals which cannot cross wide spaces of ocean, as frogs and terrestrial mammals, should not inhabit oceanic islands; and why, on the other hand, new and peculiar species . . . which can transverse the ocean, should so often be found on islands far distant from any continent."

There are few natural predators on Galápagos. The top terrestrial predator in Galápagos is not a mammal (as in most continental ecosystems), but instead is the Galápagos hawk (Figure 1.2). The lack of numerous native predators explains the temerity of animals in the archipelago. Animals appear fearless because they have no history of avoiding terrestrial predators.

The Galápagos Islands are home to species found nowhere else on Earth, many of which are closely related to each other and to species in South America. Galápagos is home to some of the highest levels of endemism on Earth. Indeed, approximately 80% of the land birds, 22% of the marine species, 97% of the reptiles, and more than 30% of the plants you will see in Galápagos are endemic. Darwin soon recognized that the wildlife of Galápagos is unique but resembled that of nearby South America. Later Darwin noted that his theory of evolution by natural selection explained this, for "species occasionally arriving long after intervals in a new and isolated district, and having to compete with new associates, will be eminently liable to modification, and will often produce groups of modified descendants." As Darwin also explained, these distinctive species often "bear so plain a stamp of affinity to those created in America," despite the fact that environmental conditions are radically different. To Darwin, this resulted from it being "obvious that the Galápagos Islands would be likely to receive colonists . . . from America" rather than elsewhere.

Whereas ecological niches in most continental ecosystems are filled by a variety of different species, ecological niches in Galápagos are filled by similar species. This phenomenon can be traced to the colonization of the islands: Because the small number of pioneer organisms that originally colonized Galápagos found empty niches and no competitors, their ancestors filled these niches by adapting to diets and lifestyles not normally available to them on the mainland. Thus, the niches filled by flycatchers, toucans, parrots, and other birds on the mainland of South America are all filled by different species of finches in Galápagos. Similarly, as Darwin noted in *On the Origin of Species* (1859), "in the Galápagos Islands reptiles . . . take the place of mammals."

Galápagos is home to flightless birds. Penguins and flightless cormorants on the western islands of Galápagos are popular with tourists

FIGURE 3.4
This vast lava flow on Santiago Island is a popular Visitor Site.

(Chapter 5). The evolution of these birds in Galápagos correlated with little competition for resources and a lack of predators; instead of investing energy in flight muscles, natural selection favored the investment of energy in other functions. Flightless birds have evolved on other islands; for example, dodos evolved on Mauritius, flightless geese evolved on Hawaii, and kiwis, kakapos, and moas evolved in New Zealand. Not surprisingly, these birds soon became extinct when predators were introduced on the islands. The variation that ultimately produced flightlessness in Galápagos and elsewhere also occurs on mainlands, but it is quickly extinguished by carnivores.

These features of Galápagos, combined with the islands' different sizes, elevations, scarcity of fresh water, and degrees of isolation, were important for Darwin's formulation of his theory of evolution by natural selection. As he later noted, "When I visited [Galápagos] during the voyage of H.M.S. *Beagle* . . . I fancied myself brought near to the very act of creation. I often asked myself how these many peculiar animals and plants had been produced: the simplest answer seemed to be that the inhabitants of the several islands has descended from each other, undergoing modification in the course of their descent; and that all the inhabitants of the archipelago were descended from those of the nearest land, namely America."

Lava

During the past 200 years, more than 50 volcanic eruptions have occurred in Galápagos (Figure 3.1). Some of these eruptions have threatened wildlife, and others have created new land. For example, the large flow of lava near Santiago's Sulivan Bay (Figure 3.4) was not present when Charles Darwin visited Galápagos in 1835, but today is a prominent feature of the archipelago.

FIGURE 3.5 Pahoehoe lava has a wavy, rope-like appearance. This photo was taken on Santiago Island (Figure 3.4).

When lava erupts from a volcano, it is fluid and begins moving. Lava is initially 700–1,200°C, but it quickly begins to cool. As the surface of the moving lava is cooled by the air or water, it becomes more viscous, thus forming a crust that solidifies into one of three forms:

pahoehoe lava (from a Hawaiian word meaning *rope-like;* Figure 3.5) is characterized by billowy, rope-like patterns that form when the lava's thin, cooling crust stops moving while the underlying molten lava continues to flow. This difference in viscosity folds the crust into crinkled, intestinal patterns. Pahoehoe lava is abundant near Sulivan Bay on Santiago Island (Figure 3.4).

a'a' lava (from a Hawaiian word meaning *hurt;* Figure 3.6) is characterized by a rough, broken, rubble-like surface formed when debris, rocks, and broken pieces of the lava's crust accumulate atop the flowing lava that is being torn apart by small explosions of gas bubbles. As a result, a'a' lava is sharp and difficult to walk over. A'a' lava occurs throughout Galápagos, and excellent examples occur at Fernandina's Punta Espinosa. Jagged a'a' lava across Isabela's Perry Isthmus (the narrow point about one-third of the way from the southern edge of the island) has blocked virtually all land animals from crossing the isthmus.

pillow lava forms when lava erupts underwater or when lava flows into water. As its name suggests, pillow lava has a rounded, pillow-like shape. Some fields of underwater pillow lava cover more area than the above-ground deposits of pahoehoe and a'a' lava. You will see pillow lava at several snorkeling sights (e.g., Sombrero Chino), as well as on uplifted islands such as North Seymour.

Extensive flows of cooling lava can produce *lava tubes* (Figure 3.7). When the surface of a flowing river of lava cools and solidifies but the underlying lava continues to move, a tube forms when the internal lava drains from the area. In some parts of Galápagos, these tunnels are more than 9 meters high and stretch several kilometers. There are several lava tubes on private land near El Chato

FIGURE 3.6
Unlike pahoehoe lava (Figure 3.5), a'a' lava has a rough appearance. This photo was taken on Islote Tintoreras near Isabela Island. The lighter areas covering the black lava are lichens, which thrive on the south-facing sides of many coastal lava fields.

and Bellavista on Santa Cruz Island. If you visit a lava tunnel, you will enter the tunnel through its collapsed roof. When you are inside, examine the sides of the tunnel; the horizontal lines parallel the direction of the lava flow.

Along coasts, lava flows have formed caves, tunnels, and grottos such as those at Santiago's James Bay. Darwin's Toilet, a pool on Santiago that fills and empties with seawater from underground tunnels, is an example, as is Los Túnelos, a series of broken, semisubmerged lava tunnels near Puerto Villamil (Isabela).

Many lava flows also have *hornitos* ("little ovens"), which are small chimneys of lava formed by gas escaping from molten lava. The escaping gas spatters lava on the nearby crust, thereby creating the hornito. Hornitos, which

FIGURE 3.7
Lava tubes such as this one at Volcán Sierra Negra (Isabela) are common throughout Galápagos. On islands such as Santa Cruz and Floreana, tourists can hike through lava tubes.

are usually less than a meter high, are common at Sulivan Bay. The escape of much larger amounts of gas from a vent forms a *spatter cone.* Lava ejected by the escaping gas accumulates in layers that form a cone. Spatter cones are common on Bartolomé.

When flowing lava contacts water, the lava quickly cools, and this cooling creates explosions that pulverize the lava into ash. The resulting *tuff cones* are made of compacted ash that, when cemented together (e.g., by heat), forms rock called *tuff.* Pinnacle Rock (Bartolomé), Darwin's Arch (Darwin), and Kicker Rock (San Cristóbal) are examples of tuff cones (Chapter 2). The northwestern side of most tuff cones in Galápagos is higher than the southeastern side because of erosion by tradewinds from the southeast (this is especially prominent at Champion, Enderby, and Cowley Islands). When rocks are ejected during a volcanic eruption, they fall back to earth and form a *cinder cone.* Cinder cones are common on Floreana Island.

Volcanic emissions of gas through Earth's crust form *fumaroles,* which are observable along the inner slopes of Volcán Alcedo and Volcán Sierra Negra. Submerged fumaroles are also observable at Roca Redonda, a popular dive-site 24 kilometers north of Isabela. This site is predicted to be the next island of Galápagos.

Beaches

Beaches are formed by the erosion of rocks on land and the deposition of marine creatures and their shells. Beaches in Galápagos have diverse compositions and colors:

Color	Composition	Examples
Black	Erosion of basaltic rock	Black Beach (Floreana) Fernandina Tintoreras (Isabela)
Golden/Tan	Pulverized coral and seashells; erosion of ochre-colored tuff cones	Darwin Bay (Genovesa)
White	Pulverized coral and seashells*	Gardner Bay (Española) Tortuga Bay (Santa Cruz)
Green/Olive	Erosion of rock containing olivine	Punta Pitt (San Cristóbal) Punta Cormorán (Floreana) Whale Bay (Santa Cruz)
Brown/Red	Erosion of scoria, a porous type of basalt	Rábida

* These beaches are usually in the path of the prevailing southeasterly tradewinds and waves.

Climate and Weather

Because Galápagos straddles the equator, temperatures are usually relatively moderate, and there is no off-season for visitors. This is because virtually all of the animals live in Galápagos year-round; that is, regardless of when you visit Galápagos, you will see iconic animals such as flightless cormorants, marine iguanas, land iguanas, penguins, Galápagos hawks, sea lions, lava lizards, tortoises, and boobies. An exception is the waved albatross; if you want to see these magnificent birds, you will need to visit Española between April and December. Unlike many places having unusual wildlife, there is no best time to visit Galápagos.

Because the Galápagos Islands are surrounded by ocean and isolated from South America by more than 900 kilometers of ocean, the climate of the archipelago is determined almost exclusively by ocean currents. These currents are, in turn, influenced by the tradewinds that push them; so, too, are the marine organisms that live in Galápagos, as well as the land-based organisms that depend on the ocean for food.

Galápagos lies at the confluence of several oceanic currents, the most influential of which are as follows (Figure 3.8):

> The cold *Peruvian Oceanic Current* from the Antarctic merges with the *Peruvian Coastal Current* (flowing along the western coast of South America) to produce the cold, nutrient-rich *Humboldt Current*.[8] This current is pushed by southeasterly tradewinds, which are strongest from June to November. Thus, during this time of year, water in Galápagos is coldest (Table 3.1). When the southeasterly tradewinds diminish from December to May, the Humboldt Current also diminishes, and waters of the archipelago get warmer. The Humboldt Current strongly influences climate in Galápagos.
>
> The warm, less saline *Panama Current* (also known as the El Niño Flow) from the northeast merges with the Humboldt Current at the equator to form the South Equatorial Current, which flows west across Galápagos. This westerly flow of water piles up water in the West Pacific, where water-levels can be up to a meter higher than in the East Pacific. When the southeasterly tradewinds diminish in May through December, the Panama Current invades Galápagos and helps produce the warm, wet season in the archipelago (see below).
>
> The cold *Cromwell Current* (also known as the Equatorial Countercurrent) originates in the western Pacific. This current flows east for thousands of miles in deep waters that are pushed to the ocean's surface along Isabela and Fernandina. This upwelling (i.e., the rise of deep water to the surface) provides nutrients for the unique marine life of Galápagos.

The intensity of these currents produces two distinct seasons in Galápagos: a warm, wet season from January to June, and a cool, dry season from July to

[8] This current was named for Prussian naturalist and explorer Alexander von Humboldt (1769–1859). Von Humboldt studied the current that cools Galápagos, but he never visited the islands.

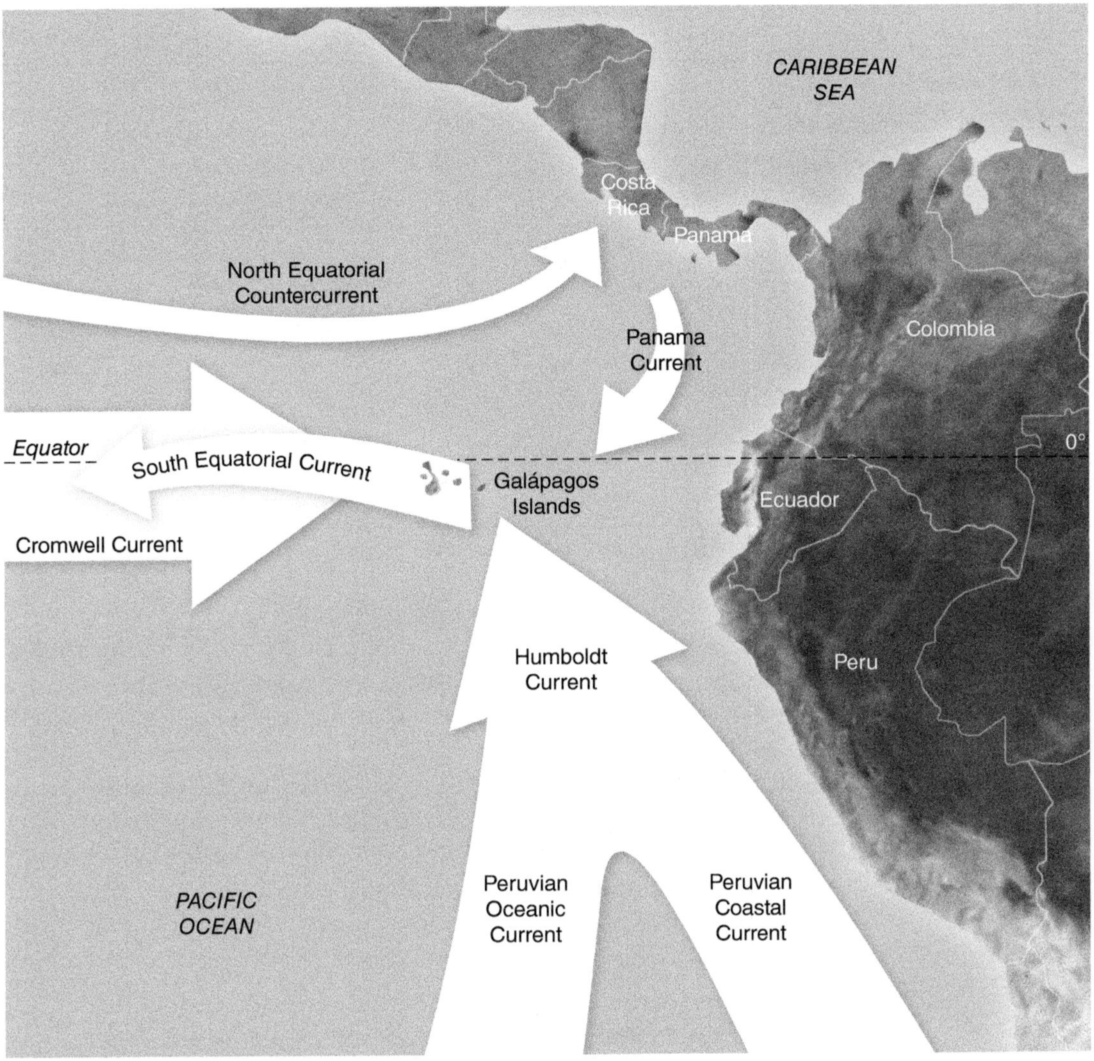

FIGURE 3.8 Galápagos is at a confluence of several major oceanic currents. Changes in the strength of these currents produce a warm, wet season from January to June, and a cool, dry season from July to December (Table 3.1).

December (Table 3.1). These seasons are not precisely fixed, and conditions during transitions from one season to the next are often unpredictable.

The *cool, dry season* lasts from July to December, during which time the cooler waters of the Humboldt Current are pushed into Galápagos by the southeasterly trade winds. Ocean temperatures drop to 19–23°C; for example, ocean temperatures in Academy Bay that average 25°C during the warm season drop to 21°C during the cool season, and land temperatures drop to 21–24°C. Because the temperature of the surface water is

TABLE 3.1 Climatic conditions throughout the year in Galápagos.

Month	Air Temp. (°C)	Water Temp. (°C)	Rainfall (mm)	Rainy Days	Clear Skies (hr d^{-1})	Seas*	Winds*
January	27	24	81	7	5.3	2	2
February	28	25	90	5	7.5	1	2
March	29	25	101	3	6	1	1
April	28	25	87	6	7.5	1	2
May	27	24	51	3	5.2	2	2
June	25	23	24	0	4.4	2	4
July	24	22	15	0	2.8	3	4
August	23	20	9	0	3.3	4	6
September	23	21	10	2	2.9	4	6
October	24	22	8	2	3.8	3	4
November	25	23	11	2	3.5	3	3
December	26	23	34	3	4	2	2

* Higher numbers indicate rougher seas and stronger winds.

cooler than the overlying air,[9] moisture accumulates in an inversion layer 200–400 meters above sea level; this is why most days of the cool, dry season are overcast. When the highlands of the tallest islands intercept this layer of moist air, mist called *garúa* forms in the highlands, thus producing lush, distinctive plant-growth in these parts of the islands (Chapter 10). However, at lower areas of islands, the climate remains cool and dry. Strong southeasterly winds (24–32 kph) produce choppy seas during the cool, dry season (especially during August and September), when underwater visibility is usually reduced from nearly 30 meters to only 8–20 meters because of phytoplankton blooms.

The *warm, wet season* lasts from January to June. During this time, the southeasterly tradewinds subside, and warm water of the Panama Current moves into the archipelago, thereby raising the average temperature of surface waters in Galápagos to near 25°C. Air temperatures also rise (to 24–29°C),[10] seas are relatively calm, there are occasional heavy rains, and skies are relatively clear (on average, the rainiest months are also the sunniest; Table 3.1). Because the Panama Current carries fewer nutrients than the Humboldt Current, the ocean water is usually clearer (i.e., visibility

[9] Although the average temperature of the ocean in Galápagos during the cool, dry season is near 21°C, the temperatures drop to 16–19°C during the cooler months of August and September in the western islands and to around 26°C near Wolf and Darwin Islands in the north of the archipelago. On average, the coldest water in Galápagos is along the western side of Fernandina and Isabela. In the southern islands, the coldest water is often along the southern side of Floreana.

[10] On average, daily land temperatures fluctuate less on the windward sides of islands than on the leeward sides. Air temperatures are warmest in February and March, when they reach 29°C.

ranges from 10–30 m) during the warm, wet season. More than 80% of the year's rain falls during the wet season, with March and April being the hottest and wettest months. Despite its name, the "wet" season in Galápagos is considerably drier than in many other places in the world.

Precipitation varies widely across the archipelago, and even across some of the individual islands. For example, although Baltra Island and the Charles Darwin Research Station (CDRS) on Santa Cruz Island are separated by only 29 kilometers, rainfall on Baltra averages only 60% of that at CDRS. This is because the highlands of Santa Cruz precipitate most of the moisture carried in the southeasterly tradewinds, thus bathing Baltra (which is north of Santa Cruz) with much drier air. The windward (i.e., southeastern) sides of islands receive more moisture than the leeward (i.e., northwestern) sides, where rain-shadows are common.

Throughout Galápagos, precipitation increases with altitude. For example, during a recent year on Santa Cruz Island, Media Luna (elevation 620 m) had 2.66 meters of rain, Devine Farm (elevation 320 m) had 1.90 meters of rain, and CDRS (elevation 6 m) had 0.47 meters of rain. Rainfall varies widely; for example, on Daphne Island, rainfall between 1976 and 1981 averaged less than 100 millimeters per year, but in 1982 it exceeded 1,300 millimeters. From 1994 to 1996, rainfall again averaged less than 100 millimeters per year, but in 1998 it exceeded 900 millimeters (and in 1999 it was less than 50 mm). These variations are most extreme during El Niño and La Niña events.

El Niño and La Niña

Every three to eight years, there are large-scale shifts in the winds and currents in the equatorial Pacific. When this occurs, the Humboldt Current weakens and does not enter Galápagos, and the Panama Current warms the archipelago and produces an *El Niño* (Spanish for "The Child," so named because it usually starts around Christmas). During an El Niño (technically called an El Niño Southern Oscillation, or ENSO), temperatures and rainfall increase dramatically, and water temperatures can reach almost 29°C. For example, the El Niño of 1953 produced 142 centimeters of rainfall at Puerto Baquerizo Moreno; just three years earlier, the annual rainfall there was only 3.55 centimeters. Similarly, the El Niño of 1987 produced more than a meter of rain at CDRS, whereas in 1985 rainfall there totaled less than 0.2 meters.

Although rains of an El Niño stimulate life on land, populations of seabirds and marine life—which depend on nutrients carried by the cooler waters of the Humboldt Current—are often devastated.[11] For example, the El Niños of

[11] Also, *Opuntia* often die during an El Niño because their roots become waterlogged and absorb so much water that their stems fill with water, become top-heavy, and fall over.

1982–1983 and 1997–1998 reduced populations of sea lions, boobies, and penguins by more than half. Similarly, the higher temperatures of surface waters during the El Niño of 1982–1983 killed 97% of the corals in Galápagos, and soon thereafter grazing by echinoderms destroyed many of the remaining corals in the islands. El Niños also diminish harvests from fisheries around Galápagos and along South America's western coast, and they have even caused extinctions. For example, as a result of the diminished production of plankton during the 1982–1983 El Niño, the endemic Galápagos damselfish (*Azurina eupalama,* a plankton feeder) disappeared and has not been seen since.

La Niña occurs when the Humboldt Current strengthens, thereby bringing more nutrients to the archipelago. When this occurs, air and water are cooler than normal, rainfall diminishes, and water temperatures drop to as low as 16°C (61°F). These conditions have the opposite effect of an El Ninõ; marine life and marine-based land-life flourish, but terrestrial productivity suffers. Plants die, seeds fail to germinate, pastures dry up, yields of rain-fed crops (e.g., corn) drop, and the reproductive success of terrestrial animals such as land iguanas and finches drops.

Daily Fluctuations of the Weather

Relative humidity and temperature follow opposing cycles during a typical day in Galápagos. At night, temperatures are coolest (e.g., 22°C) while relative humidity is highest (e.g., 90%). By early afternoon, when air temperatures peak (e.g., 28°C), relative humidity is lowest (e.g., 50%). These cycles affect life in Galápagos; for example, insects such as Darwin's darklings (*Stomion,* a flightless tenebrionid beetle) are most active during the evening and night, when temperatures are lowest and relative humidity is highest.

Sunrise occurs near 6:00 a.m., and sunset near 6:00 p.m., every day. On clear days, sunlight can be intense, with UV index readings often greater than or equal to 10 (indicating "extreme risk of harm from unprotected sun exposure").

A Year of Life in Galápagos

Many animals in Galápagos are opportunistic breeders; that is, they mate as soon as they have enough resources to do so. However, the seasonal climatic changes in Galápagos produce seasonal reproductive behaviors by many animals in the archipelago. In subsequent chapters, you will read many stories about the evolution of these animals. If you want to see particular animals (or see particular animals doing particular things), use the following guide to plan your visit:

Year-round

Great frigatebirds and magnificent frigatebirds nesting on North Seymour

Masked boobies and red-footed boobies mating and nesting on Genovesa

Penguins, flamingos, blue-footed boobies, swallow-tailed gulls, and flightless cormorants mating and nesting

January

Average temperature range: 21–29°C

Average hours of clear sky per day: 5.3

Average rainfall: 6.4 cm

Average water-temperature: 24°C

What you'll see

Highest visibility in water (through March)

Wild tortoises laying eggs

Sea turtles nesting

Lava herons nesting (until March)

Land iguanas and marine iguanas mating

Male marine iguanas on Española developing patches of red and green skin

Land iguanas on Española and Isabela begin breeding cycle

Peak season for breeding of flightless cormorants (through March)

Wet season begins

Peak season for breeding of masked boobies on Española (through March)

Land birds (e.g., finches) begin nesting after the first rain

First rains trigger flowering of plants such as palo santo and *Opuntia*

February

Average temperature range: 23–30°C

Average hours of clear sky per day: 7.5

Average rainfall: 8.6 cm

Average water-temperature: 25°C

What you'll see

Highest visibility in water (through March)

Land iguanas and marine iguanas mating and nesting

Most active time for mating of Galápagos doves and pintail ducks

Most penguins move to the cooler waters of Isabela and Fernandina

Flowering of plants such as palo santo, passion flower, and *Opuntia*

Mating season begins for flamingos on Floreana Island

Marine iguanas nesting on Santa Cruz

Lava herons nesting (until March)

Peak season for breeding of flightless cormorants (through March)

Last month to see red/green male marine iguanas on Española

Sea turtles nesting

Marine iguanas begin nesting on Fernandina

Peak season for breeding of masked boobies on Española (through March)

Wild tortoises laying eggs

Warmest water (through April)

Most penguins migrate to cooler waters of Fernandina and Isabela

Mating season begins for marine iguanas on Santa Cruz

Nesting season ends for Nazca boobies on Española

Mating season begins for black-tailed pintail ducks

Spawning aggregations of Creole "gringo" fish (through March)

March

Average temperature range: 23–31°C

Average hours of clear sky per day: 6.0

Average rainfall: 9.4 cm

Average water-temperature: 25°C

What you'll see

Sea turtles nesting

Land iguanas and marine iguanas nesting

Great frigatebirds mating and nesting on Genovesa Island

Peak season for breeding of flightless cormorants

Wild tortoises nesting

Marine iguanas nesting on North Seymour and Fernandina

Waved albatrosses return to Española at end of the month

Flowering of plants such as palo santo, passion flower, and *Opuntia*

Peak season for breeding of masked boobies on Española

Highest visibility in water

Spawning aggregations of Creole "gringo" fish

Highest amounts of rainfall in highlands (through April)

Surges of water from the north complicate landings on north-facing shores of Visitor Sites such as Bartolomé, Gardner Bay, and Puerto Egas

Warmest water (through April)

April

Average temperature range: 22–30°C

Average hours of clear sky per day: 7.5

Average rainfall: 8.4 cm

Average water-temperature: 25°C

What you'll see

- Eggs of sea turtles begin to hatch
- Highest amounts of rainfall (but it does not rain every day)
- Land iguanas and marine iguanas nesting
- Eggs of land iguanas on Isabela begin to hatch
- Waved albatrosses return *en masse* to Española and begin mating
- Great frigatebirds mating and nesting on Genovesa Island
- Wild tortoises nesting
- Hatching season ends for giant tortoise eggs in wild
- Flowering of plants such as palo santo, passion flower, and *Opuntia*
- Warmest water

May

Average temperature range: 22–28°C

Average hours of clear sky per day: 5.2

Average rainfall: 5.1 cm

Average water-temperature: 25°C

What you'll see

- Eggs of sea turtles continue to hatch at Punta Cormorán, Gardner Bay, and Puerto Egas
- Land iguanas and marine iguanas nesting
- Eggs of marine iguanas hatch on Santa Cruz
- Palo santo trees begin to shed their leaves
- Waved albatrosses mating and nesting on Española
- Great frigatebirds mating and nesting on Genovesa Island
- Wild tortoises nesting
- Peak breeding season for wedge-rumped storm petrels (through June)
- Band-rumped storm petrels begin the first of their two breeding seasons
- Sea lions breeding (through December)
- Beginning of the cool, dry season

June

Average temperature range: 20–26°C

Average hours of clear sky per day: 4.4

Average rainfall: 2.8 cm

Average water-temperature: 23°C

What you'll see

- Waved albatrosses nesting on Española
- Giant tortoises on Santa Cruz begin migrating from the highlands to the lowlands to mate and search for nesting sites
- Sea lions breeding (through December)
- Many migratory birds moving north
- Beginning of the cool, dry, garúa season
- Humpback whales moving through Galápagos
- Peak breeding season for wedge-rumped storm petrels
- Whale sharks common at Darwin and Wolf Islands (until November)
- Beginning of mating season for short-eared owls on Genovesa
- Southeasterly tradewinds begin as ocean currents intensify

July

Average temperature range: 19–24°C

Average hours of clear sky per day: 2.8

Average rainfall: 1.5 cm

Average water-temperature: 22°C

What you'll see

- Waved albatrosses nesting on Española
- Blue-footed boobies nesting
- Sea lions breeding (through December)
- American oystercatchers nesting on Santiago
- Optimal snorkeling[12]
- Flightless cormorants mating and nesting on Fernandina
- Mating season for lava lizards (until November)
- Whale sharks common at Darwin and Wolf Islands (until November)

---

[12] Snorkeling is optimal because of the increased amounts of marine life triggered by the Humboldt Current.

Dolphins and whales increasingly common near Fernandina and western Isabela

Windiest month (through August)

August

Average temperature range: 18–23°C

Average hours of clear sky per day: 3.3

Average rainfall: 0.8 cm

Average water-temperature: 21°C

What you'll see

Waved albatrosses nesting on Española

Blue-footed boobies nesting

Swallow-tailed gulls and masked boobies begin nesting on Genovesa

Sea lions breeding (through December)

Wild tortoises on Santa Cruz laying eggs

Mating season for lava lizards (until November)

Flamingos begin their mating rituals

Giant tortoises on Santa Cruz begin migrating back to the highlands

Giant tortoises begin returning to the highlands of Santa Cruz

Optimal snorkeling

Mating of Galápagos hawks on Santiago and Española

Dolphins and whales increasingly common around Fernandina

Whale sharks common at Darwin and Wolf Islands (until November)

Windiest month

Roughest seas (through October)

September

Average temperature range: 17–24°C

Average hours of clear sky per day: 2.9

Average rainfall: 0.8 cm

Average water-temperature: 21°C

What you'll see

Waved albatrosses nesting on Española

Blue-footed boobies nesting

Penguins are most active at Bartolomé

Peak season for sea lions breeding; males defending harems

- Mating season for lava lizards (until November)
- Wild tortoises on Santa Cruz laying eggs
- Optimal snorkeling
- Peak of the garúa season
- Whale sharks common at Darwin and Wolf Islands (until November)
- Tourism lowest
- Most days cool, windy, and overcast, but with little rain
- Peak breeding season for manta rays (through December)
- Roughest seas (through October)

October

Average temperature range: 18–25°C

Average hours of clear sky per day: 3.8

Average rainfall: 0.8 cm

Average water-temperature: 22°C

What you'll see

- Waved albatrosses nesting on Española
- Blue-footed boobies nesting
- Peak season for fur sea lions breeding
- Wild tortoises on Santa Cruz laying eggs
- Mating season for lava lizards (until November)
- Whale sharks common at Darwin and Wolf Islands (until November)
- Lava herons begin nesting (until March)
- Peak breeding season for manta rays and brown noddys (through December)
- Optimal snorkeling
- Roughest seas
- Dolphins and whales increasingly common around Fernandina

November

Average temperature range: 19–26°C

Average hours of clear sky per day: 3.5

Average rainfall: 1.0 cm

Average water-temperature: 23°C

What you'll see

- Sea turtles mating
- Waved albatrosses nesting on Española
- Blue-footed boobies nesting
- Breeding of brown noddy
- Band-rumped storm petrels begin their second nesting period
- Lava herons nesting (until March)
- Peak breeding season for manta rays (through December)
- Peak season for breeding of masked boobies on Española (through March)
- Peak season for sea lions breeding
- Peak season for breeding of flightless cormorants (through March)
- Optimal snorkeling
- Roughest seas (through November)
- Dolphins and whales increasingly common around Fernandina

December

Average temperature range: 20–27°C

Average hours of clear sky per day: 4.0

Average rainfall: 3.3 cm

Average water-temperature: 23°C

What you'll see

- Sea turtles begin mating
- Waved albatrosses begin leaving Española to fish at sea
- Blue-footed boobies nesting
- Peak breeding season ends for manta rays
- Sea lions breeding
- Lava herons nesting (until March)
- Many migratory birds moving south
- Peak season for breeding of masked boobies on Española (through March)
- Peak season for breeding of flightless cormorants (through March)
- Male marine iguanas on Española developing patches of red and green skin
- Land iguanas and marine iguanas begin mating
- Warm, wet season begins as the Humboldt Current diminishes
- Seas calmer

CHAPTER 4

The Islands: Where You Will Go and What You Will See

By far the most remarkable feature in the natural history of this archipelago . . . is, that the different islands to a considerable extent are inhabited by a different set of beings. I never dreamed that islands, about fifty or sixty miles apart, and most of them in sight of each other, formed of precisely the same rocks, placed under a quite similar climate, rising to a nearly equal height, would have been differently tenanted. . . .

—Charles Darwin, *Journal of Researches,*[1] 1845

ALTHOUGH GALÁPAGOS FIRST APPEARED ON A MAP IN THE 1500s (Chapter 2), none of the individual islands were named until more than a century later. The first buccaneer to name an island was William Ambrosia (Ambrose) Cowley, whose 1684 map named 16 islands of Galápagos for royalty, other potential financial supporters, and even himself.[2] In 1793, British explorer James Colnett named more islands, including some named earlier by Cowley. During the next century, British naval officers added even more names to islands to honor royalty, ships, and others. If that weren't confusing enough, in 1892 Ecuador, as part of the celebration of the 400th anniversary of Columbus's discovery of America, *again* renamed the major islands of the archipelago.[3] This explains one of the most curious—and at times confusing—features of Galápagos; namely, that most of the individual islands have multiple names. For example, Santa Cruz has had more than 10 names (including Bolívia, Chávez, Indefatigable, Valdez, San Clemente, Norfolk, Porter, and others), while a smaller island (and popular dive site) just off the west coast of Santa Cruz has had only one: Sin Nombre, meaning No Name.

When Ecuador annexed Galápagos in 1832, the most common names of the islands were those that had been assigned by English sailors, especially

[1] Darwin's *Journal of Researches* was later retitled *The Voyage of the Beagle* (see Chapter 2).
[2] Cowley Island is between Isabela and Santiago.
[3] Despite its official name—Archipiélago de Colón (Archipelago of Columbus)—Christopher Columbus was never close to Galápagos.

Cowley and Colnett (Chapter 2). In this book, we do not list or describe the origin of all of the names that have been ascribed to each island. Instead, we include only the most common Spanish and English names that appear on most maps, and which are used by residents of Galápagos (the most common names used today are listed in **bold**):

Original English Name(s)	Spanish/English Alternatives	Official Ecuadorian Name
Abingdon	Quita Sueños	**Pinta**
Albemarle	Santa Isabel	**Isabela**
Barrington	_____	**Santa Fé**
Bindloe	Diablo	**Marchena**
Charles	Marcos, **Floreana***	Santa María
Chatham	San Clemente, La Aguada	**San Cristóbal**
Culpepper	_____	**Darwin**
Duncan	Tabaco	**Pinzón**
Hood	_____	**Española**
Norfolk, Chávez	Indefatigable	**Santa Cruz**
James	**Santiago**	San Salvador
Jervis	_____	**Rábida**
Bartholomew	_____	**Bartolomé**
Narborough	_____	**Fernandina**
Seymour, South Seymour	_____	**Baltra**
Tower	Salud	**Genovesa**
Wenman	_____	**Wolf**

*Floreana was originally spelled Floriana.

VISITOR SITES

On any particular day, there are approximately 83 licensed tour boats distributed across approximately 69 Terrestrial Visitor Sites and approximately 83 Marine Visitor Sites in Galápagos. Islands such as Pinta and Pinzón are off-limits to visitors (i.e., they have no Terrestrial Visitor Sites), whereas others (e.g., San Cristóbal, Isabela) have more sites than one can see during a typical visit:

Island	Visitor Sites	
	Terrestrial	Marine
Bartolomé	2	1
Daphne Major	1	1
Daphne Minor	0	4

Darwin	0	2
Española	2	4
Fernandina	1	4
Floreana	6	6
Genovesa	2	1
Isabela	15	16
Marchena	0	5
North Seymour*	2	2
Pinta	0	4
Pinzón	0	3
Rábida	1	1
San Cristóbal	16	7
Santa Cruz	14	8
Santa Fé	1	2
Santiago**	5	7
South Plaza	1	1
Wolf	0	4

* Includes Mosquera
** Includes Sombrero Chino

At most Terrestrial Visitor Sites, you will find a trail marked with black-and-white stakes (see Figure 3.4). Visitors must stay on these trails (with their guides) to minimize their impact on the islands and wildlife. Most trails are less than a mile long, and range from smooth, flat landscapes to challenging trails over rough lava and shifting rocks. These hikes are remarkably diverse; some feature vistas, wildlife, and beautiful scenery, whereas others showcase famous historical sites. Lava lizards, iguanas, sea lions, and coastal birds are abundant at many sites, whereas other places feature unique plants and animals (e.g., waved albatrosses on Española) found nowhere else in the archipelago and, in some instances, in the world.

Many visitors tour Galápagos aboard boats (Appendix 2). According to the Galápagos National Park Service (GNPS), of the 83 licensed tour boats in Galápagos:

> 76 boats provide one-week tours that visit up to 28 Visitor Sites (usually two in the morning and two in the afternoon). Most of these boats are authorized to carry 16 passengers, plus crew.
>
> 3 boats provide two-week tours that visit up to 56 Visitor Sites (usually two in the morning and two in the afternoon).
>
> 4 boats provide daily tours to a maximum of 14 sites per week.

Some of these boats, such as the 90-meters-long *Galápagos Explorer II* and the 54-meters-long *Isabela II,* are large cruise ships that can accommodate 40–100 guests. However, more than 75% of the boats—such as the 35-meters-long *Galaxy,* the 24-meters-long *Floreana,* and the 18-meters-long *Yate Darwin*—are smaller boats that carry only 16 passengers (plus four to five crew).

Regardless of their size and number of passengers, boats are assigned an itinerary by GNPS, and that itinerary must be followed. To ensure that boats are distributed throughout the archipelago rather than crowded at only a few of the most popular sites, captains cannot depart from their itinerary. Many of the sites are near each other, whereas others are separated by hundreds of kilometers (Appendix 2). Your boat will dock near the sites, after which you will ride in a small dinghy called a *panga* to the actual site for your hike. A "wet landing" at a site means that you will disembark the panga by stepping into knee-deep water and wading ashore. During "dry landings," you will disembark onto rocks or concrete docks, and probably will not get wet.

You do not need a guide to visit some parts of Galápagos (e.g., Puerto Ayora, Puerto Baquerizo Moreno), but most Visitor Sites in the archipelago require the presence of at least one licensed guide for every 16 visitors.[4] Guides are the first line of defense for the islands; they help ensure visitors' safety, prepare visitors for their tours, identify plants and animals, enforce GNPS rules (Appendix 4), and serve as the eyes and ears for GNPS and Charles Darwin Research Station (CDRS). For example, guides have identified new species, as well as others that were thought to be extinct.

Guides are not hard to find; as of 2012, there were more than 320 guides registered with GNPS. GNPS trains and certifies guides, and their five-year licenses can be renewed after the guides take additional courses, which usually focus on safety. Guides are certified as follows:

> A Naturalist Guide I is knowledgable about Galápagos, has at least a high school diploma, has passed the Naturalist Guide I Exam, and speaks English (but not necessarily fluently). Approximately half of the certified guides in Galápagos are Naturalist Guide I.
>
> A Naturalist Guide II is knowledgeable about Galápagos, has passed the Naturalist Guide II Exam, has at least four years of experience as a guide, has at least a high school diploma (and often has taken university courses in biology and/or tourism), and is fluent in Spanish and English, French, or German. Approximately 30% of the certified guides in Galápagos are Naturalist Guide II.
>
> A Naturalist Guide III is knowledgeable about Galápagos and related issues, has at least six years of experience as a guide, has passed the Naturalist Guide III Exam, has a university degree in biology or tourism, and is fluent in English and Spanish, as well as in either French or German. Approximately 20% of the certified guides in Galápagos are Naturalist Guide III.

Dive tours must include at least one certified dive guide. In 2012, there were 38 registered dive guides in Galápagos.

[4] Although some Visitor Sites are periodically closed (e.g., Black Turtle Beach on western Isabela; the tortoise pens at CDRS), the sites we have described are current as of late-2012.

GEOGRAPHY AND THE ISLANDS' ISOLATION

In this chapter you will read about the islands, Visitor Sites, and trails that you will see in Galápagos. You will not see them all in a single visit, but your itinerary should take you to a variety of sites that showcase the archipelago's diverse wildlife and landscapes.

Although most people go to Galápagos to see the wildlife, do not underestimate the importance of the islands' geology for understanding what you see (Chapter 3). Indeed, geology constrains evolution, for speciation often involves the adaptation of life to geographic and geologic features such as temperature, altitude, and the availability of fresh water. In this chapter you will find that information for each island. We also provide measures of each island's isolation, for the isolation of Galápagos and its individual islands was important for Darwin's formulation of the theory of evolution by natural selection. Moreover, as David Lack noted, the rate of endemism of many species (including finches) on each island is directly proportional to the island's degree of isolation from the center of the archipelago,[5] and therefore is important for helping you understand what you will see during your visit to Galápagos. These measures of isolation for each of the islands are as follows:

1. The island's proximate isolation, which is the shortest distance between the island and its nearest larger neighbor. The larger neighbor must be at least twice the size of the focal island.
2. The distance from each island's geographic center to the center of the archipelago (00°32.132S, 90°31.156W). The center of the archipelago is near the tiny Isla Punta Bowditch Sur, which is along the northwestern coast of Santa Cruz (Appendix 1).

GPS coordinates for more than 350 sites are presented in Appendix 1.

Baltra

At a glance:

Other names: South Seymour
Visited by Darwin: No
Area: 27 km^2 (10.4 mi^2)
Height: 63 m
Length of coastline: 26.9 km
Age: 0.7–1.5 million years old
Inhabited: Yes; population <200 people

[5] Cocos Island, which is isolated from both South America and Galápagos, has the highest possible rate of endemism (100%) and one species of finch. That finch, *Pinaroloxias inornata,* is the only "Darwin's finch" not native to Galápagos (Chapter 5).

Proximate isolation: 361 m from Santa Cruz

Distance from center of archipelago: 29.3 km

Animals you will see: Finches, land iguanas, marine iguanas, lava lizards, red-billed tropic birds, lava herons, Galápagos herons, frigatebirds, swallow-tailed gulls, sea lions, blue-footed boobies, Nazca boobies

Plants you will see: Saltbush, prickly pear, palo santo, Jerusalem thorn, Crocker's *Scalesia,* annuals, and small vines

Most visitors first set foot on Galápagos at Baltra, a small, flat, rocky island near the center of the archipelago, just across Itabaca Channel from the northern coast of Santa Cruz. Baltra has the busiest airport in Galápagos; most flights from the mainland come to Baltra. Until 1986, Baltra was the only airport in Galápagos, but today visitors can also fly from the mainland to San Cristóbal, and from Baltra and San Cristóbal to Isabela.[6] The origin of Baltra's name is unknown.[7]

Unlike many islands in Galápagos, Baltra is not a remnant of a volcano; instead, most of Baltra was created by geologic uplift of underwater lava. As a result, Baltra is relatively flat instead of hilly like many of the other islands in Galápagos. Baltra's wildlife is less diverse than most other islands in Galápagos, and its primary vegetation is saltbush, prickly pear, and palo santo.

In 2009, GNPS annexed 73.9% of Baltra into GNP; the remaining area continues to be part of the airport and military base. Baltra's military base is closed to visitors, and there are no public accommodations or Visitor Sites on the island. That is, Baltra is not an island on which you can stay, nor is it a site you will tour. However, in 2012, GNPS was considering making at least one Visitor Site on the island.

Although Ecuador did not officially join the Allies in World War II until 1945, Baltra played a key role in the war (Chapter 2). Indeed, during World War II, Baltra was used as a U.S. Air Force base (complete with a bowling alley, beer garden, and "The Rock Street Theater") that housed soldiers who protected the Panama Canal and searched the eastern Pacific for enemy submarines. In 1946, when the war ended, the United States returned the air base and its associated facilities to Ecuador, which continues to use the island as a military base. You will see the foundations of several war-era buildings when you ride from the airport to the dock to board your tour boat or ferry to Santa Cruz.[8] Although

[6] In 2011, 78% of visitors to Galápagos arrived at Baltra's airport, and another 21% arrived at the airport on San Cristóbal (the remaining visitors arrived on private yachts). Ecuadorians comprised 35% of the visitors; U.S. residents accounted for 27%.

[7] Because it first appeared in *South America Pilot* (1928), the name "Baltra" is not—contrary to claims in several books—an acronym derived from Base Aerea para Logística y Transporte (as the island was sometimes referred to by American armed forces personnel living there during World War II; Chapter 2). The name Seymour is associated with Vice-Admiral Lord Hugh Seymour (1759–1801), a famed officer in the British Royal Navy, and his oldest son, Sir George Francis Seymour (1787–1870).

[8] Although today's tourists board their boats in Itabaca Channel between Baltra and Santa Cruz, military ships during World War II anchored in Aeolian Cove, an inlet on the western side of Baltra.

more than 2,400 U.S. military personnel and 750 Ecuadorans (hired as support staff) lived on Baltra when the base was most active (1942–45), fewer than 200 people stay on Baltra today. The base's mascot was a donkey, and the base's newspaper—the first ever published in Galápagos—was *The Goat's Whiskers*. These names are ironic, because feral donkeys and goats destroyed much of the wildlife on several islands of the archipelago (Chapter 11). Baltra's feral cats, a menace to lava lizards, land iguanas, and birds, were not eradicated until 2003.

When you land at Baltra, you will first go to the reception desks to pay your GNP entrance fee, obtain your Tourist Control Card, have your carry-on baggage inspected, and submit the customs forms you completed en route to the island.[9] (Keep all of the paperwork you are given when you arrive; you may need these forms when you depart the archipelago.) You will then find your luggage. If you booked a tour prior to arrival, a guide should be waiting for your flight. Just past the luggage pick-up area you will meet your guide, who will be holding a sign that advertises the name of your boat. Although the airport may seem chaotic, you will probably have no problem finding your guide. The Baltra airport has a reception area, bathrooms, mailbox, numerous shops and kiosks, restaurant, and a waiting area.

While you are near the terminal or standing in line at the reception desks, look around—you will likely see finches. You may also see an iguana or two outside. During Baltra's occupation during World War II, land iguanas on Baltra disappeared (Chapters 2, 6). Fortunately, however, 70 of these iguanas had been moved to nearby North Seymour in 1932. In 1980, several of these iguanas were moved from North Seymour to CDRS on Santa Cruz, and in 1991 the iguanas were reintroduced to Baltra. In 1997, scientists counted 97 iguanas on Baltra (13 of which had been born on Baltra), and in 2007 there were 420 iguanas living on the island. You may see some of these iguanas from the luggage pick-up area.

You will be on Baltra just long enough to gather your bags, find your guide, and take a bus to the docks. Most tour boats will be anchored at the dock, where you will see blue-footed boobies, sea lions, and maybe a heron at the water's edge. People going to Puerto Ayora will take a ferry across Itabaca Channel to Santa Cruz, after which they will take a half-hour ride by taxi or bus through the highlands to Puerto Ayora, the largest human settlement in Galápagos.

Mosquera Islet. Your first stop after boarding your tour boat may be Mosquera, a flat, narrow (150 m × 600 m) islet housing pelicans, herons, lava gulls, and sea lions. Like Baltra, North Seymour, Santa Fé, and Las Plazas, Mosquera was created by uplift of submerged sand and lava. You will access Mosquera's

[9] Unless they arrive on private boats, all visitors to Galápagos must go through either the Baltra or San Cristóbal airport (private boats can stay in any of the archipelago's 5 ports for up to 21 days). Unlike in past decades (Chapter 2), cargo boats cannot bring tourists to the archipelago.

white-sand beach via a wet landing, and snorkeling and swimming are the main activities. There is no trail on Mosquera; the entire islet is a Visitor Site. This is one of the best places to photograph sea lions. You may have had a good view of Mosquera during your descent to the Baltra airport.

Bartolomé

At a glance:

Other names: Bartholomew Island

Visited by Darwin: No

Area: 1.3 km^2 (0.5 mi^2)

Height: 114 m (374 ft)

Length of coastline: 6.07 km

Age: 1.5–2.0 million years old

Inhabited: No

Proximate isolation: 310 m from Santiago

Distance from center of archipelago: 28.2 km

Animals you will see: Galápagos penguins, hermit crabs, brown pelicans, green turtles, ghost crabs, lava lizards, sea lions, marine iguanas, Sally lightfoot crabs

Animals you might see: White-tipped reef sharks, sea turtles, Galápagos hawks, Galápagos snakes, spotted eagle rays, lava herons, black-tipped sharks

Plants you will see: Saltbush, lava cactus, *Opuntia,* beach morning glory, *Chamaesyce, Tiquilia, Scutia, Maytenus,* red mangroves, white mangroves, spiny plant, and mollugo. Although there is relatively little vegetation on Bartolomé, between the northern and southern beaches you will see several halophytes.

Bartolomé (Color Image 3) is a small, rugged, and desolate island on the eastern coast of Santiago (James) Island. Bartolomé, along with neighboring Sulivan Bay on Santiago Island, was named by HMS *Beagle*'s Captain Robert FitzRoy to honor Charles Darwin's friend Sir Bartholomew James Sulivan (1810–1890), the principal surveyor and second-lieutenant aboard HMS *Beagle.*[10] (Note that the correct spelling is Sulivan, not Sullivan.) Across the island are green, orange, red, and black volcanic formations.

Most tour boats anchor along the western end of Bartolomé near a 40-meter-tall, leaning spike that is one of the best-known landmarks in the archipelago: Pinnacle Rock. This large, black, eroded tuff cone formed when lava erupting from the volcano reached the ocean. When the water cooled the

[10] Some sources credit the name Bartolomé to Lt. David Ewen Bartholomew, who commanded HMS *Eurebus* in the early 1800s. However, this Bartholomew had no known association with Galápagos.

lava, the lava exploded, and particles from this explosion eventually formed the thin layers of Pinnacle Rock. You might recognize Pinnacle Rock from the 2003 movie *Master and Commander.*

There are two Visitor Sites on Bartolomé: (1) *La Playa,* a beach for snorkeling and bird watching, and (2) *Escalera,* a trail to the top of an extinct lava cone, from which you can see several nearby islands.

La Playa. Pinnacle Rock is near two beaches that are separated by a small sand dune and strip of mangroves. At the northern beach, you will have an easy wet landing on a brown sand beach; snorkeling there introduces you to rays, sharks, sea lions, and what may be your best chance to swim with Galápagos penguins (*Spheniscus mendiculus*), the only penguins that live north of the equator.

Your guide will probably ask you to leave your snorkeling gear on the north beach before you hike about 100 meters over a small sand dune through white mangroves, leatherleaf, matplant, and spiny plants to another beautiful, sandy beach on the southern side of the island. Along the way, look for *Tiquilia,* the white-flowered *Chamaesyce* (whose seeds are eaten by finches), herons, oystercatchers, and Galápagos hawks. This narrow beach, which is part of Sulivan Bay, is home to *Sesuvium,* beach morning glory, and turtle nests. You can't snorkel or swim at this beach, but you will probably see spotted eagle rays, great blue herons, and white-tipped sharks close to shore; if you visit Bartolomé between December and March, you may also see nesting sea turtles. You will probably see penguins when you return to the northern beach and take a panga ride to the west side of Pinnacle Rock. Pinnacle Rock is also a favorite site for hungry blue-footed boobies; watch them diving to catch fish. It is common to see boobies near penguins, for the penguins' pursuits render exhausted fish that are easy prey for the diving boobies (Chapter 5). The rock in the middle of Sulivan Bay is San Felipe Rock.

Escalera. You will then ride your panga to a dry landing at a rock and concrete jetty, where a boardwalk will lead you to the most famous vista in the archipelago: the view toward Santiago of Sulivan Bay and Pinnacle Rock. To best enjoy this view, you will first hike 795 meters along a wooden stairway to the 114-meter-high summit of a once-active volcano. The 30-minute ascent up the 397 steps along the eastern and southern sides of the volcano takes you past several small cones and lava tubes; the landscape is stark, and visitors often report feeling like they are walking on the moon. This hike is strenuous only because it is steep; there are no slippery rocks to climb, nor are there any pools of water to wade through. If you get tired, you can stop and enjoy the stark, ethereal landscapes along the boardwalk, as well as examine the nearby pioneer plants (e.g., lava cactus, *Tiquilia, Chamaesyce*) along the slope. Be sure to stay on the boardwalk; this will give you safe footing and reduce erosion of the ash. Near Bartolomé's summit, it is often windy, so take a windbreaker. On clear nights, the beacon atop Bartolomé can be seen from several miles away.

Your ascent will lead you past several overlooks. At the first overlook, you will see matplant, *Brachycereus,* and the paths of lava flows down the north side of the volcano. At the second overlook, you can examine spatter cones. The third overlook is between two spatter cones; from here you will have a beautiful view of Daphne Major and Daphne Minor. You can also find large blocks of lava nearby that are surprisingly easy to lift because of their many gas bubbles.

The panoramic view from atop the volcano is the most photographed (and most famous) vista in the archipelago. It is breathtaking—it includes lava formations, cinder cones ejected by the volcano, Santiago Island, James Bay to the west, and towering Pinnacle Rock. To the southeast are Baltra, Santa Cruz, and North Seymour, and to the southwest is Rábida. Tens of islets and rocks protrude from the Pacific's surface. At shops in Galápagos, you can buy postcards of the view to send to your friends, telling them, "I was here."

Daphne Major

At a glance:

Other Names: Daphne Island

Visited by Darwin: No

Area: 0.33 km^2 (0.12 mi^2)

Height: 123 m

Length of coastline: 2.14 km

Age: 1.8 million years old

Inhabited: No

Proximate isolation: 7.6 km from Santa Cruz

Distance from center of archipelago: 20.8 km

Animals you will see: Blue-footed boobies, Nazca boobies, red-billed tropicbirds, finches

Plants you will see: Palo santo, chamaesyce, *Opuntia,* Galápagos purslane, gray matplant, puncture weed

Approximately 10 kilometers west of Baltra (and directly north of Santa Cruz) is Daphne Major. This island, which is the size of about 80 football fields, was formed by a now-extinct volcano, and retains its distinctive volcano-like shape. The smaller, older, and heavily eroded Daphne Minor (whose area is only 0.07 km^2) is 6 kilometers northeast of Daphne Major.[11] Both of these islands were named in honor of HMS *Daphne,* an 18-gun ship that policed the Pacific and Indian Oceans in the 1800s.

Although GNPS recently opened Daphne Major to limited access by the public, these visits are restricted to only one visit per month by approved boats

[11] Daphne Minor is also referred to as Daphne Chica.

(typically special charters) carrying fewer than 12 passengers. The dry landing on Daphne Major can be especially challenging, and it is best done in calm seas. (The site of the dry landing is marked by GNPS black-and-white stakes.) You will then climb a steep 350-meter trail along the eastern side of the island to the 123-meter-high summit. Craters atop the cone house hundreds of blue-footed boobies; red-billed tropicbirds nest in crevices, and Nazca boobies live along the crater's rim. Although visitors are not allowed inside the crater, this island is one of the best places to photograph blue-footed boobies. The palo santo (*Bursera malacophylla*) on Daphne Major is endemic to Daphne Major, Daphne Minor, North Seymour, and Baltra.

Despite its limited access, Daphne Major is one of the most famous islands in the Galápagos, for it is the site of the decades-long studies of finches by Peter and Rosemary Grant and their colleagues (Chapter 5). Although your captain will probably not dock near Daphne Major, he or she may circumnavigate the island to give you a look at its plants and animals. You will see Nazca boobies, blue-footed boobies, red-billed tropicbirds, and eroded tuff formations.

Darwin

At a glance:

Other names: Culpepper Island

Visited by Darwin: No

Area: 1.1 km^2 (0.4 mi^2)

Height: 168 m

Length of coastline: 5.41 km

Age: 0.4–1.6 million years

Inhabited: No

Proximate isolation: 175.2 km from Pinta

Distance from center of archipelago: 294.3 km

Animals you will see: Galápagos sharks, reef fish, shearwaters, fur sea lions, swallow-tailed gulls, Nazca boobies, dolphins, sea turtles, frigatebirds, manta rays, vampire finches, sooty terns, scalloped hammerhead sharks, marine iguanas, and sea lions. There are no terrestrial reptiles on Darwin Island.

Animals you might see: Whales, whale sharks

Darwin Island, the northernmost island in Galápagos, is named in honor of Charles Darwin, despite the fact that Darwin never visited the island.[12] The island is an eroded volcano that, like Wolf Island to the south, has vertical cliffs that make landings difficult; this explains why the first visitors to the island did not set foot on Darwin Island until 1964, when park officials landed on

[12] In 1684, Ambrose Cowley had named this island after Thomas, Lord Culpepper (1635–1689), the governor of colonial Virginia from 1677–1683.

the island via helicopter. The island's peak reaches only 168 meters above sea level, but the volcano itself rises more than 1,000 meters above the ocean floor. Several smaller volcanoes nearby—some only 30,000 years old—do not reach sea level. Darwin, which is 160 kilometers north of Isabela, is 3 hours (by boat) north of Wolf Island, its nearest neighbor.

If you visit Darwin Island, you will be there to dive (there are no trails on the island, and landing on the island is prohibited). The best sites for diving are along the southeastern and northeastern sides of the famed "Darwin's Arch," a small tuff-arch that sits atop a wide rock plateau just off the southeastern coast of Darwin Island (Appendix 5). The diversity of life here is stunning. Hammerhead sharks (that are often spooked by divers' bubbles) will venture close to you, and sometimes swim over the reef where they are cleaned by king angelfish. Along the barnacle-covered wall you will find rainbow chubs, Moorish idols, creole fish, green spotted eels, striped snappers, coronet fish, trumpet fish, parrotfish, flounders, and large schools of bigeye jacks. Whale sharks are common around Darwin Island between June and October. Sooty terns breed only on Darwin Island, where surface waters are often rough, and currents can change quickly.

Darwin Island, which is more than 400 kilometers north of Española, is affected relatively little by the Humboldt Current. As a result, waters around Darwin are warmer than those of the central islands (24–28°C during December–April, and 22–26°C during May–November). There are also more corals around Darwin Island than around the southern islands of Galápagos. Darwin and Wolf Islands house the largest colonies of Nazca boobies, which you may see in feeding frenzies with sharks and dolphins.

Given their remote location, both Darwin and Wolf Island are vulnerable to unregulated fishing and tourism. To improve its ability to patrol and manage the area, GNPS in 2008 established a permanent floating base at Wolf Island.

Española

At a glance:

Other names: Hood Island

Visited by Darwin: No

Area: 61 km^2 (23 mi^2)

Height: 213 m

Length of coastline: 42.8 km

Age: 3.3–4.0 million years old

Inhabited: No

Proximate isolation: 45 km from San Cristóbal

Distance from center of archipelago: 132 km

Animals you will see: Hood (Española) mockingbirds, Audubon's shearwaters, ghost crabs, warbler finches, small ground finches, large cactus finches, lava herons, Sally lightfoot crabs, Galápagos sea lions, Galápagos doves, Galápagos hawks, Nazca boobies, marine iguanas, lava lizards, red-billed tropicbirds, blue-footed boobies

Animals you might see: Blue-footed boobies, nesting marine iguanas, waved albatross (April–January), American oystercatchers, yellow-crested night herons, swallow-tailed gulls

Plants you will see: Saltbush, mesquite, Galápagos carpetweed, *Parkinsonia,* Galápagos shore petunia, *Opuntia,* puncture weed, Galápagos lantana, beach morning glory. Most of the vegetation on Española is typical of transitional and arid zones.

Española is the southernmost and oldest remaining island in Galápagos.[13] Parts of the small, relatively flat island were formed by uplift, and others by volcanic activity. Española (Hood) Island is named in honor of Spain and after Viscount Samuel Hood (1724–1816), a famed British admiral who served in the American War of Independence.

Española's remote location in Galápagos helps explain why it has so many endemic species (e.g., lava lizards, mockingbirds). You will also see brightly colored marine iguanas and waved albatrosses, whose feeding grounds are near the mainland of Ecuador and Peru. Española's iguanas are the only iguanas in Galápagos that change color during their breeding season. Adult marine iguanas on Española normally have a reddish color, but during mating season they also develop green patches of skin. Española's tortoises live in the highlands, so you will not see them at either of the island's Visitor Sites.

Gardner Bay. Along the northeastern shore of Española is Gardner Bay, whose turquoise waters wash a long, beautiful, tan-colored sandy beach that offers excellent swimming and snorkeling. The crescent-shaped, 1.3-kilometer-long beach, which faces the islet of Gardner-near-Española (as opposed to Gardner-near-Floreana), is all you will find at this Visitor Site; there are no inland trails. When you arrive you will disembark your panga via a wet landing and spend most of your time along the beach. You will see many sea lions (some may swim with you), as well as Galápagos hawks, Galápagos doves, three endemic species of finches, extroverted Hood (Española) mockingbirds, marine iguanas, oystercatchers, and sea turtles just offshore. Plants common here include saltbush, prickly pear, Jerusalem thorn, and puncture weed.

Gardner Islet. Snorkeling is also good at nearby Gardner Islet, where caves harbor numerous invertebrates such as variegated sea stars and cushion sea stars. You will meet sea lions and schools of large fish, including yellow-tailed surgeonfish, bumphead parrotfish, and king angelfish.

---

[13] There have been many islands in Galápagos much older than Española, but erosion and plate tectonics have sunk these islands below sea level (Chapter 3).

Tortuga Rock. More excellent snorkeling is just offshore at Tortuga Rock, an islet that looks like a turtle—that is, a *tortuga*—coming up for air. If you swim to Tortuga Rock (it's about a 10-minute swim from the beach), you will pass over several large rocks littered across a sandy bottom. Watch for stingrays resting in the sand. Along the side of Tortuga Rock facing the beach, you will find calm waters housing surgeonfish, wrasses, parrotfish, Moorish idols, and—if you're lucky—the well-camouflaged scorpionfish. The opposite side of Tortuga Rock (i.e., the side farthest from shore) has deeper water and stronger currents. There, you may see white-tipped reef sharks and manta rays.

Punta Suárez. Punta Suárez, which is located along the northwestern tip of Española and about an hour-long boat ride from Gardner Bay, offers some of the most diverse and abundant wildlife (especially birds) in the archipelago. The Punta Suárez trail, which begins near a small lighthouse and monument, is one of the most enjoyable and memorable hikes in Galápagos. The rate of endemism at Punta Suárez is among the highest of all Visitor Sites in Galápagos; this site is a great place to see Galápagos doves, mockingbirds, and finches. Common plants here include saltbush, mesquite, *Lycium, Sesuvium,* and puncture weed.

The 1.7-kilometer hike at Punta Suárez will lead you from one end of the island to the other and back, and begins with a dry landing at a concrete jetty. Here, you may be greeted by bodysurfing sea lions. The trail then leads you south to a small, white-sand beach, beyond which you will see many lava lizards (the largest of the seven species endemic to Galápagos), inquisitive mockingbirds, swallow-tailed gulls, Galápagos doves, blue-footed boobies, Nazca boobies, red-billed tropicbirds, Galápagos hawks, three species of finches, and red-and-green-colored marine "Christmas" iguanas. This will likely be your best chance to photograph many of these animals, especially Nazca boobies along the gray cliffs. (These wave-eroded cliffs reveal the layers of lava that formed the original shield volcano.) If any of the Nazca boobies or blue-footed boobies have abandoned their eggs, you may see the eggs being scavenged by mockingbirds (the largest species of mockingbird in Galápagos). The cliffs, which are about 20 meters above the crashing waves, provide a beautiful background for photos. Along the trail by the cliffs you will see Española's famed blowhole ("El Soplador"), where water sprays more than 15 meters into the air. You will then ascend the trail to open regions that cover the island.

If you visit Española between April and January, you will then be treated (after about 300 m of hiking) to a breeding colony of waved albatrosses.[14] These

[14] Contrary to many claims, the birds you will see along the trail do not comprise the world's entire population of waved albatrosses. Indeed, most of the waved albatrosses in Galápagos nest at Punta Cevallos on the southern side of Española, and another tiny population of these birds nests on Isla de la Plata, a small island that is a half-hour boat ride off the coast of mainland Ecuador. Isla de la Plata, which is part of Parque Nacional Machalilla (the only coastal national park in mainland Ecuador), is often referred to as "Poor Man's Galápagos" because it showcases blue-footed boobies, Nazca boobies, red-footed boobies, *Opuntia,* sea lions, and other wildlife often associated with Galápagos. Parque Nacional Machalilla, like Galápagos, is a UNESCO World Heritage Site.

large, beautiful birds, which arrive *en masse* in April to begin breeding, may be displaying their ritualized fencing-like mating behavior, caring for their eggs, or nurturing their large, down-covered chicks (Chapter 5). You may also see their graceful take-offs from the nearby cliffs (i.e., "The Albatross Runway"), or their awkward, clumsy landings. If you visit Española between January and March, you may see only one or two albatrosses. The population of waved albatrosses along the trail has decreased in recent years (the number of abandoned eggs there has also increased), presumably because of human impact.

Just past the albatross's nesting grounds the trail becomes more challenging and is strewn with volcanic boulders. You will be surrounded by tall vegetation, and a nearby open area (by an especially large rock) is home to a small colony of Galápagos hawks. On the ground you might see lava lizards and Galápagos doves, which have distinctive blue circles around their eyes.

You will then walk along the beach, where you will see blue-footed boobies, Sally lightfoot crabs, sea lions, marine iguanas, and mockingbirds. In the bay you might see sea turtles gulping air. The hike at Punta Suárez will take about three hours.

Fernandina

At a glance:

Other names: Narborough Island

Visited by Darwin: No

Area: 642 km^2 (248 mi^2)

Height: 1,463 m

Length of coastline: 114.6 km

Age: 700,000 years old (and still forming)

Inhabited: No

Proximate isolation: 4.02 km from Isabela

Distance from center of archipelago: 112.5 km

Animals you will see: Galápagos sea lions, Galápagos hawks, lava lizards, marine iguanas, Galápagos penguins, striated herons, flightless cormorants, Sally lightfoot crabs, yellow-crowned night herons, great blue herons, lava herons, pelicans, fur sea lions, yellow warblers, and American oystercatchers

Animals you might see: Whales, dolphins, blue-footed boobies, Galápagos snakes, octopus

Plants you will see: Mangroves (red, black, and white), lava cactus, saltbush, and Galápagos shore petunia. Because of Fernandina's recent eruptions, the diversity of plants on the island is relatively low. Vegetation, which is arid and sparse, is concentrated in *kipukas,* which are areas untouched by recent lava flows. Humid-zone plants grow in the southern and eastern highlands, and the coast houses several forests of mangroves.

Fernandina is an island west of Isabela named for King Ferdinand II of Aragon (1452–1516), who—as the husband (and second cousin) of Queen Isabella—sponsored Columbus's voyage to the New World. In 1684, Cowley originally named the island Narborough to honor British Admiral Sir John Narborough (1640–1688), who had been appointed a commissioner of the navy four years earlier. The island is dominated by the 6.5-kilometer-wide, 270-meter-deep, and 1,463-meter-high Volcán La Cumbre and surrounding lava fields. The only documented collapse of a caldera in Galápagos occurred in 1968 on Fernandina.

Fernandina is the westernmost, most volcanically active (it sits atop the Galápagos Hotspot; see Chapter 3), and youngest of the central islands in Galápagos. The island, which is separated from Isabela by the 4.8-kilometer-wide Bolivar Channel, is still being formed; at least 11 eruptions have occurred since 1813, and fumaroles of steam from the volcanoes are often visible from passing boats. Like the volcanoes of Isabela, Fernandina's Volcán La Cumbre is shaped like an inverted soup bowl, with a large, central caldera. Snorkeling along Fernandina's eastern shore is some of the best in the archipelago; you will see many types of fish, as well as penguins and flightless cormorants.

Marine life around Fernandina is especially rich because of upwelling by the Humboldt and Cromwell Currents (Chapter 3), and includes penguins, flightless cormorants, dolphins, and the largest population of whales (orca and humpback) in the archipelago. Because of its remote location, Fernandina is one of the least-visited and most pristine (i.e., least disturbed) islands in Galápagos. Most lava on Fernandina is a'a' lava; pahoehoe lava is largely restricted to vents along the coastal plain (Chapter 3). Many geologists visit Cabo Douglas (at the northwestern tip of Fernandina) to study volcanic activity in the archipelago.

Punta Espinosa. Fernandina has only one Visitor Site: Punta Espinosa. Punta Espinosa is a rocky peninsula of lava, extending eastward from the base of Volcán La Cumbre along the northeast edge of Fernandina (i.e., just across from Isabela's Tagus Cove). On your way to shore, watch for sea turtles. On the southern side of Punta Espinosa is a small, mangrove-lined cove whose narrow strip of black lava and sand is accessed via a dry landing onto either a dock or slippery volcanic rocks.[15] You will hike east 300 meters past green mangroves to a brown beach and lava area where hundreds of marine iguanas gather for mating during the first months of the year. These iguanas are not as colorful as those on Española, but their numbers are impressive. Indeed, there are more marine iguanas (as many as 3,000 iguanas per square kilometer) on Fernandina than on any other island; you will have to be careful not to step on them or

[15] In 1975, the landing area was uplifted by almost a meter, thereby making the dry-landing site inaccessible to boats it was built to serve. This pier is now used only for dry landings during high tides.

their nests (which can be identified as holes or depressions in the sand). The marine iguanas at Punta Espinosa are among the largest in Galápagos. Fernandina also has a large population of land iguanas, but their favorite nesting area (along the rim of the volcano) has been disturbed by recent eruptions (some of the northern lava flows are less than 200 years old). Punta Espinosa is one of the best places to see lava cactus (*Brachycereus nesioticus*), a pioneer plant that grows in lava fields (Color Image 43).

You will hike past some whale bones gathered by park guides; these bones stand out prominently on the black lava. The trail then splits. The trail to the right leads you past some iguana burrows and along beaches and tidal pools, where you will see more marine iguanas. You will also see sea lions, pelicans, and frigatebirds, after which you will see the world's only species of flightless cormorants near the tip of the site. These curious birds nest on rocks just above the high-tide line. This may be the only place you see these memorable birds (they live only in the western islands of Galápagos).

You will pass a small cove and again pass the whale bones as you hike 0.7 kilometers along a different trail through a field of pahoehoe and a'a' lava. If you visit during low tide, you will see rays, sea lions, sea turtles, and other animals frolicking in the tide pools. If your hike ends with snorkeling, you might also see sea turtles, penguins, flightless cormorants, and marine iguanas eating algae. Your hike at Fernandina—which many visitors consider the most beautiful place in Galápagos—will take two to three hours.

Floreana

At a glance:

Other names: Charles Island, Santa María

Visited by Darwin: Yes

Area: 174 km^2 (67 mi^2)

Height: 640 m (2,099 ft)

Length of coastline: 61.4 km

Age: 0.7–1.5 million years old

Inhabited: Yes; population = 160

Proximate isolation: 49.2 km from Santa Cruz

Distance from center of archipelago: 84.3 km

Animals you will see: Galápagos sea lions, finches, willets, black-necked stilts, ruddy turnstones, white-cheeked pintail ducks, hermit crabs, common noddies, yellow warblers, herons, rays, storm petrels, lava lizards, Galápagos flycatchers, brown pelicans, marine iguanas, sea turtles, and tortoises from other islands. Although Darwin in 1835 commented that tortoises were abundant on Floreana (e.g., ships were taking as many as

200 tortoises per day), the Floreana race of giant tortoises was extinct by 1846.

Animals you might see: White-tipped reef sharks, stingrays, flamingos, medium tree finch

Plants you will see: Mangroves, leatherleaf, *Parkinsonia,* mesquite, *Scalesia,* croton, palo santo, saltbush, Galápagos lantana, beach morning glory, passionflower, and Galápagos cotton

Floreana (Santa María) is an isolated, rectangular (11 × 16 km) island west of Española and south of Santa Cruz. Santa María was the name of one of Columbus's boats, and the more common name Floreana honors Juan José Flores, the first president of Ecuador and "The Founder of the Republic."[16] Floreana is an active volcano; in July, 1813, Captain David Porter saw the volcano erupt, and the island's highlands are dotted with several young pyroclastic cones. Floreana, which was the first island of Galápagos colonized by humans (Chapter 2), has lost more endemic species than any other island in Galápagos.[17] Floreana was the only inhabited island when Darwin was in Galápagos.

Although Floreana today is home to only about 160 people, it is famous for its many legends, scandals, unexplained deaths, and landmarks (Chapter 2). It also has scientific significance: Charles Darwin's visit to Floreana was his most important stop during the voyage of the *Beagle,* for it was there that Darwin first suspected that each island of Galápagos had unique mockingbirds. (It was mockingbirds, not finches, that helped Darwin understand that distinct species live on different islands; Chapter 2.) Unfortunately, these mockingbirds are now gone from the island.[18] Floreana was also where Darwin was first told that tortoises on each island have unique shells (Chapter 2).

Floreana is a four-hour boat ride from Santa Cruz and the same distance west of Española. Floreana, which consists of several eroded volcanic hills, is one of the few islands in Galápagos with a reliable supply of fresh water; there is a spring in the lush southeastern highlands. (This water and Floreana's nutrient-rich soil help explain Floreana's long history of human colonization; see

[16] Flores (1800–1864) served as Ecuador's first supreme chief and president from 1830 to 1834 and 1839 to 1845. He died of uremia in 1864 while under house arrest during the presidency of Gabriel Garcia Moreno (who in 1875 was assassinated in Quito). Numerous books and charts have claimed that the island christened by English buccaneer William Ambrose Cowley as "King Charles the Second's" Island was present-day Floreana. However, Cowley never came close to Floreana. Instead, the island that Cowley named "King Charles the Second's" Island is present-day San Cristóbal.

[17] In 2007, GNPS began preliminary work on Project Floreana, a five-year, $1.7-million project aimed at stabilizing Floreana's economy and environment, managing its conservation effectively, and protecting the island against invasive species.

[18] The Charles mockingbird, with its distinctive beak, was abundant when Charles Darwin visited Floreana in 1835, and prompted Darwin to collect mockingbirds from other islands that he visited. However, this mockingbird has since been driven to extinction on Floreana itself by introduced predators, and today lives only on Champion (700 m from Floreana) and Gardner (8 km from Floreana) Islands. Gardner is the only place in Galápagos where the three species of boobies nest side-by-side. A 2007 survey reported 38 Charles mockingbirds on Champion Island and 98 Charles mockingbirds on Gardner Island (Chapter 5).

Chapter 2). Floreana's highest peak is Cerro Pajas (640 m), a cinder cone that is populated by the invasive shrub *Lantana camara.*

Approximately 65% of Floreana's residents live in Puerto Velasco Ibarra, a small town near Cerro Pajas on the island's west coast that has one hotel, one school, and one church. The western side of the island features bare lava flows and a black beach. In recent years, several cabins for tourists have opened at the southern end of Black Beach.

Punta Cormorán. Along the northern coast of Floreana is Punta Cormorán, a steep lava cone named for a ship in the U.S. Navy (and not for the islands' famous flightless bird). Your tour of Punta Cormorán begins with a wet landing on a beach made of volcanically derived brown sand and olivine crystals rich in silicates of iron and magnesium. These translucent crystals give the beach a greenish tinge. You will likely encounter sea lions and introduced wasps (be careful!) in vegetation near the landing site.

About 710 meters from the beach the trail splits. Hiking to the right takes you along the saltbush-lined beach, after which you will turn inland to see the bushy cut-leaf daisy (*Lecocarpus pinnatifidus,* which lives only on Floreana). About 100 meters inland you will find a hyper-saline lagoon that is surrounded by hills covered in palo santo. This lagoon often houses flamingos, but they are usually too distant for good photographs. If you visit between February and April, you might see the flamingos' memorable courtship rituals. Other common birds here include stilts, oystercatchers, willets, white-cheeked pintail ducks, sanderlings, phalaropes, and western sandpipers.

The other path features inland vegetation, including cut-leaf daisy, leather leaf, palo verde, passion flower, *Nolana galapagensis,* chala, algarrobo, passion flower, palo santo, *Scalesia vellosa, Lecocarpus,* and *Lantana* (endemic *Lantana* is now threatened by an introduced *Lantana camara;* see Chapter 10). After hiking for about 350 meters, you will approach powdery white-sand "Flour Beach" made of pulverized bits of coral; you might see turtles and stingrays feeding near the shore. If you look west from Flour Beach, you will see flat Champion Island and the more-distant (and pointed) Enderby Island. If you wade into the water, be sure to shuffle your feet to urge the stingrays to leave.

Devil's Crown. About 250 meters north of Punta Cormorán is Devil's Crown (Figure 4.1), a submerged volcanic cone that erosion has transformed into a semicircle of jagged peaks (hence the name "crown").[19] This is a Marine Visitor Site; it has no trails, but it is one of the best snorkeling sites in the archipelago. Look for hawkfish, king angelfish, yellowtail grunts, white-tipped sharks, wrasses, sea cucumbers, sea stars, tiger snake eels, and eagle rays. Be careful and heed the advice of your guide—currents are often strong and unpredictable.

[19] Devil's Crown is also referred to as Onslow Island.

FIGURE 4.1 Devil's Crown is a popular dive site and landmark in Galápagos. The "crown" is the circular jagged peak of a submerged volcanic cone just north of Floreana Island. Although currents here are often strong, many people consider Devil's Crown to be the premier site for snorkeling in the archipelago.

When snorkeling you might see some hammerhead sharks far below. Don't worry—they will not bother you, and they will probably leave when they sense your boat's presence. Coral is abundant at Devil's Crown, especially inside the crown, where you might also see surgeonfish, wrasses, angelfish, blennies, parrotfish, and other colorful fish. On the rocks you will see pelicans, herons, lava gulls, frigatebirds, Galápagos shearwaters, swallow-tailed gulls, and red-billed tropicbirds, as well as *Opuntia* and candelabra cactus.

Post Office Bay. Post Office Bay at the northern end of Floreana (and just west of Punta Cormorán) is another place where you can retrace Charles Darwin's footsteps in Galápagos. Unlike most other Visitor Sites in Galápagos, Post Office Bay is visited for its human history (Chapter 2). After a wet landing on a brown sand beach, a trail leads you to the famed "Post Office" (Figure 4.2). Feel free to leave postcards (no postage necessary!) for friends, and pick up cards for people living near your home; when you get home, either mail the cards or—better yet—deliver them to their destination.[20] When you go ashore, avoid wearing bright colors or strongly scented sunscreen; introduced wasps (*Polistes versicolor*) in this area are attracted to strong scents and bright colors. Common plants here include Galápagos cotton, mesquite, palo santo, algarrobo, *Parkinsonia, Lantana,* and saltbush.

After visiting the Post Office barrel, some tours continue inland on the trail to the remains of Casa Matriz, a commercial fishing company (Chapter 2). This Norwegian business began in 1927, but was abandoned soon thereafter; the company's

[20] Over the years, we have tracked hundreds of cards sent from mainland Ecuador and the Galápagos to the United States. The average times for their trips from various locations has been as follows: Baltra—42 days (range: 14–74); Santa Cruz—53 days (range: 18–67); Isabela—37 days (range: 17–46 days); San Cristóbal—38 days (range: 13–67); Floreana's Post Office Barrel—39 days (range: 6–884); Guayaquil—14 days (range: 10–18); Quito—15 days (range: 10–19). As of 2012, stamps for postcards were $2 each; cards mailed at Post Office Bay require no postage.

FIGURE 4.2 The famed Post Office is a popular Visitor Site on Floreana Island. The Post Office remains functional; letters to anywhere in the world can be left for the next passing ship. In the background is a trail leading to vats and other remains of Casa Matriz, a short-lived fishing business established on Floreana in 1927 (see Figure 2.6). The Post Office is the only site in Galápagos where graffiti is still acceptable. In 2012, GNPS placed a new post office barrel at Post Office Bay.

giant, rusting vats still litter the area (Figure 2.6). About 100 meters beyond the barrel is a lava tunnel; if you enter the tunnel, be sure to take a flashlight. You will descend about 6 meters, after which you will hike to the tube's other end. Your walk through the tube and back will take about 45 minutes. The trail to the lava tunnels was once the trail to Floreana's highlands, but that trail now begins at Puerto Velasco Ibarra. While you are at the Post Office Barrel, the crew of your boat may be playing soccer at the soccer field just north of the landing area.

Champion Island. Less than a kilometer from Floreana's northeastern shore is Champion Island, a small, circular, low-lying tuff cone covering about 0.1 km^2. This islet is one of the last two habitats of the critically endangered Charles mockingbird (*Mimus trifasciatus*). Champion Island, which is renowned for its snorkeling and scuba diving, is also home to penguins, sea turtles, Galápagos sharks, eels, mantas, sea stars, wrasses, and other marine creatures.

Mirador de la Baronesa. Mirador de la Baronesa ("The Baronesa's Viewpoint"; Color Image 4) is a tuff cone between Punta Cormorán and Post Office Bay.

Plants you will see here include *Cordia leucophylyctis, Lantana peduncularis,* and *Plumbago scandens.* In addition to providing a beautiful view of La Lobería and Floreana, this site is historically important: Baroness Eloisa von Wagner (Chapter 2) came here to monitor approaching ships. During your climb to the overlook, you will pass a solar-powered beacon marked "Post Office" with GPS coordinates ("Lat. 01°14.5'S, Long. 90°29.5W"). Views from Mirador de la Baronesa are often spectacular at sunset.

La Lobería. Near the Mirador de la Baronesa is La Lobería, where you will see sea lions, sea turtles, *Opuntia,* red mangroves, and marine iguanas. At low tide, you might see sea lions with their pups in the tide pools. A small colony of penguins lives here; you may see them late in the day.

Puerto Velasco Ibarra and Wittmer Pension. Not far from Mirador de la Baronesa is Wittmer Pension and the small village of Puerto Velasco Ibarra at Black Beach. These sites are near the island's original settlement (Chapter 2). Your visit will begin with a dry landing at a trail leading to the village of about 160 residents. The hotel is named for Margret (often misspelled Margaret) Wittmer, a pioneer who arrived in Galápagos with her husband and stepson in 1932 (Chapter 2). Staff at Wittmer Pension, now operated by Margret Wittmer's daughter and grandchildren, will stamp your postcards. Wittmer's book *Floreana* (originally titled *Postlagernd Floreana,* after Floreana's famous post office barrel) describes Margret's early days on Floreana, as well as a series of mysteries, disappearances, and unexplained deaths of other settlers (Chapter 2).

Asilo de la Paz. From Wittmer Pension it is easy to go 8 kilometers into the highlands and visit Asilo de la Paz (Haven of Peace),[21] where you will find the island's primary source of fresh water. This so-called Governor's Drip (its flow is more like that of an open faucet) was often visited by pirates, whalers, and others to collect water; nearby you will find a cave used by pirates, Patrick Watkins, Dr. Ritter and Dore Strauch, and the Wittmer family when they settled on Floreana (Figure 4.3; Chapter 2). Near Asilo de la Paz is a corral built by GNPS where you can see tortoises from several different islands in the archipelago. This pen is merely a tourist attraction; there is not a tortoise-breeding program on Floreana.

Cerro Alieri. Just east of Puerto Velasco Ibarra is the 340-meter-high Cerro Alieri. Vegetation at this site is impressive; of the 48 different species here, approximately 56% are native, and 33% are endemic. Plants of interest here include the threatened *Linum cratericola* (once believed to be extinct, and today having only about 40 individuals remaining), *Darwiniothamnus tenufolius,*

[21] The source of the fresh water at Asilo de la Paz (Asylum of Peace) is about the size of a long panga.

FIGURE 4.3
This cave in Floreana's highlands was home to buccaneers, pirates, and the Wittmers. The cave includes a fireplace and benches carved into the lava.

Lippia salicifolia, Lecocarpus pinnatifidus, and *Scalesia pedunculata.* Your visit to Cerro Alieri will take about two hours.

Genovesa

At a glance:

Other names: Tower Island

Visited by Darwin: No

Area: 14 km^2 (5.4 mi^2)

Height: 64 m

Length of coastline: 23.1 km

Age: 700,000 years[22]

Proximate isolation: 46.8 km from Marchena

Distance from center of archipelago: 113.6 km

Animals you will see: Red-footed boobies, swallow-tailed gulls, Nazca boobies, marine iguanas, wedge-rumped storm petrels, wandering tattlers, Audubon shearwaters, lava herons, large ground finches, warbler finches, white-cheeked pintail ducks, blue-footed boobies, black-crowned night herons, fur sea lions, great and magnificent frigatebirds, Galápagos sea lions, Galápagos doves, Galápagos owls, and Genovesa mockingbirds (the smallest of the Galápagos mockingbirds).

Animals you might see: red-billed tropicbirds, yellow-crowned night herons, short-eared owls, large cactus finches, whimbrels, semi-palmated plovers, and ruddy turnstones. Marine iguanas here are the smallest in Galápagos (some of the largest are on Fernandina).

Plants you will see: Mangroves, palo santo, *Opuntia,* Galápagos cotton, *Cordia,* saltbush, Galápagos shore petunia, Galápagos croton, and lava cactus. Most vegetation of Genovesa is arid-zone vegetation.

---

[22] Much of the lava on Genovesa is so low in potassium that it is difficult to estimate its age with potassium-argon radiometric methods.

Genovesa, which was described by William Beebe as "dead cinders of the world's end," is a flat, low-lying, horseshoe-shaped island named for Genoa, Italy, the supposed birthplace of Christopher Columbus. The 3.2-kilometer-wide island, which is located in the northeastern part of the archipelago, is the remaining rim of a giant, submerged volcanic crater. The crater of the shield volcano in the center of Genovesa is 60 meters deep and 600 meters across, and at the bottom of the crater is a 350-meter-wide lake. Most boats do not visit Genovesa because it is somewhat remote—it takes six hours to reach Genovesa from Itabaca Channel, and the seas are often rough. Unlike the central islands of the archipelago, lava on Genovesa is not from the Galápagos Hotspot, but instead resembles that produced by the mid-oceanic ridge of the Galápagos Spreading Center.

Genovesa, like Española in the south, is a refuge for seabirds; this explains why guides often refer to Genovesa as "Bird Island." On Genovesa you will see Galápagos doves, storm petrels, four species of finches, lava gulls, red-footed boobies, noddy terns, red-billed tropicbirds, and nesting frigatebirds. There are no land reptiles on Genovesa. Genovesa's isolation, as well as its lack of fresh water, explain the absence of introduced species on this island.

Darwin Bay. Your boat will anchor along the northeastern side of Darwin Bay, which is an ocean-filled caldera of an extinct (and mostly submerged) volcano. The caldera opens to the sea to the south via a shallow channel; boat captains navigate this channel and enter the eastern side of Darwin Bay by lining up the two metal towers on the bay's northern shore.[23] Cliffs bordering the landing site are adorned by *Cordia* and graffiti left by earlier visitors (e.g., *Gracious* in 1958, *Velero III* in 1932). A wet landing on the tan coralline beach leads to nests of great frigatebirds and swallow-tailed gulls with their squeaking chicks. A flat trail then leads you past tidal lagoons and mangroves. Near the beginning of the 0.8 km trail you will find red-footed boobies nesting in red mangroves, saltbush adorned with nests of frigatebirds, blue-footed boobies, yellow warblers, night and yellow-crowned herons, whimbrels, wandering tattlers, and more swallow-tailed gulls. Be careful here—some birds may be nesting on the trail. You will see many instances of red mangroves colonizing the bare lava. Red-footed boobies thrive here, presumably because there are no predatory hawks (Chapter 5).

Just past the intertidal zone, you will find more red-footed boobies as well as nests of great frigatebirds, whose males will likely be displaying their red, puffed-up gular sacs during the mating season (February to June; Color Image 17). Some frigatebirds will likely be chasing other birds in hopes of stealing their food. Marine iguanas at Darwin Bay are small (largely because of the reduced amounts of algae around Genovesa), but this area is excellent for snorkeling with marine iguanas as they eat. *Opuntia* around Darwin Bay has short, soft spines, a trait associated with the absence of cactus-eating herbivores on the

[23] The entrance to Darwin Bay—a 12-meter-wide, 40-meter-deep channel—can be challenging, and ships unaware of the shallow waters in this area may run aground. For example, in 2010, *Espondelos* crashed at the entrance to Darwin Bay; its 16 passengers were saved by the passing *Monserrat.* Near a cliff by the nearest tower is a plaque commemorating two scuba divers lost at sea in Darwin Bay.

FIGURE 4.4 Prince Philip's Steps is a popular Visitor Site on Genovesa that showcases myriad birds and scenic landscapes. To access the trail, visitors must climb this steep staircase.

island. The last part of the trail consists of sharp a'a' lava that leads you to cliffs showcasing a beautiful view of the bay. If you have time, snorkel along the western side of Darwin Bay, where you may see hammerhead sharks, moray eels, and manta rays. Common plants here include yellow cordia, Galápagos croton, *Sesuvium, Bursera, Rhizophora,* and *Cryptocarpus.*

Prince Philip's Steps. Prince Philip's Steps (El Barranco; Figure 4.4), which is near the tip of the southeastern arm of Darwin Bay, is an area that you may first navigate by panga to see the frigatebirds, red-billed tropicbirds, and fur sea lions along the plateau's cliffs. Among the branches of the palo santo trees protruding from the cliffs you will see red-footed boobies, and on the ground you will see nests of Nazca boobies.

You will then dry-land near the base of a steep, rocky staircase (with a handrail) commemorating Prince Philip's visit to Galápagos (in 1964 and again in 1981), and then enjoy a steep, 25-meter climb to the top of the staircase. At the top, you will find nests of Nazca boobies, red-footed boobies, and four species of finches: sharp-beaked ground finches, warbler finches, large ground finches, and large cactus finches. You will then smell the sandalwood aroma from a forest of palo santo that houses more red-footed boobies; this is one of the best places in Galápagos to photograph the faces and feet of these birds. Your 1.5-kilometer hike then leads you to a lava field that is home to flocks of storm petrels that live in cracks in the lava (Figure 4.5);[24] you might also see a short-eared owl, which eats the petrels. You will also see muyuyo (*Cordia lutea*), palo santo, Galápagos croton, *Chamaesyce,* and lava cactus (*Brachycereus*) at this site. Your visit to Prince Philip's Steps will take about three hours.

Isabela

At a glance:

Other names: Albemarle Island

Visited by Darwin: Yes

---

[24] Recent surveys estimate that there are more than 200,000 pairs of petrels on Genovesa.

FIGURE 4.5 The trail from Prince Philip's Steps (Figure 4.4) leads to a lava field that houses thousands of storm petrels. Here you may also see short-eared owls, which prey on the petrels.

Area: 4,588 km^2 (1,771 mi^2)

Height: 1,707 m

Length of coastline: 617.4 km

Age: 700,000 years old

Inhabited: Yes; population = 2,200

Proximate isolation: Not applicable

Distance from center of archipelago: 74.8 km

Animals you will see: Penguins, flightless cormorants, marine iguanas, black-necked stilts, American oystercatchers, finches, pelicans, Sally lightfoot crabs, land iguanas, lava herons, striated herons, hermit crabs, yellow warblers, paint-billed crakes, pink flamingos, common stilts, short-eared owls, ruddy turnstones, wandering tattlers, white-cheeked pintails, gallinules, rays, Galápagos tortoises, Galápagos hawks, doves, sea lions, yellow-crowned night herons, vermilion flycatchers, blue-footed boobies, sting rays, green sea turtles, creole fish,[25] and great blue herons. Isabela's El Muñeco—which forms the tip of the seahorse-shaped island's nose—is

[25] Creole fish are also called gringo fish; the pink color of their backs reminds locals of sunburned tourists.

the only place in the Northern Hemisphere where you can see penguins in their native habitat. Isabela is home to 60% of the flora and fauna in Galápagos.

Animals you might see: Whales (e.g., orca, blue, sperm, humpback), mangrove finches[26]

Plants you will see: *Opuntia,* palo santo, *Parkinsonia,* beach morning glory, lava cactus, candelabra cactus, poison apple, saltbush, Galápagos cotton, Galápagos *Lantana,* Darwin's daisy, Galápagos tomato, soapberry, *Scalesia,* guayabillo, pega-pega, Galápagos acacia, and Galápagos croton. Vegetation in the highlands along the southern part of Isabela is dense.

Isabela, which is named for Queen Isabella of Spain (1451–1504),[27] is the largest island in Galápagos; it is more than four times larger than Santa Cruz, the second-largest island of the archipelago. Isabela includes the 1,707-meter-high Volcán Wolf, the highest peak in the archipelago,[28] as well as several other calderas up to 10 kilometers across. This area once consisted of six separate volcanoes, but today the 130-kilometer-long, seahorse-shaped island consists of merged lava erupted from the six volcanoes.[29] With the exception of the highly eroded Volcán Ecuador (at the westernmost tip of Isabela), all of Isabela's volcanoes remain active. Isabela, whose white beaches, palm trees, mangroves, lagoons, and cliffs are reminiscent of prototypical Pacific islands, is the only island in Galápagos crossed by the equator. Approximately 1.1% of Isabela is used for agriculture.

Some of Isabela's most popular Visitor Sites are along the western side of the island, and are reached only after a long navigation (Appendix 2). The western side of the island, where upwelling by the Cromwell Current produces nutrient-rich waters, is also the best place to see whales in Galápagos. Several Marine Visitor Sites around Isabela feature sharks, manta rays, and many other animals. More than 70% of the surviving giant tortoises in Galápagos live in northern Isabela, and

[26] In 2006, GNPS officials determined that the remaining 200 mangrove finches were living at only three sites on Isabela.

[27] Isabella of Castile, Queen of Spain, helped finance Columbus's voyages in the 1490s. Ambrose Cowley named the island for Christopher Monck (1653–1688), the Second Duke of Albemarle. Christopher's Point, a promontory at the southwestern end of the island, also honors Christopher Monck, who was a supporter of buccaneers. Christopher became chancellor of Cambridge University, which Charles Darwin attended from 1828 to 1831 in hopes of becoming a clergyman. Although the island was not officially named Isabela until 1892, the name Isabela appeared on maps as early as 1794.

[28] Volcán Wolf at the northeastern end of Isabela is named for geologist Theodor Wolf, who visited Galápagos in the 1870s to study the archipelago's volcanoes (Chapter 2). Volcán Wolf has a large caldera (5 km × 7 km) and the second-deepest caldera (600 m) in Galápagos. Volcán Wolf has erupted at least 10 times since 1797. Although the sides of Volcán Wolf are steep and made mostly of a'a' lava, feral goats were reported on Wolf in 1995. Volcán Wolf, along with Volcán Ecuador, straddles the equator.

[29] From north to south, these volcanoes are Volcán Wolf (1,707 m), Volcán Ecuador (775 m), Volcán Darwin (1,280 m), Volcán Alcedo (1,097 m), Volcán Sierra Negra (1,490 m), and Volcán Cerro Azul (1,689 m). A unique subspecies of tortoise evolved with each volcano; 35% of all of the tortoises in Galápagos live at Volcán Alcedo, and another 15% live at Volcán Darwin and Volcán Wolf.

more than 90% of the penguins in Galápagos live along western Isabela and Fernandina, where sea surface temperatures are lowest in the archipelago.

Whalers, pirates, and others who visited Isabela left behind donkeys and goats that were used as sources of milk and fresh meat. By the 1990s, the goat population on Isabela exceeded 135,000. In 2006, park officials began using dogs, sterilized "Judas" goats with locator collars, and helicopter-based hunting to eradicate these goats from northern Isabela (Chapter 11). These efforts also eliminated 1,520 donkeys from northern Isabela—especially around Volcán Alcedo (the last 99 of which were killed from helicopters)—and 1,102 donkeys from southern Isabela. As of 2011, park officials were trying to kill the last 200 donkeys remaining in southern Isabela.

Puerto Villamil. Puerto Villamil, a small fishing village on the southern edge of Volcán Sierra Negra, is the third-largest town in Galápagos (population 2,200).[30] Puerto Villamil was founded in 1897 by Antonio Gil for the production of lime (made by burning coral harvested from off of Isabela). There was a nearby coffee plantation, and another village—Santo Tomás, named for the nearby 1,490-meter-high volcano now named Sierra Negra (where sulfur was mined from the western side of the crater)—was a few miles inland.[31] Today, the residents of these two towns are supported by fishing and tourism. When a small airport opened on Isabela in 1996 for inter-island flights, tourism began increasing rapidly. Although Charles Darwin's name and image are common on Santa Cruz and San Cristóbal (Appendix 3), they are conspicuously absent in Puerto Villamil.

Tortoise Breeding Center. About 1.5 kilometers outside of Puerto Villamil is the GNPS Arnaldo Tupiza Tortoise Breeding and Rearing Center (est. 1993), where workers breed endangered giant tortoises of southern Isabela (as of late 2012, the center housed 751 tortoises). Surrounding the breeding center are several lagoons that harbor mangroves, flamingos, white-cheeked pintails, whimbrels, common stilts, and other birds. The breeding center also includes gardens showcasing native plants such as Galápagos cotton, manzanillo, spiny plant, Darwin's daisy, *Opuntia,* yellow cordia, and palo santo (Chapter 10).

Wetlands. A 1.5-kilometers path leads visitors from the breeding center to the Wetlands, which are north of town. The Wetlands includes a series of trails that lead to several Visitor Sites showcasing plants, animals, and beautiful vistas; for example, *Cerro Orchilla* is a hill from which you will have a stunning view

[30] Puerto Villamil is named for General José Villamil, a native of Louisiana who moved to Ecuador when the United States purchased Louisiana. Believing that the Galápagos archipelago would be a prized possession, Villamil urged Ecuador's president Juan José Flores to give him exclusive rights to colonize and govern Charles Island (later renamed Floriana, and now spelled Floreana; Chapter 2).

[31] The Sulfur Mine Visitor Site is on the western side of Isabela's Volcán Sierra Negra, about 10 kilometers from the "El Cura" control site. The trail to the Sulfur Mine is lined with guava, an invasive species. You can rent horses at "El Cura" to ride to the rim of Volcán Sierra Negra.

of Puerto Villamil, Volcán Sierra Negra, and Cerro Azul; *El Estero* is a 227-meter mangrove-lined path of lava that leads to a picnic area popular with local residents; and *Poza Redonda* and *La Poza Escondida* are short paths leading through mangroves to recreational areas. A 75-meter path leads to *Túnel del Estero,* a staircase inside a lava tunnel that flowed out under the ocean. The rocky and scenic *Playa de Amor* is a popular site for local residents; and *Pozas Verde* and *Los Tunos* are paths that lead to lookouts.

Wall of Tears. Approximately 6 kilometers west of Puerto Villamil is the Wall of Tears (El Muro de las Lágrimas), a 9-meter-high, 6-meter-wide, 160-meter-long wall of lava stones built by prisoners at one of Isabela's infamous penal colonies (Figure 2.11). Near the wall are the foundations of several buildings used to house the colony's wardens, as well as a large tank that was used to store water.[32] A staircase will lead you to the top of the wall, from which you can see the remnants of a U.S. military base from World War II. The trail back to Puerto Villamil showcases several other Visitor Sites (see above), several scenic views along the coast, and a cemetery that includes the grave of Puerto Villamil's founder, Antonio Gil (for whom the town's sand-covered, palm-lined main street is named). Also here are the graves of several officers of the Isabela prison, including policeman Oliverio Camino, whose tombstone tells visitors that Camino was "murdered in the Alemania Camp in 1950." Most prisoners were buried in a crater near the camp.

Punta Vicente Roca. Volcán Ecuador is a small, extinct volcano (height 610 m) on the northwestern tip of Isabela. This volcano, which is near the shoreline, is younger than Isabela's other volcanoes and is marked by a rocky point called Punta Vicente Roca; this area forms the lower lip of the mouth of seahorse-shaped Isabela. The two coves here are not official Visitor Sites, but snorkeling and a tour aboard your panga are worth the effort; the upwelling of currents near here produces diverse marine life. At Punta Vicente Roca, which is on the southwestern edge of Volcán Ecuador,[33] you will see blue-footed boobies, Nazca boobies, marine iguanas, yellow warblers, flightless cormorants, swallow-tailed gulls, rays, sea lions, fur sea lions, and brown noddy terns—and maybe dolphins and whales. Just north of Punta Vicente Roca you might see fins of curious-looking giant sunfish (*Mola mola*), which can weigh more than 500 kilograms. If you are going to kayak during your stay in Galápagos, the protected coves here are good places to practice.

---

[32] The Wall of Tears was built between 1946 and 1949, after which the penal colony was closed. The buildings at this site (an area called El Provenir, meaning "the future") were then torn down and moved with prisoners to other penal colonies on Isabela (e.g., at Alemania and Santo Tomás). A plaque commemorating the closing of Isabela's penal colonies is on the abandoned church behind the new church on the east side of Puerto Villamil's town square. Many sources inaccurately claim that the Wall of Tears was closed in 1959, but this was when all of the archipelago's remaining prisons were closed. The Wall of Tears closed in 1949.

[33] Half of Volcán Ecuador has already slid into the ocean, thus giving the volcano its amphitheater-like shape. If you round Cape Berkeley (the "nose" of seahorse-shaped Isabela), you will enjoy a beautiful cutaway view of the volcano and its caldera.

Islote Tintoreras. Just south of Puerto Villamil is Islote Tintoreras, a group of islets that can be reached with a 10-minute panga ride. The turquoise water here is usually calm and features sea lions, marine iguanas, rays, and other wildlife. When the tide is low, you can often see reef sharks in the small channels of the islet. A 1.4 kilometer trail through lichen-encrusted a'a' lava (Figure 3.6) showcases marine iguanas, sea lions, red and white mangroves, and spiny plants.[34] At the first beach you encounter along the trail, you will also see Galápagos clubleaf (*Nolana galapagensis*), a rare endemic plant. The second beach along the trail showcases mangroves. Because of the sharks, visitors are usually not allowed to swim at Islote Tintoreras. Your hike at Islote Tintoreras will take about 1.5 hours.

La Calera. Features similar to those of Islote Tintoreras are found at La Calera, a site that is a five-minute panga ride from the pier at El Embarcadero east of Puerto Villamil. You will hike 200 meters through an intertidal zone to a calm area ideal for snorkeling when the tide is low. La Calera is an ideal place to snorkel with sea turtles and to watch marine iguanas feeding underwater. This site was used by the island's early settlers to repair boats and prepare materials to build homes (some of these materials continue to litter the site). A boardwalk from El Embarcadero will lead you to *Concha de Perla,* a shallow, mangrove-lined lagoon used by local residents for swimming and snorkeling.

Mirador del Mango. Approximately 8 kilometers northwest of Puerto Villamil is Mirador del Mango, a former gravel mine used to build roads in Isabela. The road to Mirador del Mango features dry-zone and transition-zone vegetation (Chapter 10). This site, which is named for the prominent mango tree nearby, features a lookout from which you can see southern Isabela and part of the island's agricultural area. Prisoners and others carrying supplies between the coast and farms in the highlands rested under the mango trees here.

Tagus Cove. South of Punta Vicente Roca along Isabela's western coast (and across Bolívar Channel from Fernandina) is Tagus Cove (Color Image 5), a narrow channel named for HMS *Tagus* that moored here in 1814 (while pursuing the USS *Essex;* Chapter 2) to collect tortoises. Tagus Cove, which is west of Volcán Darwin, is a young, partially breached pyroclastic cone.

Your tour of Tagus Cove[35] will probably start with a panga tour below the cliffs, during which you will see marine iguanas, blue-footed boobies, Nazca boobies, flightless cormorants, and penguins living along the ledges of lava near the shore. Tagus Cove recalls the recent history of Galápagos (Chapter 2), for its steep cliffs display graffiti left by passing boats dating back to the 1800s; the

[34] Lichens grow primarily on the southern surface of the a'a' lava because the prevailing winds that bring moisture to the area come from the south.

[35] When Darwin visited Galápagos in 1835, Tagus Cove was known as Banks Cove. Darwin devoted one-third (i.e., 36 pages) of his geological notes while in Galápagos to Tagus Cove, despite being there for only one day.

oldest legible carving here was left under a ledge near the landing site by the crew of the *Phoenix* in 1836. (Note: Posting graffiti is now prohibited throughout Galápagos, except at Post Office Bay; Figure 4.2.) Near here, the Cromwell Current raises nutrients that often attract whales and dolphins; in December and January, you may snorkel with dozens of breeding sea turtles. Water temperatures are usually colder here than anywhere else in the archipelago.

A dry landing at Tagus Cove leads you to a 2-kilometers hike (one-way), one of the longest in the archipelago (aside from the ascents of volcanoes). After climbing 129 wooden steps through palo santo and spiny plant to an overlook, you will then hike through more palo santo and *Scalesia* to a peak that provides a stunning view of Tagus Cove, Darwin Lake (a 9-m-deep saltwater lagoon that is above sea level), a lava flow from Volcán Darwin, and Volcán Wolf. Volcán Darwin, the fifth-largest volcano in Galápagos, has a large caldera that is nearly 200 meters deep and 5 kilometers across. There have been no observed eruptions of Volcán Darwin, but several relatively young flows of lava decorate the caldera and edges of the volcano. Common plants here include Galápagos cotton, castela, yellow cordia, chala, palo santo, and saltbush.

Urbina Bay. South of Tagus Cove at the base of Volcán Alcedo on Isabela's western coast is Urbina Bay, a nesting site for flightless cormorants, penguins, and brown pelicans. After a wet landing on a white-sand beach, your hike along a looped trail will feature land iguanas, marine iguanas, blue-footed boobies, pelicans, and—if you're lucky—Galápagos penguins. In 1954, an eruption of Volcán Alcedo drove the coastline more than a kilometer out to sea while raising the coastline about 4 meters. The coral and marine remnants that were raised above sea level by that eruption are still eerily visible here. Giant tortoises live nearby during the wet season, but migrate to the highlands during the dry season. At Urbina Bay you will also see Galápagos hawks and Galápagos cotton, an endemic plant that may have been brought to the islands by Incas (or that floated to Galápagos from Peru). Other common plants at Urbina Bay include spiny plant, palo santo, and thin-leafed Darwin's shrub. If you snorkel here, you may see spiny lobsters (*Panulirus penicillatus*). The trail at Urbina Bay is long (3.2 km), so bring extra water for your hike.

Punta Tortuga. Just north of Tagus Cove is Punta Tortuga, a beach surrounded by mangroves at the base of Volcán Darwin. Here you will find the endemic mangrove finch, which is one of two tool-using finches. This site, which was the first official Visitor Site created on the western side of Isabela, offers a great view of Fernandina's Volcán La Cumbre.

Volcán Sierra Negra. Volcán Sierra Negra (also called Volcán Santo Tomás), the oldest, largest, and one of the most active volcanoes on Isabela, is in the southeastern part of Isabela between Volcán Alcedo and Volcán Cerro Azul. The oval caldera of Volcán Sierra Negra is more than 9 kilometers across, and touring the volcano and surrounding area takes most of a day. Your tour will begin in Puerto Villamil, where you will take an 18-kilometers van ride to Santo Tomás, from which you can ride horses or hike the remaining 9 kilometers to the caldera. There, you will walk around part of the 10-kilometer-diameter rim and then go northeast across a lava field to Volcán Chico, which is a group of steaming craters that last erupted in 1979. During your walk you may see whimbrels, egrets, Galápagos hawks, vermilion flycatchers, finches, and Galápagos martins; common plants here include acacia, yellow cordia, Galápagos croton, candelabra cactus, *Scalesia,* and *Tournefortia.* Although you may also enjoy a stunning panoramic view of Isabela, garúa often diminishes the view from the rim of the caldera.

Volcán Alcedo. Just north of the 10.8-kilometer-wide Perry Isthmus (the narrowest point on Isabela, about one-third up from the southern tip of the island) is Volcán Alcedo. This symmetric, large, and shallow (270 m) volcano, which is located between Sierra Negra (to the south) and Volcán Darwin (to the north), last erupted between 1946 and 1960, and is one of the oldest volcanoes on Isabela. The caldera of Volcán Alcedo, along with that of Sierra Negra, is the only caldera in Galápagos that people can visit. Approximately 5,000 tortoises live near Volcán Alcedo; you will find their paths along the rim of the volcano. Volcán Alcedo is often closed to tourists.

Reaching the caldera of Volcán Alcedo requires a long hike, special permission from the GNPS, and a thorough quarantine procedure. On your way to the rim, you will see pega-pega (in the transition zone) and guayabillo (in the wet zone; Chapter 10); other plants here include palo santo, yellow cordia, Galápagos croton, lance-leafed Darwin's shrub, prickly pear, small-headed *Scalesia,* spiny plant, and cat's claw. The view from atop the rim is spectacular; the caldera is almost 5 miles across and includes several steaming fumaroles, the largest of which hikers call "The Geyser." Giant tortoises here had been endangered by thousands of introduced goats, which had stripped the area of much of the vegetation needed as food by the tortoises. GNPS's Project Alcedo has been killing the goats (often with gunners aboard helicopters) since 1995. By 2006, goats and donkeys in northern Isabela had been almost eradicated, and populations of native plants and animals were recovering.

Although the view from the summit of Volcán Alcedo is stunning, getting to the summit can be arduous—it's a long climb over sometimes rugged land. Your day will start before sunrise with a wet landing at Shipton Cove on Isabela's eastern coast. If you take a bus to the highlands, you will still have a 2-hour hike to the rim of the volcano. If you don't take the bus, you will hike several miles to the base of the volcano, after which you will hike on a trail of white pumice along the volcano's eastern flank.[36] This hike is not overly challenging, but

[36] In Galápagos, pumice is found only at Isabela's Volcán Alcedo. However, because it floats, it often washes ashore on other islands, especially after the rainy season.

you will finish with a steep ascent along an overgrown trail to the rim. The rim is 10 kilometers across, and the caldera is the third-largest in the world (after Africa's Ngorongoro and the Canary Islands' Las Cañadas caldera). You will get to the rim at about noon, eat lunch (no campfires allowed!), and hike another few miles along the rim. There you will see tortoises, steaming fumaroles, Galápagos martins, vermilion flycatchers, Galápagos hawks, and other wildlife. The view is spectacular.[37] If you hike another hour, you will reach the sometimes-active Volcán Chico. This surreal area will likely be one of the highlights of your trip. It will take you several hours to get back to the beach, after which you will probably visit the tortoise-breeding station operated by GNPS (see above). You may also see a saltwater lagoon that houses flamingos and several shore birds.

Punta Albemarle. At the northern tip of Isabela is Punta Albemarle, which forms the "ears" of the seahorse-shaped island. Rough seas and challenging navigations are why relatively few boats visit this site. However, the water here is laden with plankton, and you will have one of your best chances to see whales while you are near Punta Albemarle. If you go ashore, the 490-meter (one-way) hike will show you pahoehoe lava (bordered by mangroves), large marine iguanas, and flightless cormorants. Near this site was a U.S. radar base during World War II; you may be able to see a tower and several abandoned water tanks from your boat.

Punta Garcia. Across Isabela Channel from Santiago Island is Punta Garcia. This site, which features a'a' lava, is reached with a dry landing that is sometimes difficult. Looming above the barren area is Volcán Darwin. When flightless cormorants disappeared from this site, visitation declined. Aside from Puerto Villamil, this is the only site on Isabela reachable without sailing around Isabela's west coast. Common plants here include palo santo, white and red mangrove, and radiate-headed *Scalesia.*

Roca Redonda. An especially interesting site is Roca Redonda, an isolated, flat-topped islet 24 kilometers north of Isabela. This islet extends only 45 meters above sea level, but it is the tip of a huge mountain that is 3 kilometers high and 20 kilometers wide at its base. The southeastern side of the steep-sided islet is one of the few places on Earth where scuba divers can safely see gas bubbling into the water from underwater fumaroles. Hammerhead sharks are common here. Visitors are not permitted to land at Roca Redonda, but from your boat you will see why Herman Melville described the islet as "the aviary of Ocean."

Elizabeth Bay. Along Isabela's western coast is Elizabeth Bay, a Marine Visitor Site. This site has no trails, and landings are forbidden, so your exploration

[37] Some tour operators have permission to camp atop the volcano, in which case you will need to pack your own gear, as well as your trash when you return. If you are with such a group, be sure to rise early to see sunrise—it's one of the most spectacular sites in the archipelago.

will be via panga. You will see mangrove trees colonizing the lava, along with green turtles, rays, sharks, and several different kinds of shore birds. In the red mangrove trees are nests of pelicans, frigatebirds, finches, warblers, herons, and other birds. The shallow water near the lava-lined entrance is usually exceptionally clear, thereby creating an excellent window to view marine life. Common plants at Elizabeth Bay include mangroves, palo santo, and prickly pear cactus. Elizabeth Bay was a popular site for whalers; when whales came to the bay to calve, whalers would blockade the entrance with their ships and, as Herman Melville noted, "so had the Leviathans neatly in a pen."

Las Marielas. Near the entrance to Elizabeth Bay is Las Marielas, a group of rocky islets that houses colonies of penguins (water in the western parts of Galápagos is usually cooler than elsewhere in the archipelago, thus explaining the penguins' occurrence in this region). These penguins—sometimes with gray, downy chicks—are often visible early in the morning and near dusk, especially in the garúa season. If you visit Las Marielas and Elizabeth Bay between November and February, you will also probably see mating green turtles. Upwelling near Las Marielas and Elizabeth Bay supports many flightless cormorants and some of the largest marine iguanas in Galápagos.

Punta Moreno. Southwest of Elizabeth Bay (and north of Volcán Cerro Azul[38]) is Punta Moreno, whose shores and lagoons are often frequented by white-cheeked pintails, common gallinules, flamingos, paint-billed crakes, penguins, marine iguanas (some of the largest in Galápagos), stilts, brown pelicans, and great blue herons. After a dry landing, you will enjoy a three-hour hike that showcases numerous lava fields interspersed with ponds and lagoons that house carpenter bees, flamingos, finches, Galápagos doves, mockingbirds, white-cheeked pintail ducks, flightless cormorants, and other wildlife. Common plants here include lava cactus, palo santo, *Scalesia,* candelabra cactus, white mangrove, mollugo, and thorn bush. From Punta Moreno you will have spectacular views of Volcán Cerro Azul, Volcán Sierra Negra, Volcán Alcedo, and Volcán La Cumbre on Fernandina.

Marchena

At a glance:

- Other names: Bindlow
- Visited by Darwin: No
- Area: 130 km^2 (50 mi^2)
- Height: 343 m (1,125 ft)

---

[38] Cerro Azul, which is home to the largest tortoises in all of Galápagos, is on the southwestern corner of Isabela; it has erupted at least nine times, the most recent of which occurred in 1998. Its caldera, which measures only 4 kilometers × 3 kilometers, is one of the smallest in the western Galápagos, but its depth (650 m) is among the largest in the western Galápagos. Cerro Azul is the youngest volcano on Isabela; Isabela's other volcanoes, from youngest to oldest, are Wolf, Fernandina, Darwin, Alcedo, and Sierra Negra.

Length of coastline: 51.4 km

Age: 600,000 years

Inhabited: No

Proximate isolation: 55.8 km from Santiago

Distance from center of archipelago: 96.6 km

Animals you will see: Galápagos hawk, sea lions, and the endemic Marchena lava lizard

Marchena is a large shield volcano, of which only the top 343 meters protrude above sea level. In 1684, Cowley named this island Bindloe, probably to honor Captain Robert Byndloss, the brother-in-law of Admiral Sir Henry Morgan ("King of the Buccaneers"). Others argue that Bindloe honors Captain John Bindloe. The name Marchena honors Fray Antonio Marchena, a confidant of Christopher Columbus who helped Columbus and his son at the Convent of Rábida in 1484. Marchena, the seventh-largest island in Galápagos, is the largest island in the archipelago to have no Terrestrial Visitor Sites. Visitors are not permitted ashore at Marchena, and the island is not on the route for most live-aboard dive boats.

Marchena, which is rather desolate, has no fresh water and has never been settled. Consequently, its plants and wildlife have never been disturbed by introduced animals. There is no evidence that tortoises ever lived on Marchena. The volcano's elliptical caldera (7 km × 6 km) is almost full of lava, some of which has spilled over the sides. Although Marchena has many young lava flows, the only observed eruption occurred in 1992. Most of Marchena is covered by black lava; there are also several fumaroles. Marchena achieved worldwide fame in 1934 when the desiccated bodies of Rudolph Lorenz and Trygve Nuggeröd were discovered on its beach (Figure 2.8).

North Seymour

At a glance:

Other names: Seymour Island

Visited by Darwin: No

Area: 1.9 km^2 (0.73 mi^2)

Height: 28 m (98 ft)

Length of coastline: 5.5 km

Age: 1.3 million years old

Inhabited: No

Proximate isolation: 1.46 km from Baltra

Distance from center of archipelago: 30.6 km

Animals you will see: Great and magnificent frigatebirds, finches, common noddies, lava gulls, pelicans, lava lizards, green sea turtles, swallow-tailed gulls, Galápagos sea lions, land iguanas, Galápagos doves, marine iguanas, blue-footed boobies, Nazca boobies, red-billed tropicbirds, pelicans

Animals you might see: Galápagos snakes

Plants you will see: Palo santo, saltbush, *Sesuvium,* arrayancillo, yellow cordia, *Maytenus,* spiny plant, puncture weed, *Opuntia,* scorpion plant, *Parkinsonia,* Galápagos croton, leatherleaf, Galápagos carpetweed

North Seymour is a small, boulder-strewn island just north of Baltra that was formed by geological uplift, thus explaining the island's low, flat profile and shoreline cliffs. The island is named for Lord Hugh Seymour (1759–1801), a senior officer in the British Royal Navy.

Given its close proximity to Baltra, North Seymour is where many visitors to the Galápagos take their first hike. There is one looping trail that starts along the south side of North Seymour, which is crowded with life. The island is relatively flat, so the 1.8-kilometer walk is not difficult. You will have a dry landing (this can be challenging if there are waves) onto reddish-brown rocks littered with sea lions and swallow-tailed gulls. Just above the landing site you will see a small grove of palo santo trees, which are leafless except during part of the wet season. Then you will begin a two-hour hike that features one of the archipelago's largest nesting sites for blue-footed boobies and magnificent frigatebirds. Watch your step—some of the boobies nest on the trail. (You will find few other sites in Galápagos better suited for photographs of blue-footed boobies.) The island's plants are sparse and typical of the arid zone; prickly pear cacti are common along the shores and cliffs. You may also see male magnificent frigatebirds with their inflated pouches.[39]

North Seymour's land iguanas—the largest in Galápagos—were introduced here from nearby Baltra in 1932 by California oil tycoon Allan Hancock aboard *Velero III.*[40] (Hancock believed that the iguanas would better survive on North Seymour than Baltra, where introduced goats were already common.) This introduction was fortuitous, for Baltra's land iguanas disappeared during the wartime occupation of the island (1942–1946; see Chapter 2). Thereafter, the introduced iguanas on North Seymour were repatriated to Baltra. Today, there are growing populations of land iguanas on Baltra and North Seymour.

North Seymour is one of the premier dive sites in the central islands of Galápagos. Currents are strong, but you will likely see hammerhead sharks, sea lions, green sea turtles, garden eels, and countless tropical fish.

---

[39] The word *magnificent* here is part of the common name for these birds. That is, *magnificent* is not simply a descriptive adjective; it distinguishes magnificent frigatebirds from the other species of frigatebird in Galápagos, the great frigatebird.

[40] Hancock donated the La Brea Tar Pits to Los Angeles (with the stipulation that admission would be free to the public) and developed Rancho La Brea into what is most of Hollywood and the Wilshire District of Los Angeles.

Pinta

At a glance:

Other names: Abingdon Island

Visitor Sites: None

Visited by Darwin: No

Area: 60 km^2 (23.1 mi^2)

Length of coastline: 39.0 km

Height: 762 m

Age: 800,000 years old

Proximate isolation: 28.8 km from Marchena

Distance from center of archipelago: 128.1 km

Animals you will see: Marine iguanas, fur sea lions, Galápagos hawks, finches, pelicans, swallow-tailed gulls, sea lions, tortoises, dolphins. Tortoises that once lived on Pinta (and elsewhere in Galápagos) were harvested by whalers, fishermen, and naval crews, and goats introduced by fishermen in 1958 hastened the decline of the population. Lonesome George, the last surviving tortoise from Pinta and an icon of conservation in Galápagos for more than 40 years, died in captivity at CDRS in 2012 (Chapters 2, 6).

Plants you will see: *Opuntia,* saltbush, palo santo, Galápagos cotton, Galápagos passionflower

Pinta, which was named for Columbus's boat *Pinta,* is an elongate shield volcano that has erupted once in the past century. The name Abingdon honors James Bertie (1653–1699), the First Earl of Abingdon. Pinta is home to one-third of all native species of plants in Galápagos.

Although Pinta has no Terrestrial Visitor Sites, there are several landing areas around the island. However, to go ashore, you will need a special permit issued by GNPS. The island's steep slopes make visits challenging. Much of the lava on Pinta contains unusually large, white crystals of plagioclase, a rare, white compound that is the basis for this lava being called abingtonite. (Abingtonites also occur on nearby northern islands of Galápagos.) Pinta is the only island in Galápagos that is north of the equator and has a moist highland.

Goats were introduced on Pinta by fishermen in 1958. When the goats were finally eradicated from Pinta, park officials realized that the island's plants—which had earlier been controlled by tortoises—were growing excessively. In 2010, as part of "Project Pinta," park officials released 39 hybrid tortoises onto Pinta; these were the first tortoises there since Lonesome George was removed almost 40 years earlier (Chapters 2, 6). The reintroduction of these tortoises is a temporary solution to restore ecological balance on Pinta; the tortoises are sterile and therefore will not repopulate the island.

Pinzón

At a glance:

Other names: Duncan, Deane

Visited by Darwin: No

Area: 18 km^2 (7 mi^2)

Height: 458 m (1,503 ft)

Length of coastline: 17.0 km

Age: 0.7–1.5 million years old

Inhabited: No

Proximate isolation: 10.4 km from Santa Cruz

Distance from center of archipelago: 18 km

Animals you will see: Sea turtles, eagle rays, marine iguanas, sea lions, finches, lava lizards, Galápagos hawks, dolphins

Plants you will see: *Scalesia, Opuntia*

Pinzón is named for Martín and Vicente Pinzón,[41] who were captains of the *Pinta* and *Niña* boats of Columbus's armada to the New World. Cowley originally named this island after shipbuilder Sir Anthony Deane (1638–1721), a protégé of English politician Samuel Pepys (1633–1703).[42] Pinzón's name of Duncan Island honors British admiral Adam Duncan (1731–1804), the First Viscount Duncan.[43] Because of its harsh landscape and abundant *Opuntia,* Pinzón is also called Devil's Island. In 1942, a U.S. LB-30A bomber crashed on Pinzón, killing 10 of its crew. The wreckage was rediscovered in the mid-1970s when CDRS and GNPS were trying to eradicate introduced rats from the island. Some of the wreckage still litters the island.

When the CDRS and GNPS were formed, surveys of the archipelago's tortoise populations revealed that only about 200 tortoises remained on Pinzón. Moreover, all of these tortoises were old adults; *all* of the eggs and hatchlings for many decades had been killed by black rats, which were introduced to Pinzón in the late 1800s. To save the tortoises from extinction, scientists took tortoise eggs collected on Pinzón to the CDRS breeding center, and 20 tortoises were repatriated to Pinzón five years later. During subsequent decades, more than 500 additional adult tortoises were repatriated to Pinzón as park officials tried to eradicate the rats. Although there were temporary increases in

---

[41] Martín Alonso Pinzón (1441–1493) captained *La Pinta,* and his brother Vicente Yáñez Pinzón (1462–1514) captained *La Niña.* Another brother, Francisco Martín Pinzón (1445–1502), was master of *La Pinta.*

[42] In 1679, Deane and Pepys were falsely accused of treason (for trying to overthrow Charles II) and jailed in the Tower of London, but they were eventually released. Pepys's diary, which he kept from 1660 to 1669, described several important events (e.g., the Great Plague, the Great Fire of London) and was an important source for the English Restoration.

[43] In 1797, Duncan (1731–1804)—in one of the most important battles in naval history—defeated the Dutch fleet at Camperdown in the French Revolutionary War.

tortoise populations and other animals (e.g., lava lizards), the eradication program ultimately failed, and today the population of tortoises on Pinzón is again threatened by black rats. Until the rats are eradicated, park officials continue to collect tortoise eggs on Pinzón and hatch them at the CDRS breeding center on Santa Cruz. Tortoises are returned to Pinzón when they are about four years old and are no longer prey for the rats.

Pinzón is a central, cliff-bound island that has no Terrestrial Visitor Sites. To go ashore, you will need a permit from GNPS. (These permits are typically reserved for scientists.) At Pinzón you will see sea lions, sea turtles, Galápagos hawks, finches, lava lizards, and marine iguanas. Pinzón has rocky shores and heavy surf.

The only Marine Visitor Site near Pinzón is No Name Island, which is just southeast of the island. Pinzón is famous for scuba diving that features hammerhead sharks, Galápagos sharks, and rays.

Rábida

At a glance:

Other names: Jervis Island

Visited by Darwin: No

Area: 4.9 km^2 (1.9 mi^2)

Height: 367 m (1,204 ft)

Length of coastline: 10.5 km

Age: 0.7–1.5 million years old

Inhabited: No

Proximate isolation: 4.46 km from Santiago

Distance from center of archipelago: 25.1 km

Animals you will see: Flamingos, Galápagos sea lions, white-tipped reef sharks, blue-footed boobies, Galápagos hawks, great blue herons, lava lizards, common stilts, Nazca boobies, vermilion flycatchers, white-cheeked pintail ducks, brown noddies, marine iguanas, pelicans, and mockingbirds. Feral goats, which had led to the extinction of geckos, land iguanas, and rice rats on the island, were eradicated on Rábida in 1972. Today, Rábida is goat-free, but introduced rats plague animals, as well as the endemic and critically endangered shrub, *Galvezia leucantha.* There are no land iguanas or tortoises on Rábida.

Plants you will see: Black mangroves, saltbush, palo santo, *Maytenus,* spiny plant, leatherleaf, Galápagos croton, muyuyo, Galápagos cotton, Galápagos tomato, and *Opuntia*

Rábida, a small, dry island near the geographic center of Galápagos, is about 5 kilometers south of Santiago and a 1.5-hour sail from Sombrero Chino.

Rábida is named for the convent of Rábida, which cared for Columbus's son Diego while Columbus sailed for the New World.[44] The island consists of several steep, eroded cinder cones, lava flows, and reddish cliffs. Except for a beach at the Visitor Site, most of Rábida's coastline is rocky.

Your visit to Rábida begins with a wet landing on the island's famous rust-colored beach. The sand's color, which results from the high iron content of the lava, provides a unique backdrop for photographs of the area's many sea lions. In past years, pelicans often nested in saltbush and black mangroves along the beach, but these birds are now rarely present. If pelicans are there, watch for adults feeding their chicks; this may be the only site where you will see pelicans' nesting and family life (you may also see corpses of baby pelicans, many of whom die in their first year of life). East of the beach you will see a picturesque hillside covered with palo santo and soft-spined *Opuntia*. Colors along the beach are most intense in late afternoon.

Rábida's 1.1-kilometer trail takes you to a saline lagoon, at which you might see yellow-crowned night herons, Galápagos hawks, Galápagos doves, yellow warblers, stilts, white-cheeked pintail ducks, brown pelicans, boobies, and finches. Flamingos were common here until 1997, but they are seldom present now. The trail includes a beautiful overlook at the east end of the beach. If you snorkel along the east end of the beach at Rábida, you will see a rich diversity of fish. Rábida's vegetation typifies that of the arid zone.

San Cristóbal

At a glance:

Other names: Chatham Island

Visited by Darwin: Yes

Area: 557 km^2 (215 mi^2)

Height: 896 m

Length of coastline: 156.1 km

Age: 2.4–4.0 million years old

Inhabited: Yes; population = 7,000

Proximate isolation: 66.6 km from Santa Cruz

Distance from center of archipelago: 124.7 km

Animals you will see: Great frigatebirds, Galápagos sea lions, pelicans, lava lizards, yellow warblers, small ground finches, small tree finches, vermilion flycatchers, tortoises, Sally lightfoot crabs, red-billed tropicbirds, Galápagos sharks, white-tipped reef sharks, red-footed boobies, San Cristóbal mockingbirds, marine iguanas, dolphins, swallow-tailed gulls

[44] The name Jervis honors British Admiral John Jervis (1735–1823), who later became Earl of St. Vincent after he defeated the Spanish fleet in 1797.

Animals you might see: Great blue herons, sea turtles, blue-footed boobies, Nazca boobies, rays, yellow-crowned night herons. Black flies (*Simulium bipunctata*), introduced in 1990, now plague farmers and livestock living at altitudes above 300 meters.

Plants you will see: Palo santo, *Miconia,* matazarno, Galápagos cotton, mollugo, pega-pega, *Scalesia,* Galápagos croton, saltbush, Galápagos lantana, Darwin's daisy, *Calandrinia galapagos, Lecocarpus darwinii, Lignum vitae*

San Cristóbal, the easternmost island in Galápagos, is named for Saint Christopher, the patron saint of sailors. The name Chatham probably honors either John Pitt, the Second Earl of Chatham, or William Pitt the Younger (1759–1806), who was prime minister and Lord Warden of the Confederation of Cinque Ports (near the English Channel).[45] San Cristóbal, the fifth-largest island in the archipelago, was the first island in Galápagos visited by Charles Darwin, who came ashore with Captain Robert FitzRoy and several shipmates at Frigatebird Hill on September 16, 1835 (Chapter 2).

San Cristóbal consists of two joined volcanoes, both of which are still considered active. The northeastern part of San Cristóbal was formed by eruptions only a few centuries ago, whereas the southwestern half of the island is a symmetric shield volcano formed 600,000 to 2,400,000 years ago. San Cristóbal, which has the only permanent stream of fresh water in Galápagos that reaches the coast,[46] is also home to El Progreso, the oldest surviving settlement in the archipelago (established 1869; see Chapter 2). Approximately 15% of San Cristóbal's land is used for agriculture. The abundance of fresh water made San Cristóbal the most populous island in Galápagos until the late 1950s.

Puerto Baquerizo Moreno. Puerto Baquerizo Moreno (population 6,500), which is at the southwestern tip of San Cristóbal, is the archipelago's second-largest town. As the provincial capital of Galápagos, Puerto Baquerizo Moreno houses governmental offices and Ecuador's Second Zone Naval Base. San Cristóbal's airport, one of the two major airports that serve Galápagos from the mainland, processes more than 45,000 tourists each year. Prices at San Cristóbal are usually cheaper than at Puerto Ayora. There are four excellent areas for surfing near town as well as several of the few campgrounds in the islands (get information at the GNPS Information Site just outside of town). Although Puerto Baquerizo Moreno is smaller than Puerto Ayora, it is Baquerizo Moreno that has the Charles Darwin International Convention Center and the archipelago's only two stoplights.

[45] Some sources claim that James Colnett's naming of Chatham in 1793 (Chapter 2) honors William Pitt the Elder (1708–1778), the First Earl of Chatham. Other sources claim that Colnett named the island to honor British soldier John Pitt (1756–1835), the Second Earl of Chatham, the oldest son of William Pitt the Elder, and an older brother of William Pitt the Younger. Chatham was also the name of HMS *Chatham,* which visited Galápagos several times in the late 1700s.

[46] This stream flows into the ocean from a 9-meter-high cliff at La Honda, which is 20 kilometers east of Puerto Baquerizo Moreno. The *Beagle*'s Captain FitzRoy collected water there for two days before leaving Galápagos.

Puerto Baquerizo Moreno includes several tributes to Charles Darwin (Appendix 3), including a bust of Darwin just outside the entrance of the Second Zone Naval Base. The bust at the Naval Station (Figure 2.9) is not far from another waterfront bust of Darwin in a small plaza along Malecón Charles Darwin.[47] You will find several displays of the islands' endemic plants at both ends of Malecón Charles Darwin, as well as near the memorial to fishermen near Oro Beach. On your way to the naval station, a shop will remind you when Darwin visited the islands (*Darwin Shop, Since 1835 . . .*).

Interpretation Center. Most visits to Puerto Baquerizo Moreno include a visit to the Interpretation Center, which is about 1.5 kilometers northeast of town. This open-air center describes the history of Galápagos, as well as the archipelago today. The Interpretation Center, which opened in 1998, can be reached on foot. You do not need a guide to visit the center.

Frigatebird Hill. When you leave the Interpretation Center, follow the 3-kilometer trail that takes you to Frigatebird Hill (Cerro Tijeretas).[48] The cove below this hill is where Charles Darwin first set foot on Galápagos in 1835 (Color Image 1); there is a large statue of Darwin (holding a book titled "Galápagos") near this cove. Aside from North Seymour, Frigatebird Hill is the only site at which you can see both species of frigatebirds in Galápagos (i.e., great frigatebirds and magnificent frigatebirds) in the same colony. Many people who visit Frigatebird Hill finish their tour with a cooling swim at Playa Mann, a popular beach across the street from Universidad San Francisco de Quito's Galápagos Academic Institute for the Arts & Sciences. Don't stray far; currents at nearby Love Beach and Punta Carola (marked by a lighthouse) are often strong. Not far from Frigatebird Hill, a fishing company named Sociedad Pasquera de Galápagos, also known as La Predial, operated from 1952 to 1960 (Chapter 2). Along the trails at Frigatebird Hill you will find palo santo, yellow cordia, Galápagos croton, Galápagos cotton, mollugo, and *Piscidia*.

Playa Ochoa. If your tour began in San Cristóbal, your first (or last) stop might be Playa Ochoa. This small, 160-meter-long beach, which is about 30 minutes by boat from Puerto Baquerizo Moreno, is accessed by an easy wet landing. You

[47] Many visitors to Galápagos encounter their first statue of Charles Darwin on San Cristóbal, near where Malecón Charles Darwin intersects Avenida 12 de Febrero. This bust of Darwin rests atop a compass-like, eight-pointed base on which Darwin faces east. Given this elaborate base and the bust's orientation, many visitors assume that it is significant that Darwin faces east. Apparently, it's not significant; there is nothing mentioned about the bust's orientation in the records associated with the monument. Moreover, other busts and statues of Darwin located elsewhere in Galápagos do not all face east. For example, the bust outside the nearby Naval Station on San Cristóbal (Figure 2.9) faces west, the statue at the base of Frigatebird Hill faces west, the bust in El Progreso faces east, the bust in Puerto Ayora faces west, and the bust in Guayaquil (a duplicate of the one near the Naval Station) faces east. More information about the many monuments to Charles Darwin in Galápagos is in Appendix 3.

[48] To get a panoramic view of Darwin's landing site, climb the 100 steps to the first overlook of the bay. The second overlook, which is another 100 steps up the hill, leads to a trail that takes you along the top of the hill. This last vista from Frigatebird Hill will lead you to Playa Baquerizo, at which local residents often fish for squid and other invertebrates.

will be greeted by sea lions, ghost crabs, hermit crabs, finches, and the tough, long-lived matazarno trees (*Piscidia carthagenensis*). A trail will lead you to a tidal lagoon where you will see various birds, including the rare San Cristóbal mockingbird. You can also snorkel at this beach. Stories abound of treasures buried in this area, but no treasures have ever been found.

Punta Pitt. Here, on the northeastern side of San Cristóbal, you will land at a sandy cove along a 90-meter-long olivine beach and hike a 1.4-kilometer trail that takes you to the top of a tuff cone. This is one of the only sites in Galápagos where all three types of boobies gather to breed; you will also see swallow-tailed gulls, storm petrels, and both great and magnificent frigatebirds here. Common plants at Punta Pitt include stickleaf, cut-leaf *Scalesia,* and Galápagos lantana. Unless you visit Genovesa, this will probably be your best chance to photograph red-footed boobies. Your tour of Punta Pitt will take about two hours.

Jardín de las Opuntias. Not far from Puerto Baquerizo Moreno is Jardín de las Opuntias, which is named for the prickly pear cactus (*Opuntia megasperma*) that is common in the area. This site has a trail 1.6 kilometers long (round-trip) that showcases vegetation of the dry zone and coastal zone. Also near Puerto Baquerizo Moreno is Manglecito, a 475-meter-long beach that is lined with mangroves. A lagoon here features white-cheeked pintail ducks and other wildlife.

La Lobería. About 10 minutes from San Cristóbal's airport is La Lobería, a coralline beach inhabited by sea lions (and popular with local surfers). Along this site's 900-meter trail you will see palo santo, morning glory, cacho de chivo (eaten by several species of finches), mangroves, yellow warblers, frigatebirds, marine iguanas, chala, and *Tiquilia.* This site is outside of GNP, but park officials help maintain the area.

Galapaguera Natural. Along the northern coast of San Cristóbal is Galapaguera Natural. You will make a wet landing at a coralline beach that is a nesting site for sea turtles from January to May. This site, which was opened to visitors in 1989, is also home to *Sesuvium,* Galápagos milkweed, palo santo, matazarno, mockingbirds, lava lizards, geckos, and tortoises, but the population of these tortoises has been damaged by feral cats, goats, and donkeys. Galapaguera Natural is a 5-kilometer hike (one-way) from the beach at which you will land. Wasps (*Polistes versicolor*), which were introduced to Floreana in the late 1980s, are abundant at this site, especially during the wet season.

La Galapaguera de Cerro Colorado. Approximately 20 kilometers east of Puerto Baquerizo Moreno is La Galapaguera de Cerro Colorado, a GNPS facility where you will see giant tortoises in the wild.[49] This 12-hectare facility (est. 2003)

[49] Although tortoises in southern San Cristóbal were driven to extinction by whalers, there are 1,200 tortoises in northeastern San Cristóbal. However, these tortoises are increasingly threatened by feral goats, rats, and cats.

FIGURE 4.6 Kicker Rock is a famous landmark northeast of Puerto Baquerizo Moreno on San Cristóbal. Kicker Rock is the remains of a tuff cone that rises 148 meters above sea level.

includes a 900-meter trail that showcases an interpretation center, a lagoon, the Jacinto Gordillo Tortoise Rearing Center, and several plants endemic to San Cristóbal, including Galápagos acacia, lechoso, cat's claw, matazarno, romerillo, and Galápagos croton. After the walk, you may be able to swim, possibly with green sea turtles.

Kicker Rock. About 1.5 hours northeast of Puerto Baquerizo Moreno is spectacular Kicker Rock (Figure 4.6). If you approach Kicker Rock from the south, you will see that erosion has molded the tuff cone into the shape of a Sphinx or sleeping sea lion ("Leon Dormido"); others view its shape as resembling a shoe (and thereby the origin of the name "Kicker Rock"). If you approach Kicker Rock from the north, you will see that the rock is split and forms a channel that can be navigated by pangas and small boats. Kicker Rock is inhabited by Nazca boobies, blue-footed boobies, and frigatebirds. Strong currents often make snorkeling along Kicker Rock exciting. You may see spotted eagle rays, seahorses, barracudas, Galápagos sharks, turtles, and possibly hammerhead sharks. There are no landing sites on Kicker Rock; it is a Marine Visitor Site.

Cerro Brujo. Adjacent to Kicker Rock on San Cristóbal's northwestern shore is Cerro Brujo, an eroding tuff cone that was one of the first sites visited by Charles Darwin (and Robert FitzRoy) during his visit to Galápagos aboard the

Beagle.[50] (*Brujo* is Spanish for *sorcerer* or *wizard,* a reference to Cerro Brujo being shaped like a witch's hat.) Decades ago, a nearby lagoon was used as a source of salt for preserving tortoise meat.

You will have a wet landing on a beautiful white coral beach, where you can walk 2 kilometers south and—depending on the season—see swallow-tailed gulls, sea lions, marine iguanas, blue-footed boobies, and brown pelicans. At several places you will see a'a' lava, and along the beach you will find a variety of halophytes. Near here, you can also take a panga ride along beautiful cliffs and through a lava tunnel. Your tour of Cerro Brujo will take about three hours, including your panga ride.

El Progreso. About 6 kilometers east of Puerto Baquerizo Moreno is El Progreso. This small village (population 520) is at the base of Cerro San Joaquin, the highest peak on San Cristóbal. Although El Progreso was a notorious penal colony in the late 1800s (Chapter 2), today it is a calm farming community. El Progreso, which was established in 1869, is the oldest surviving settlement in Galápagos.

El Progreso includes several Visitor Sites, most of which memorialize the area's agricultural development by Manuel J. Cobos (Figure 2.4; Chapter 2). Beneath the welcome sign at the edge of town you will find remains of the machinery used at Cobos's sugar mill, and if you turn right when entering town you will see Cobos's large, now-empty tomb. A few blocks east of Cobos's tomb is Escuela Carlos Darwin, where you will find a bust of Charles Darwin (Appendix 3). If you then go north for two blocks and turn left (i.e., west) on Avenida Victor Maldonado, at the end of the street you will see the ruins of Cobos's home on the north side of the street, and the ruins of his servants' quarters on the south side of the street. In the middle of town is the largest tree in Galápagos, which includes a tree house in which you can spend a cramped but memorable evening. Guava on San Cristóbal is reputed to have come from three plants brought to the area by Cobos.

La Soledad. Just north of El Progreso is the small community of La Soledad. Nearby is Mirador de la Soledad, a lookout that also houses a church and an abandoned restaurant. From Mirador de la Soledad you can see Isla Lobos, Kicker Rock, and other parts of southwestern San Cristóbal.

Laguna El Junco. About 10 kilometers beyond El Progreso is Laguna El Junco (Figure 4.7), the largest freshwater lake in Galápagos.[51] The lake is named for junco (*Eleocharis geniculata*), an introduced sedge common in the area. From

---

[50] The intersections of Charles Darwin Avenue and 12 de Febrero in Puerto Baquerizo Moreno and Puerto Ayora (Figure 2.1) are popular photographs with biologists, most of whom mistakenly believe that 12 de Febrero honors Charles Darwin's birthday (February 12, 1809). It doesn't; instead, it honors February 12, 1832, the date that the Ecuadorian government of General Juan José Flores annexed Galápagos (Chapter 2).

[51] The lake covers 60,000 square meters, has a diameter of 270 meters, has a maximum depth of 6 meters (which drops to 1 m during droughts), and holds 360,000 m^3 of water. Fish and nitrates, which were absent in a 1966 study of the lake, were in the lake by 1977. A common fish in the lake is mullet, probably introduced by local residents. Analyses of sediments of Laguna El Junco show that the lake hasn't dried during the last 10,000 years (i.e., since the end of the last Ice Age).

Coffee at El Cafetal

Most agriculture in the southern highlands of San Cristóbal can be traced to Manuel J. Cobos, a brilliant and cruel businessman who built a farming empire there in the late 1800s (Chapter 2). Cobos focused on growing sugar cane, but in 1879 he produced his first commercial crop of coffee at Hacienda El Cafetal (The Coffee Plantation). This plantation, located about three kilometers southeast of El Progreso, thrived until Cobos was murdered in 1904 by disgruntled workers (Chapter 2). Thereafter, harvests at El Cafetal decreased, and by 1929—with Cobos's heirs facing enormous debt—the plantation had been abandoned.

The increased demand for specialty coffees in the 1980s prompted the Gonzalez-Duche family to purchase El Cafetal in 1990 and begin harvesting coffee beans from Cobos's original plants. Today, El Cafetal covers 800 hectares, approximately 350 of which are devoted to growing Cobos's Arabica bourbon coffee. From December to January, approximately 120 workers (half of whom are residents of San Cristóbal) hand-pick 200 metric tons of the organically grown beans. The most prized beans come from the 60% of the crop that grows in the shade of other trees (especially guava). The harvested beans are soaked in water for 12 to 14 hours, and those that sink are sun-dried and sent to Guayaquil for roasting and packaging. (A small portion of the coffee grown in Galápagos is roasted, packaged, and sold locally.) Beans that float are made into instant coffee or are used to make less-expensive "blends" of Galápagos coffee. Efficient workers at El Cafetal can earn more than $100 per day.

The road from El Progreso to El Cafetal is rough in some places, and some parts are difficult to travel on after rains. However, you can buy coffee from El Cafetal in Puerto Baquerizo Moreno at Mockingbird Bar Cafeteria, which is at the intersection of Avenidas Española and Ignacio de Hernandez.

FIGURE 4.7 San Cristóbal's Laguna El Junco, the largest freshwater lake in Galápagos, is the collapsed caldera of an extinct volcano. The rain-fed lake, which is 650 meters above sea level, provided water for many early settlers of Galápagos (e.g., Manuel Cobos's agricultural empire) and the U.S. military base on Baltra (see Figure 2.10).

January to April (i.e., the rainy season), the site is muddy, and fog often impedes views of the lake. On your way to and from Laguna El Junco, you will see different types of vegetation as well as white-cheeked pintails, several species of finches, stilts, common gallinules, and frigatebirds drinking and bathing in the lake. A southbound trail from El Junco will take you to a facility where volunteers eradicate highly invasive plants (e.g., guava and blackberry, which are now competing with *Miconia*) and replace them with native species.

San Joaquin Hill. The 920-meter-long trail around Laguna El Junco showcases several beautiful views of San Cristóbal, including San Joaquin Hill. Climbing this peak, which is best visited by truck or bus, shows you the island's many vegetation zones. Bring rain gear. The round-trip visit to San Joaquin Hill and Laguna El Junco will take two to three hours. Just east of San Joaquin Hill you will see the San Cristóbal Wind Park (Chapter 11).

Isla Lobos. About 40 minutes by boat northeast of Puerto Baquerizo Moreno is Isla Lobos. This small, rocky island is seldom visited because its dry landing and 700-meter-long trail can be challenging. Moreover, its blue-footed boobies are seasonal, and some visitors will miss them entirely. At Isla Lobos, you will see black mangroves, Galápagos cotton, and black rats. If you stop at Isla Lobos, your visit will last about an hour. You will find many sea lions here at "Sea Lion Island."

Santa Cruz

At a glance:

Other names: Indefatigable Island

Visited by Darwin: No

Area: 986 km^2 (381 mi^2)

Height: 864 m (2,834 ft) = Cerro Crocker

Length of coastline: 170.5 km

Age: 0.7–2.3 million years old

Population = 14,000

Proximate isolation: 27.6 km from Isabela

Distance from center of archipelago: 20.6 km

Animals you will see: Vermilion flycatchers, Galápagos hawks, white-tipped reef sharks, Galápagos doves, petrels, whimbrels, pelicans, manta rays, golden rays, giant tortoises (wild and in captivity, including "Super Diego" [Figure 2.12]), great frigatebirds, marine iguanas, Sally lightfoot crabs, blue-footed boobies, black-necked stilts, striated herons, lava herons, great blue herons, lava gulls, yellow warblers, lava lizards, mockingbirds, American oystercatchers, cactus finches, ground finches, land iguanas (in captivity at the CDRS), and smooth-billed anis.

Animals you might see: Woodpecker finches, vegetarian finches, Galápagos rails, short-eared owls, and dark-billed cuckoos. Populations of Galápagos petrels (*Pata pegada*) are now threatened by black rats.

Plants you will see: *Miconia, Scalesia,* beach morning glory, saltbush, poison apple, spiny bush, candelabra cactus, *Opuntia,* mangroves, and Galápagos acacia. Santa Cruz is the only island that has six vegetation zones, which range from the littoral and arid zones (where *Opuntia* and palo santo are abundant) to mist-shrouded areas inhabited by ferns, *Scalesia, Miconia,* and moss-crusted trees (Chapter 10).

Santa Cruz, located across Itabaca Channel from Baltra and home to the town of Puerto Ayora, is the center of Galápagos tourism. The island is a large, circular shield volcano and is the second-largest island in the archipelago (Isabela is the largest). From Santa Cruz, you can charter small boats to several nearby islands in the archipelago. Santa Cruz means "holy cross" in Spanish, and the name "Indefatigable" honors HMS *Indefatigable,* a battleship that visited Galápagos in 1815.

Santa Cruz has two distinct geologic areas: the older peninsula along the northeastern coast is uplifted submarine lava and tuffs, whereas the larger (and younger) part of the island was formed by several lava flows and volcanic cones. Although lavas on the northeastern side of Santa Cruz are from 1 to 2.3 million years old, the pahoehoe lava along the crest of the island is probably only a few thousand years old. Approximately 11% of Santa Cruz is used for agriculture (of which more than 70% is pastures). In and around Puerto Ayora you will see noddies, yellow warblers, great blue herons, and finches.

Puerto Ayora. Most tours visit Puerto Ayora, the largest town (population = 12,000) in Galápagos. Puerto Ayora, which is located on the southern coast of Santa Cruz, is the economic center of the archipelago; it is home to several Visitor Sites, CDRS, GNPS offices, and more tourist facilities and amenities than any other island in Galápagos. Puerto Ayora is also where most tour boats replenish their supplies. In some of the shops along Charles Darwin Avenue, Otavaleño Indians from the Santa Cruz highlands sell rugs, tapestries, and sweaters. The southern end of Charles Darwin Avenue ends at a commercial dock, where supplies are unloaded for the island's residents, visitors, and businesses.

Boats visiting Puerto Ayora anchor in Academy Bay, named for the 1905–1906 expedition to Galápagos by the California Academy of Sciences aboard the boat *Academy* (Chapter 2).[52] Puerto Ayora's cobblestoned main street, Charles Darwin Avenue, and adjoining side streets feature numerous bars, restaurants, tour operators, hotels, pharmacies, dive shops, Internet cafes, churches, banks, taxis, and shops selling T-shirts and postcards. Although prices are higher along the waterfront in Puerto Ayora than in Puerto Baquerizo Moreno (San Cristóbal), the selection in Puerto Ayora is better. You can also restock your supply of snacks, toiletry items, and liquor in shops in Puerto Ayora, as well as take numerous "I Was Here" pictures beside giant statues of tortoises, albatrosses, and land iguanas. The economy of Puerto Ayora is based on tourism, boat-building, and fishing. Much of Santa Cruz can be seen independently or with a guide, and many of the central archipelago's Visitor Sites can be visited on day trips from Puerto Ayora.[53] Many of the sites mentioned here can be reached on foot or by taking a taxi. Water taxis will take you to Angermeyer Point, which is the gateway to Playa de los Alemanes and Las Grietas.

At the fish market on Charles Darwin Avenue, you can watch local residents buy a variety of fish from local fishermen, as well as pelicans, sea lions, and other animals hoping to get a free meal (Figure 5.13). If you look east from the fish market, you will see small, flat Islote Caamaño,[54] which houses sea lions year-round. In the distance, between Islote Caamaño and the northeastern side of the bay, is Santa Fé. At the market is also a monument to fishermen.

A variety of important sites along Charles Darwin Avenue are unmarked. For example, the beach behind Pacifico Bank was a popular place to collect lobsters, the Artesanal Market was a herding area for goats and cows, and the

[52] The 266 giant tortoises collected at Galápagos during the expedition included the last tortoise on Fernandina. Because the San Francisco Earthquake of 1906 (which occurred while *Academy* was in Galápagos) destroyed the buildings that housed the California Academy of Science, the specimens collected at Galápagos became the foundation for the new, rebuilding Academy.

[53] The best places to get answers to your tourism-related questions are at the Cámara de Tourism (Charles Darwin Avenue at Charles Binford Avenue) and the GNPS office (near the CDRS).

[54] Islote Caamaño (often misspelled Camaño) is named for José Plácido Caamaño (1837–1901), who was president of Ecuador from 1833 to 1888.

large windmill marks the town's original source of fresh water (the pump there is now nonfunctional). Before it was renovated in 2001, the public plaza was a popular hangout for grifters, criminals, and prostitutes.

Charles Darwin Research Station. Headquarters of CDRS and GNPS are a 1.5-kilometer walk eastward along Charles Darwin Avenue from the municipal dock at Academy Bay. Both of these organizations help preserve and conserve wildlife of the archipelago. Along the way to these sites, you will see many different plants (including *Opuntia,* candelabra cacti, *Parkinsonia,* spiny plant, Galápagos acacia, yellow cordia, saltbush, *Maytenus,* and mangroves), finches, yellow warblers, and possibly some lava herons and marine iguanas. You will also pass the local cemetery (on your left) and the GNPS Store (on your right). Proceeds from the store's sales help support the work of the Park Service.

CDRS is on every tour of Santa Cruz; it is a popular Visitor Site and home of the archipelago's conservation and preservation programs, which are performed by GNPS. If your tour doesn't take you to the highlands, CDRS will probably be your only chance to see giant tortoises. About 150 meters past the GNPS checkpoint is rocky Playa de la Estación, a small beach used by local residents for recreation. There, you can swim with turtles and sea lions, and see marine iguanas, parrotfish, small rays, and damselfish. But be careful; there was a shark attack here in 2012. From the beach you can also enjoy scenic views of Puerto Ayora and Angermeyer Point.

At CDRS you will see conservation in action; its many research programs include the Fausto Llerena Tortoise Rearing Program. You will see hatchlings, baby tortoises, and juvenile tortoises being raised for reintroduction on various islands of the archipelago,[55] as well as the Van Straelen Interpretation Center, which features exhibits describing the natural history of the archipelago. The Tortoise Breeding and Rearing Program began in 1970 with the rescue of the 14 remaining tortoises (two males and twelve females) of Española's distinctive saddleback subspecies of tortoise. Since that time, CDRS and GNPS have repatriated more than 1,700 of the Española subspecies of giant tortoises to Española; hundreds more have been repatriated to other islands (Chapter 11).[56] Adult tortoises at CDRS are fed three times per week.

While at CDRS, be sure to visit the Nature Trail, which tours the grounds and includes numerous stops documenting local vegetation. The trail will lead you to the CDRS Store; your purchases there will support work at CDRS.[57] These stores sell a variety of books, bags, clothing, postcards, and other items, as well as the ubiquitous T-shirts.

---

[55] A five-year-old tortoise, which can withstand feral rats, pigs, and dogs, is about 12 inches in diameter.

[56] Now that introduced predators have been exterminated on Española, tortoises are repatriated to the island when they are only 1.5 to 2.0 years old.

[57] The CDRS store grosses about $500,000 per year, and its profits are about half of that amount.

Las Grietas. A half-hour walk from Puerto Ayora is Las Grietas (The Crevices), where fissures have created two giant lava piers separated by a canal of water.[58] Lagoons along the trail feature many migratory and coastal birds, including whimbrels, lava gulls, ruddy turnstones, black-neck stilts, and white-cheeked pintail ducks. In the fissures, you will see two layers of water: the surface water is fresh water that filters down from the nearby ravine, and the deeper water is salt water entering from the ocean. You do not need a guide to visit Las Grietas, but be careful if you go near the ravines. If you feel adventurous, you can join the local children as they jump from the cliffs into the cool, calm water. The swimming and snorkeling are excellent.

Playa de los Perros. Playa de los Perros is a 10-minute boat ride from Puerto Ayora. After a dry landing, a 5-meter ladder will lead you to an easy, 500-meter hike along a trail to a 75-meter-long white sand beach inhabited by marine iguanas, sea lions, and crabs. The trail features finches and vegetation of the arid zone, including chala (*Croton scouleri*), muyuyo (*Cordia lutea*), mangroves, and *Opuntia.* Once you're in the canal, you will find blue-footed boobies, pelicans, great frigatebirds, and brown noddies. At low tide, you can often see reef sharks.

Tortuga Bay. Tortuga Bay, the longest and most famous white-sand beach in Galápagos, is a 50-minute (5 km) walk from Puerto Ayora. To get there, walk west on Charles Binford Avenue on the marked path, go up the steps, and pass the Miguel Cifuentes Arias Community Center and GNP post;[59] you will soon reach a beautiful, wide, white-sand beach. Resist the urge to swim here—currents are often dangerously strong. Just around the rocky point and its *Opuntia* forest you will find a calm, mangrove-lined inlet that is ideal for swimming with rays, sea turtles, and many species of fish. At Tortuga Bay you will see marine iguanas, pelicans, mangroves, lava herons, stilts, whimbrels, and maybe some white-tipped reef sharks and flamingos; common plants include saltwort, *Opuntia,* Heller's *Scaesia,* and spiny plant. You do not need a guide to visit Tortuga Bay, but there are no commercial facilities or drinkable water there. Be careful not to step on turtles' nests.

Bahia Ballena. Bahia Ballena (Whale Bay), a cove along the western coast of Santa Cruz, has brown olivine sand and is at the base of Whale Bay Hill. The beach here is steep, and wet landings are often challenging. Along the trail you may find ceramic artifacts left by human settlers near the beach in 1846. (You cannot remove any of these artifacts.) Bahia Ballena was where pirates and whalers brought tortoises before loading them onto their boats. Your hike around Whale Bay Hill will take about 45 minutes and will introduce you to endemic carpenter bees, spiny plants, Galápagos hawks, finches, and smooth-billed anis.

---

[58] Begin the walk at the trailhead originating at Playa de los Alemanes.

[59] These are the closest bathrooms and the last place to buy water on your way to Tortuga Bay.

Tortoises are sometimes resting under *Acacia* along the trail. You may also find a skull of a goat, an invasive species that was once abundant on the island (Chapter 11). A dense forest of palo santo surrounds Whale Bay Hill.

El Mirador de los Túneles. El Mirador de los Túneles, which is not managed by GNPS, is a partially collapsed lava tube near Puerto Ayora. Vegetation at El Mirador de los Túneles is typical of the arid zone (e.g., matazarno, palo santo), and you will likely see finches near the tube's entrance (and white owls inside). You do not need a guide to visit this site.

Highlands. Santa Cruz is the only island in Galápagos having a road across its interior, thus giving visitors access to the island's highlands. Along Avenida Baltra,[60] which crosses Santa Cruz from north to south, you will see the vegetation change from the arid vegetation along the coast to the wet forests and fields of the highlands. From the village of Bellavista (formerly called Fortuna), which is 7 kilometers north of Puerto Ayora, you can go east or west 1 to 2 kilometers to visit lava tubes. These tunnels, some of which are on private land not regulated by GNPS, are easy to walk through. Landowners often charge a small fee (that may have been included in your tour fee) to tour the lava tubes. Some of the cool, wet tubes are easily accessed by wooden stairways. Wear nonslip footwear.

The highlands of Santa Cruz were some of the earliest regions to be cultivated; the contrasts you see here between the natural areas and the adjoining agricultural areas (cleared for cattle and plantations) will make obvious the impacts humans have had on the island. You do not need a guide to visit these sites, but it is advisable to have one. The highlands are usually lush, especially during garúa, at which time you may encounter tortoises in the fog. All of the tortoise-viewing farms and visiting sites have restrooms, souvenir shops, and snack bars.

Media Luna. About 5 kilometers northwest of Bellavista is a crescent-shaped volcanic cinder cone named Media Luna, which is at the upper edge of the *Miconia* zone (Chapter 10). Here, you might see the endemic Galápagos rail (*Laterallus spilonotus*) nesting under the area's dense vegetation. The trail from Bellavista to Media Luna takes you through agricultural land used to grow papaya, avocado, lemons, cattle, coffee, sugarcane, and wild guava, beyond which you will see *Miconia* and giant ferns. Along this path you will probably also see the paint-billed crake, vermilion flycatcher, Hawaiian petrels, woodpecker finches, and large tree finches. On a clear day, from Media Luna you can see Puerto Ayora.

Cerro Crocker. About 3 kilometers north of Media Luna is Cerro Crocker (height = 864 meters),[61] the highest point on Santa Cruz. The shallow, volcanic

---

[60] In Puerto Ayora, Avenida Baltra is also called Avenida P. Julio Herrera. The road from Puerto Ayora to Itabaca Channel was paved in 1996.

[61] Cerro Crocker is named for California philanthropist and yachtsman Charles Templeton Crocker (1884–1948).

FIGURE 4.8
In the highlands of Santa Cruz are Los Gemelos, a pair of sinkholes created by collapsed lava. These sinkholes are an excellent place to see the endangered *Scalesia pedunculata* (Figure 10.6).

soil of Cerro Crocker is covered by a forest of ferns, shrubs, and *Miconia* trees, as well as by nests of Galápagos petrels, Galápagos rails, and other birds. You will also see introduced plants such as cascarilla and guava, which have spread from the agricultural areas. Animals and fires originating on nearby farms have damaged populations of native vegetation. Guides are not required here, but they are recommended. If it's wet, your hike here will be muddy.

El Garrapatero. About 20 kilometers northeast of Puerto Ayora is El Garrapatero. This site includes a beautiful 1.5-kilometer-long white sand beach and, about 90 meters behind this beach, a lagoon that hosts flamingos and white-cheeked pintail ducks. You can camp here if you have permission to do so from GNPS. The path leading to the beach is lined by plants typical of the arid zone, including palo santo, manzanillo, chala, spiny plant, candelabra cactus, mesquite, and arrayancillo.

Los Gemelos. Up the road in the moist highlands are Los Gemelos (The Twins), a pair of 30-meter-deep sinkholes created when caverns of flowing lava collapsed (Figure 4.8). On the steep sides of the craters you will see the layers of lava left behind by numerous lava flows. The craters, which are 2.5 kilometers past Santa Rosa (and 13 km past Bellavista), are on both sides of the road and hidden by vegetation, so watch for the signs. These craters are surrounded by an endangered endemic forest of *Scalesia pedunculata* (Figure 10.6). Along the way, watch for beautiful crimson-feathered vermilion flycatchers, as well as tree finches, warbler finches, short-eared owls, tool-using woodpecker finches, Galápagos hawks, Galápagos doves, and other birds. By the side of the road you might see white cattle-egrets and short-eared owls looking for food. Also nearby may be smooth-billed anis, which are large black birds introduced to rid cattle of ticks, but which may now be outcompeting endemic finches for food. Los Gemelos are in the *Miconia* zone (Chapter 10) and among the lichen- and moss-covered *Miconia* and guayabillo are the endemic *Darwiniothamnus, Cyathea,* and *Passiflora.* The road between Puerto Ayora and Los Gemelos is lined with invasive elephant grass.

Los Túneles. Not far from Los Gemelos are Los Túneles, the largest lava tunnels in Galápagos. The most famous of these tubes is the 800-meter-long "Tunnel of Love," named for the heart-shaped hole in its roof. It takes 15 to 20 minutes to walk through these tunnels. Bring a flashlight. El Mirador de los Túneles, which is between Puerto Ayora and Bellavista, features a shorter tube.

El Chato Tortoise Reserve. A downhill (and often muddy) 5-kilometer trail from Santa Rosa leads to El Chato Tortoise Reserve, where you will see giant tortoises in the wild.[62] The trail goes through private property and, at the park's boundary, splits into two paths: the 3-kilometer path to the right climbs *Cerro Chato,* and the 2-kilometer path to the left leads to *La Caseta,* where you can camp. About halfway along the trail to La Caseta is a small pond in which you will find tortoises during the rainy season. Bird watching is excellent along both trails; you will see yellow warblers, finches, white-cheeked pintail ducks, Galápagos rails, paint-billed crakes, and vermilion flycatchers. Vegetation typifies the humid *Scalesia* zone; unlike palo santo in drier zones, trees in the *Scalesia* zone are green and crusted with moss and lichens. You will also find invasive blackberry, guayava, and wasps. Feral pigs, who dig up and eat tortoise eggs, are kept in check here by park officials. You do not need a guide to visit Cerro Chato or La Caseta, but one is recommended; visitors often get lost here.[63] Although La Caseta was closed in 2011 to protect endangered plants, it may be reopened. A visit to La Caseta takes about three hours.

Although wildlife in the Santa Cruz highlands is diverse, the highlight of your visit to the Tortoise Reserve may be seeing the dome-backed Santa Cruz tortoises in the wild. You may spot some of them even before you get to the reserve, or you may hear them hiss as they pull their heads into their shells as you approach. Bring a rain jacket and sturdy shoes (that you don't mind getting dirty) on your hike in the Tortoise Reserve, for it is often wet and muddy. You may want insect repellant as well, for protection against biting ants.

Las Bachas. Along the northern shore of Santa Cruz is Las Bachas, which is about an hour-long boat ride from Puerto Ayora. This site, which you will access via a wet landing, consists of two coral-white sandy beaches having a combined length of about 1 kilometer. These beaches are among the main nesting sites of Galápagos green turtles (*Chelonia mydas*) in Galápagos. When the eggs hatch, frigatebirds gather to catch and eat the hatchlings. Near the beach you will find saltbush, spiny bush, black and white mangroves, stilts, leatherleaf, and *Opuntia,* and inland you will find herons, common stilts, white-cheeked pintail ducks, flamingos, and other birds in lagoons. Given its close proximity to Baltra, Las Bachas is often the first or last site visited by tours. However, many

[62] Many tours also visit tortoises in the wild at Rancho Primicias, a farm bordering the reserve.
[63] For example, 23-year-old Israeli tourist Guy Nachmoni disappeared in Chato in 1991 and his corpse was not found until five months later. In Chato you will see the memorial to Nachmoni's death.

boats skip Las Bachas; although its protected cove has calm water popular for sunbathing and swimming, there is relatively little to see if you snorkel there. The foundations of the two piers (built for barges) on the beaches at Las Bachas are the remains of the area's occupation during World War II (Chapter 2).[64]

Conway Bay. Along the northwestern shore of Santa Cruz is Conway Bay, an excellent site for watching sea lions. This site, which is near the Islet of Eden, also features marine iguanas. Feral dogs wiped out the area's original population of land iguanas; the ones you will see have been repatriated by CDRS. Other wildlife here includes common stilts, white-cheeked pintail ducks, lava lizards, lava herons, and yellow warblers. Conway Bay is seldom visited; if you go, you will probably have the entire beach to yourself.

Cerro Dragón. Cerro Dragón (Dragon Hill), a low-lying hill along the northern shore of Santa Cruz, is named for the land iguanas that often gather nearby. In the 1970s, GNPS moved the entire population of land iguanas (plus some soil) from Cerro Dragón to Venice Islet to protect the iguanas from feral dogs that were abundant on the island. After the dogs were eradicated (GNPS paid hunters $1 per dog), some of the iguanas were repatriated to Cerro Dragón in 1990. In 1993, after these iguanas had reestablished themselves, the site was reopened to tourists. In 2012, feral donkeys were still present near Cerro Dragón.

Cerro Dragón is reached by a wet or dry landing, depending on the tide. You will find a rocky, white sandy beach from which a dusty, rocky trail leads to a hyper-saline lagoon often frequented by common stilts, herons, pintail ducks, flamingos, and other birds. The 1.6-kilometer trail then leads you to the top of Dragon Hill, where you will enjoy beautiful views of Rábida, Santiago, and Santa Cruz. Along the trail to Dragon Hill, which has been used since the 1800s, you will see Galápagos cotton and leatherleaf. Invasive wasps have become increasingly abundant here, so avoid wearing bright colors or using strongly scented colognes.

Caleta Tortuga Negra. A popular final stop by many Galápagos tour boats is Caleta Tortuga Negra (Black Turtle Cove), a shallow, mazelike tidal lagoon along the northern shore of Santa Cruz near Las Bachas (just west of Canal de Itabaca between Santa Cruz and Baltra). The cove is surrounded by red, white, and black mangroves, which house pelicans in their branches and oysters among their roots. There are no landing sites at Caleta Tortuga Negra; instead, you will tour this site in your panga with its motor off so you will not disturb the cove's many turtles, spotted eagle rays and yellow cow-nosed rays (often swimming

[64] During World War II, the U.S. military used barges to transport troops and supplies (especially fresh water from San Cristóbal) between islands, and the word *bachas* refers to how local residents pronounced the English word *barges*. Contrary to many reports, the floating piers are no longer visible at Las Bachas; all you will find of the piers are rapidly eroding stakes embedded in the sand.

in a diamond formation), great blue herons, lava herons, yellow warblers, pelicans, and white-tipped reef sharks. The cove extends almost a kilometer inland. If you visit between September and February, you will likely see turtles mating in the clear water. Rocks near the eastern entrance to Black Turtle Cove are popular with blue-footed boobies early in the day.

Santa Fé

At a glance:

Other names: Barrington Island

Visited by Darwin: No

Area: 24 km^2 (9 mi^2)

Height: 244 m

Length of coastline: 22.8 km

Age: 2.7–3.9 million years old

Inhabited: No

Proximate isolation: 16.65 km from Santa Cruz

Distance from center of archipelago: 60.1 km

Animals you will see: Marine iguanas, land iguanas (endemic), great frigatebirds, ghost crabs, Galápagos hawks, Galápagos rice rats, Galápagos sea lions, ground finches, cactus finches, manta rays, Audubon shearwaters, Galápagos flycatchers, yellow warblers, and swallow-tailed gulls

Animals you might see: Galápagos snakes

Plants you will see: Saltbush, Galápagos *Lantana,* leatherleaf, yellow cordia, spiny plant, *Maytenus,* palo santo, and Galápagos croton. Vegetation is arid and dominated by cacti. *Opuntia* on Santa Fé are the tallest cacti in the world.

About 26 kilometers southeast of Santa Cruz is Santa Fé, which is named for a city in Spain. (Barrington honors British Rear Admiral Samuel Barrington [1729–1800]). For decades, Santa Fé was believed to have been formed only by uplift of underwater lava formations, but recent evidence suggests that some of the island's landscape has also been influenced by terrestrial volcanic activity. Because there is no fresh water on Santa Fé, the island has been relatively unaffected by humans and introduced animals.

Your visit to Santa Fé will begin with a wet landing at a sheltered white-sand beach of Barrington Bay, which is on the northeastern side of the island. After being greeted by sea lions on the beach and *Opuntia* on the nearby cliffs, you will walk along one of two trails. A short 300-meter trail takes you west to a coastal forest of giant *Opuntia* (some more than 10-meter tall) that includes "The Cactus Arch." (Figure 4.9). Along the cliffs you will also see swallow-tailed

FIGURE 4.9 Santa Fé is home to some of the tallest *Opuntia* in Galápagos, including the famed "Cactus Arch" shown here. These cacti are approximately 13 meters high.

gulls, red-bill tropicbirds, and shearwater petrels, and enjoy one of the most scenic views of the archipelago.

The longer, 1.4-kilometer trail leads you to land iguanas in the island's interior. These large (some more than a meter long), yellow land iguanas live nowhere else in the world. You will also see lava lizards, sea lions, and marine iguanas. Santa Fé is also home to one of the two surviving species of endemic rats in Galápagos. If you snorkel at Santa Fé, watch for jellyfish.

Santiago

At a glance:

Other names: James Island, San Salvador

Visited by Darwin: Yes

Area: 585 km^2 (226 mi^2)

Height: 905 m

Length of coastline: 132.2 km

Age: 0.7–2.0 million years old

Inhabited: No[65]

[65] Although Santiago had been colonized repeatedly (Chapter 2), no one was living on the island when Ecuador created GNP. As a result, all of Santiago is part of the park and cannot be inhabited.

Proximate isolation: 16.86 km from Santa Cruz

Distance from center of archipelago: 60.1 km

Animals you will see: Fur sea lions, sea lions, whimbrels, marine iguanas, Galápagos hawks, Sally lightfoot crabs, yellow-crowned night herons, striated herons, pelicans, vermilion flycatcher, black-necked stilts, white-cheeked pintail ducks, lava lizards, mockingbirds, ruddy turnstones, semi-palmated plovers, American oystercatchers, snails and nudibranchs (in tidal pools), great blue herons, finches, blue-footed boobies

Animals you might see: Flamingos, Galápagos doves, marine turtles, dolphins, sharks, lava herons, Galápagos snakes, penguins

Plants you will see: Palo santo, Galápagos croton, *Opuntia,* mangroves, muyuyo, Galápagos acacia, poison apple, pega-pega, saltbush, saltwort, *Sesuvium*

Santiago is a large, barren, volcanic island northwest of Santa Cruz. Buccaneer William Ambrose Cowley originally named this island Duke of York Island to honor Charles II's brother James II, Duke of York, but when James ascended to the throne (as King James II) in 1685, Cowley renamed the island "by the grace of God, King James the Second's Island."[66] Santiago is the Castillian name for St. James.

Santiago, the fourth-largest island in Galápagos (after Isabela, Santa Cruz, and Fernandina), consists of two merged volcanoes—a low volcano on the southeastern end of the island, and a typical shield volcano (Cerro Cowan) on the northwestern end of the island.[67] Santiago's volcanic landscape includes numerous tuff cones along the southern, eastern, and western sides of the island. There are also young lava flows along the low-lying (<250 m) southern and eastern sides of Santiago.

In 1813, Captain David Porter of the US *Essex* released some goats at James Bay to graze on the island. Not surprisingly, Santiago's plants and animals have been significantly altered by descendants of these and other introduced species. While charting the islands, Porter observed an eruption on Floreana and also noted that tortoises on different islands had different shells. When Darwin visited Santiago two decades later, he saw pigs on the island. Darwin had trouble setting up his camp because of all of the iguanas; these iguanas were later almost wiped out by feral cats and dogs.

---

[66] The name "San Salvador" honors the first island discovered by Columbus in the Caribbean.

[67] Cerro Cowan (905 m) is named for James S. Cowan, a 21-year-old U.S. Navy lieutenant on the US *Essex* who was killed in a duel by shipmate Lieutenant John Marshall Gamble on Santiago (James) Island on August 12, 1813. Cowan's grave "among some green bushes near the beach" of James Bay has been sought by the U.S. Navy and many others, but never found (Chapter 2). A carving on the wall of the ravine along the western side of Cerro Cowan by Cameron Hathawson of the "Ship Halard" (made on April 4, 1804) is the oldest known graffiti in Galápagos.

Puerto Egas. James Bay, which is on the western side of Santiago, offers access to several Visitor Sites. Puerto Egas,[68] an abandoned salt-mining operation, is reached by a wet landing on the steep, dark-sand beach of James Bay; this is the most popular Visitor Site in this area. The dark lava shoreline here includes several pools, caves, and inlets, the most popular of which are the fur sea lion grottos (also look for hermit crabs, barnacles, and fish such as the endemic four-eyed blenny). The 800-meter hike through arid vegetation to these grottos ends at a black lava flow where you will see fur sea lions, Sally lightfoot crabs, night herons, and marine iguanas. This area, with its crystal-clear blue water and crashing waves, is one of the most scenic places in Galápagos. You will also see Darwin's Toilet, a grotto that fills and empties as waves hit the coast. Common plants on this hike include *Scutia,* acacia, muyuyo, palo santo, *Psidium, Opuntia,* and *Tiquilia.* You will soon note that fur sea lions are not as comfortable with humans as are sea lions.

A 2-kilometer trail leads past Hector Egas's failed salt-mining operation and follows part of a path that wagons formerly used to haul salt from the area.[69] Although you can't go into the crater of the salt mine, a 3.5-kilometer trail (one-way) will lead you on an easy hike to and along its rim. Along the shore you will see barnacles, hermit crabs, Galápagos hawks, smooth-billed anis, and yellow-crowned night herons. Nearby are the ruins of buildings and ironworks of the abandoned salt-mining operation. Between the landing site and these buildings is a soccer field for crews of tour boats. This is one of the few sites outside of towns at which crews and tourists can go ashore and play soccer (another is near Post Office Bay on Floreana). Your visit to Puerto Egas will take about three hours.

Volcán Sugar Loaf. Southeast of your landing site at Puerto Egas is the most dominant feature of the James Bay and Puerto Egas area: Volcán Sugar Loaf (395 m). Near here you will see finches, Galápagos doves, and lava lizards, and atop the peak you will have incredible views of the island. To the east and north is a vast lava flow that Charles Darwin wrote about during his visit in 1835. In the open area at the end of the trail you may see Galápagos hawks. Your round trip to Volcán Sugar Loaf will take about an hour.

Buccaneer Cove. Less than an hour's sail north of James Bay along the northwest corner of Santiago is Buccaneer Cove, a popular haunt of pirates, whalers, and

[68] Puerto Egas (South James Bay) is named for Darío Egas Sanchez, who owned a now-abandoned salt-mining operation there in the 1920s. Egas's heirs reopened the mine in 1963, but just days before they were to send their first shipment to the mainland, a new government lowered the price of salt, and the new company soon went bankrupt. Buildings from these settlements are visible on the south side of James Bay along the beach; nearby are goat pens and a large tank that Egas and his 17 workers used to store fresh water (Galápagos hawks often perch there). Salt was mined from a volcanic cone on the eastern side of the island. This cone, which is called Minas de Sal, is home to flamingos and other wildlife; it was visited by Charles Darwin, but today is off-limits to visitors.

[69] In *The Voyage of the Beagle,* Charles Darwin wrote about the submerged "beautifully crystallized white salt" in this circular lake just south of James Bay. He also saw a "skull lying among the bushes" of a boat captain who had been murdered here by his crew. While near James Bay, Darwin found parts of a marmalade jar in pahoehoe lava. The jar, which was manufactured in 1684, is the basis for lava here being named the Marmalade Pot Flow and indicates that the volcano erupted between 1684 and 1835.

other visitors (including Charles Darwin) for centuries. The popularity of this site is due to its excellent snorkeling, relatively calm waters, a nearby source of fresh water, and its cliffs, which are made of tuff formations that house numerous birds. Pay special attention to the colorful strata and the tuff cone along the southern rim, and the lava, cinders, and caves along the northern rim. Common birds at the cove include boobies, noddy terns, yellow warblers, Galápagos snakes, herons, Galápagos hawks, Galápagos doves, and swallow-tailed gulls. The beach's reddish-purple sand creates a sharp contrast with the hundreds of seabirds at this site. Few boats stop at Buccaneer Cove because the wet landing can be challenging in active seas. Moreover, the site's trail doesn't showcase many unique plants or animals. However, the snorkeling is excellent; you will see sea lions, fur sea lions, and rays.

Playa Espumilla. About 5 kilometers from Puerto Egas is the gold-colored sand of Espumilla Beach. After a wet landing, a 2-kilometer trail will lead you through mangroves and palo santo past a lagoon often inhabited by white-cheeked pintail ducks, Galápagos hawks, vermilion flycatchers, herons, whimbrels, common stilts, and flamingos. Along the way you will see coves housing the Southern Hemisphere's smallest pinniped—the Galápagos fur sea lion. Be careful as you walk—sea turtles dig their nests along the edge of the mangroves. Common plants here include arrayancillo, palo santo, mangroves, pega-pega, Galápagos acacia, saltbush, and manzanillo.

Sulivan Bay. Sulivan Bay, which is across a narrow channel from Bartolomé, is southeast of Santiago and features a vast lava field that stretches to the sea. You will examine this giant sheet of black pahoehoe lava, which was formed in 1897 (i.e., it was not present when Darwin visited the area in 1835), during your hike along the 1.5-kilometer trail. The lava has cooled into countless interesting designs (Figure 3.5). The youngest lava is black; in some areas, it surrounds peaks of older, reddish lava. Along the trail, which loops around an iron-oxidized cinder cone, you will see imprints of tree trunks in the lava, lava bubbles, and pioneering plants such as Galápagos carpetweed (*Mollugo*) and lava cactus (*Brachycereus*). Other plants here include matplant, palo santo, *Maytenus,* and mullugo. Wildlife here includes Galápagos hawks, penguins, and herons. The hike here takes about 1.5 hours.

Sombrero Chino. About 150 meters from the southeastern coast of Santiago is the aptly named Sombrero Chino (Chinese Hat, area $<0.25\ km^2$), a cinder cone reached by a wet landing onto a crescent-shaped white-sand beach. The island's downward-facing hat-like shape (Color Image 2) is best appreciated when approached from the north. Sombrero Chino was formed by relatively recent volcanic activity; many lava tubes go down the side of the islet to the coast. The island's 700-meter (round-trip) trail, which leads you to crashing waves at a rocky point, showcases volcanic rocks and lava fields. Wildlife on Sombrero Chino includes sea lions, lava lizards, yellow warblers, Galápagos hawks, and marine iguanas; you might also see penguins on the nearby shore of Santiago, and sea lion mothers teaching their pups to swim in the tidal pools. Common plants here include *Sesuvium, Opuntia,* lava cactus, puncture weed, Galápagos purslane, saltwort, and saltbush. Candelabra cactus, which is abundant on nearby Santiago, is virtually absent on Sombrero Chino.

Most visits to Sombrero Chino culminate with snorkeling just off Santiago. Because of the protection by Santiago, the water here is relatively calm and snorkeling is exceptionally good. Don't miss this one—you might see grazing marine iguanas, sharks, Galápagos penguins, and mating eagle rays.

South Plaza

At a glance:

Other names: Plaza Island

Visited by Darwin: No

Area: 0.13 km^2 (0.05 mi^2)

Height: 25 m (82 ft)

Length of coastline: 2.54 km

Age: 3.4–4.2 million years old

Inhabited: No

Proximate isolation: 302 m from Santa Cruz

Distance from center of archipelago: 39.9 km

Animals you will see: Land iguanas, red-billed tropicbirds, frigatebirds, blue-footed boobies, Galápagos doves, Galápagos sea lions, Audubon's shearwaters, finches, short-eared owls, herons, Nazca boobies, Sally lightfoot crabs, yellow-tailed mullets, marine iguanas, lava lizards, swallow-tailed gulls

Animals you might see: Orcas (hunting sea lions)

Plants you will see: *Sesuvium edmonstonei, Portulaca howellii, Opuntia,* puncture weed, *Maytenus,* and leatherleaf. Most plants on South Plaza are cacti and scrub of the arid zone.

Las Plazas are two small, crescent-shaped islands (facing each other like parentheses) that are a few hundred meters off the eastern coast of Santa Cruz. Because both of these islands were formed by uplift, they are relatively flat and slanted, much like a tilted table. Although North Plaza is closed to the public, South Plaza (the larger of the two islands) is a favorite site because of its spectacular 30-meter-high cliffs. South Plaza, which is only 1 kilometer long and 130 meters wide, is named for General Leónides Plaza Gutiérrez, a former president of Ecuador.[70] South Plaza can be visited during a day trip from Puerto Ayora.

[70] Gutiérrez (1865–1932) was president of Ecuador in 1901–1905 and 1912–1916. His son, Galo Plaza Lasso (1906–1987), served as Ecuador's president from 1948 to 1952.

Tour boats visiting South Plaza (Color Image 6) dock in the sandy, narrow channel separating South Plaza and North Plaza. Your tour begins with a dry landing at a concrete dock usually inhabited by sea lions (your guide may need to clear the way for you). Near the dock you will see a Galápagos monument and several prickly pear cacti (*Opuntia*). Yellow land iguanas camp beneath these cacti, and cactus finches will be in the cacti or nearby vegetation. In recent years, a population of hybrid iguanas has been discovered on South Plaza; these hybrids, which are a cross between a land and a marine iguana, have stripes and a facial structure intermediate between that of the marine iguana (flat-faced) and land iguana (elongated snout). The hybrids are presumed sterile (Chapter 6).

Covering South Plaza are extensive growths of the succulent ground-matted ice-plant *Sesuvium edmonstonei,* which gives the island a reddish color during the dry season (after rain, *Sesuvium* is green). In the wet season, the yellow blooms of *Portulaca oleraceae* (another succulent) are a favorite meal for the island's land iguanas. With the red *Sesuvium,* green cacti, and the channel's turquoise water, the colors of South Plaza can be overwhelming. GNPS officials banned snorkeling at Plazas several years ago after swimmers were attacked by sharks.

The looping 1-kilometer trail on South Plaza leads you to steep cliffs along the south side of the island. There, you will see many birds, including swallow-tailed gulls (with their beautiful red-ringed eyes), acrobatic Galápagos shearwaters, blue-footed boobies, Nazca boobies, frigatebirds, and fast-moving red-billed tropicbirds. In water below the cliffs, you will often see schools of yellow-tailed mullet. Pull your hat on tight and be careful—it is often windy along the cliffs, and people have fallen to their deaths here.[71]

Near the end of the walk along the cliff you will see "The Hospital," an area of basalt that has been rubbed smooth by generations of male sea lions recuperating from mating battles lost to other males. (About 1,000 sea lions live on South Plaza; if you have time, watch them climb up from the water to their rest areas atop the cliff.) The trail then crosses to the northern shore and ends at the landing site. Your tour of South Plaza will take about 1.5 hours.

Wolf

At a glance:

Other names: Wenman Island

Visited by Darwin: No

Area: 1.3 km^2 (0.5 mi^2)

Height: 253 m

[71] Victims here have included tourists as well as experienced visitors. For example, in 1998, Pepe Salcedo—captain of the yacht *Sulidae*—fell from the cliff. He fractured his skull and died the next month.

Length of coastline: 6.20 km

Age: 0.9–1.6 million years old

Inhabited: No

Proximate isolation: 140.4 km from Pinta

Distance from center of archipelago: 256.7 km

Animals you will see: Vampire finches, lobsters, yellowtail surgeonfish, frigatebirds, red-billed tropicbirds, butterfly fish, trumpet fish, morays, Galápagos sharks, manta rays, sea turtles, reef sharks, scalloped hammerhead sharks, Nazca boobies, red-footed boobies, fur sea lions, frigates, marine iguana, dolphins, swallow-tailed gulls

Animals you might see: Whales, whale sharks

Plants you will see: *Scalesia, Opuntia,* Galápagos croton

Wolf Island, along with Darwin Island, forms the northwestern corner of Galápagos. Wolf is 353 kilometers from San Cristóbal, 257 kilometers from Santa Cruz, and 34 kilometers from Darwin. Cowley named this island Wenman after Philip, third Viscount Wenman (sometimes misidentified as Wainman). The island's name was later changed to honor German geologist Theodor Wolf (1841–1924), who studied the geology of Ecuador (Chapter 2). Wolf Island is a classic seabird island; you will see thousands of frigatebirds, boobies, and other birds along its cliffs and plateaus.

Neither Wolf nor Darwin has a Terrestrial Visitor Site; if you visit either of these islands, you are there to scuba dive. Diving here is among the best in the world. At Wolf's numerous dive sites—for example, The Caves, The Pinnacle, and North Islet—you will see large schools of hammerhead sharks, cow-nosed rays, and Galápagos sharks. Diving at many of Wolf's sites (e.g., The Elephant) is spectacular but challenging because currents are often strong. The northern end of Wolf Island is honeycombed with caves and tunnels.

Wolf Island is a 14-hour navigation from Puerto Ayora. Visiting GNPS rangers stay at Wolf Island two to three weeks at a time. Surface waters around Wolf and Darwin Islands are the warmest in Galápagos; during May through November they are 22–24°C, and in December through April they warm to 24–29°C. Wolf's vampire finches, a subspecies of the sharp-beaked finch (*Geospiza difficilis*) endemic to Wolf and Darwin Islands, feed partly on blood pecked from other birds.

CHAPTER 5

Birds

When you have the opportunity to observe these finches at first hand, make sure that you appreciate the role that they have played in the history of human thought.

—Fitter, Fitter, and Hosking, *Wildlife of Galápagos,* 2007

ALTHOUGH GALÁPAGOS OFFERS MANY REWARDS—THE FASCINATING geology of relatively young volcanoes, southern constellations, and a human history that is too odd to be fiction—most visitors go to Galápagos to see the fascinating plants and animals that tell stories of colonization, evolution, and conservation. Animals in Galápagos are famous for their fearlessness, some showing so little concern for humans that they will walk, crawl, or slide right over your feet en route to their destination. The flora of the islands is remarkable as well, for most of the flowering plants have small, drab flowers that are self-pollinating, having colonized the islands before the arrival of pollinating insects. By 1994, the Galápagos National Park Service (GNPS) had near-complete lists of the vascular plants and vertebrate animals on the 77 largest islands, which comprise more than 99.9% of the total land area of the archipelago. Our knowledge of other groups of organisms (e.g., nonvascular plants, fungi), however, remains incomplete.

Many of the plants and animals of Galápagos are endemic, meaning that they are found nowhere else on Earth except Galápagos—and in some instances, only on particular islands in Galápagos. These plants and animals include 28 endemic species of birds, more than 50 endemic species of fish, 17 endemic species of reptiles, more than 100 endemic species of algae, 7 endemic species of lava lizards, 12 endemic species of sharks, 5 endemic species of rays, more than 200 endemic species of vascular plants, and only 6 endemic species of terrestrial mammals (2 species of sea lions, 2 species of bats, and 2 species of rice rats). Many of the animals are widespread—for example, laid-back marine iguanas sprawl on every island—whereas other animals live on only one island during particular times of the year (e.g., virtually all waved albatrosses live only on Española from March to January). As Charles Darwin noted, "the natural history of this archipelago is very remarkable; it seems to be a little world within itself."

The following chapters—Birds, Reptiles, Mammals, Fish, Invertebrates, and Plants—are devoted to the fauna and flora of the islands. We have not attempted a complete list of all the animals and plants in the islands; after all, the beetles alone would require their own large book. Instead, we have focused on the native and endemic icons that fill the must-see lists of most visitors to Galápagos.

BIRDS

For many visitors to Galápagos, birds are the primary attraction, and it is easy to see why. The islands are home to many iconic species—such as blue-footed boobies, penguins, and albatrosses—that people can identify without ever visiting Galápagos. Many of these birds are associated with memorable stories that may or may not be true: great frigatebirds pirate the food of other birds; waved albatrosses mate for life; the finches gave Darwin his ideas about evolution; and Nazca booby chicks kill their siblings. In our discussion of 43 iconic or endemic bird species, we will clarify which of these stories are true, and which are conditional or even false.

Who Are the Birds?

We identify birds by the presence of feathers. Whether these feathers are used in flight or not, they are key to understanding birds. Feathers, along with hollow bones, a reduction in redundant organs (like ovaries and testes), and efficient respiratory systems are unifying features of birds, and they are all adaptations that make flight possible.

Evolutionarily, birds are grouped with the *amniote* lineage of the vertebrates; that is, their embryos are encased in an *amniotic egg,* which has four membranes (the amnion, the yolk sac, the allantois, and the chorion) that surround the developing chick. Specifically, birds are in the *diapsid* group of Reptilia that includes the crocodiles, lizards, and snakes (yes, birds are feathered reptiles)—the same group that includes the now-extinct dinosaurs.

It is difficult to imagine a better place to study avian evolution than Galápagos. In the archipelago's birds, we see speciation in action (e.g., in sharp-beaked ground finches), myriad effects of competition for limited resources (e.g., competitive displacement in ground finches), intersexual selection (e.g., the blue-footed booby courtship dance, the male frigatebird's red pouch), the consequences of genetic drift (e.g., in Floreana mockingbirds), and the existence of evolutionary relics, or vestigia (e.g., the stunted wings of the flightless cormorants). The birds of Galápagos can teach us a lot about evolution.

Birding in Galápagos

One striking thing about birds of Galápagos is that there isn't a lot of biodiversity—that is, the *number* of species you encounter is not large. This makes bird identification relatively straightforward, especially if you have tried identifying birds in Ecuador's extremely diverse Amazon basin or Andean mountains. What makes birds of Galápagos so remarkable is their high degree of *endemism*—at least 28 are found *only* in Galápagos and appear to have evolved in the archipelago in the last few million years. Thus, many of the birds you encounter are new species, many are still undergoing rapid evolutionary changes, and some are not found anywhere else on Earth.

Avian Conservation

One negative aspect to endemism and restricted ranges (e.g., flightless cormorants, which occur only on the east coast of Fernandina and the west coast of Isabela, have a total population size of approximately 1,500 individuals) is that endemic species are vulnerable to unpredictable environmental changes. In the descriptions that follow, you will note that restricted ranges and smaller breeding populations may be especially vulnerable to changing weather conditions (such as El Niño events; Chapter 3), the effects of introduced predators (e.g., cats and rats; Chapter 11), and pathogens (e.g., *Plasmodium* and *Toxoplasma*), and habitat destruction from human activities. This young ecosystem, undergoing rapid evolutionary changes and changing in response to human impacts, ensures that the Galápagos species you see now differ from those seen a generation ago, or those to be seen in the near future.

Galápagos Finches

Of the 23 endemic land birds of Galápagos, 14 are finches.[1] For the most part, these birds are what ornithologists would call the "little brown jobbies" of Galápagos: not only are they unremarkable in size or color, there is considerable morphological overlap between *sympatric* (or co-occurring) species, and their songs are neither melodic nor memorable. Identifying specific finches can be difficult.

The 14 species of Galápagos finches are all endemic; an allied species occurs on Cocos Island 1,000 kilometers to the northeast. Galápagos finches are categorized in five genera: *Geospiza* (ground finches), *Platyspiza* (vegetarian finches), *Camarhynchus* (tree finches), *Cactospiza* (tool-using woodpecker and mangrove finches, also called cactus finches), and *Certhidea* (warbler-like

---

[1] Some texts refer to 13 species of Galápagos finches. However, with the recent division of the warbler finches into two species—*Certhidea olivacea* and *Certhidea fusca*—we recognize 14 species. Prior to 2000, all warbler finches were considered *C. olivacea*.

finches). In addition to their substantial morphological overlap, these finches have similar life histories. They all build small, domed nests with a small hole in the side for passage, they all incubate eggs for approximately 12 days, and their chicks all leave the nest after about two weeks of parental care.

Any introductory biology text, or any treatment of birds in Galápagos, addresses the significance of these finches to our understanding of the mechanisms of evolution. In the 1980s and 1990s, research emphasized how character displacement reduces competition between species, and how Galápagos finches are a model for understanding speciation through adaptive radiation. This work is discussed below, as are some of the recent and ongoing studies of these birds. Finches have played, and continue to play, an important role in studies of speciation, sexual selection, coevolution, and conservation.

It seems likely that a finch population from the mainland began the adaptive radiation that has produced today's Galápagos finches. But how many individuals would it take to found a successful colonizing population? Vladimir Vincek and colleagues modeled various scenarios based on the current distribution of major histocompatibility genes and concluded that "the founding population consisted of no fewer than 30 individuals," emphasizing that this figure is likely an underestimate of the actual number of founding individuals. Recent evidence, however, challenges the notion of a South American ancestor for the finches of Galápagos; genetic analyses suggest that the island finches belong to a group of birds (who also build dome-shaped nests) that radiated from the Caribbean, and that includes the grassquits of the genus *Tiaris*.

The truth about "Darwin's finches"

On the Origin of Species, in which Charles Darwin makes the case for his theory of evolution by natural selection, does not mention finches, nor are finches discussed in Darwin's field notebooks from the voyage. However, although the finches of Galápagos have become aligned with Darwin in a way that belies their initial [in]significance, their *contemporary* influence on evolutionary biology cannot be overstated.

In 1936, English ornithologist Percy Lowe's (1870–1948) paper "The Finches of the Galápagos in Relation to Darwin's Conception of Species" introduced the term *Darwin's finches*. And in 1938, Julian Huxley—grandson of Darwin's "bulldog," Thomas Henry Huxley, and the then-secretary of the Zoological Society of London—asked English schoolteacher and amateur ornithologist David Lack to study Galápagos finches for an entire breeding season. Lack and five colleagues collected birds for one month on San Cristóbal and three months on Santa Cruz. In all, Lack collected 31 live finches spanning four species, and measured more than 7,800 birds. This work ultimately allowed Lack and his colleagues to document natural selection in the wild acting through interspecific competition, and to show that differences in bill size among finches are adaptations to specific food niches.[2] Lack's book, *Darwin's Finches* (1947), popularized the term "Darwin's finches" and made famous the birds to which Darwin had paid little attention more than a century before. After reading Lack's book, evolutionary biologist E. O. Wilson proclaimed that Darwin's finches "shout the truth of evolution." Textbooks began including the finches as evidence for Darwin's theory, and the finches soon became an icon of evolution. As biologist Ernst Mayr noted in 1973, "there is no modern textbook of zoology, evolution or ecology which does not include an account of [Lack's] work."

In 1973, decades after Percy Lowe assigned the name "Darwin's Finches," Peter and Rosemary Grant began their study of evolution in Galápagos finches on Daphne Major.[3] By 1977 the Grants and their students had banded more than half of the island's birds, and thereafter this percentage remained near 100%. Their work has documented evolution in action, and their results have been unequivocal. As they noted in 1977, "our data, and those of [Robert] Bowman, indicate an adaptive significance of the beak differences." The Grants's more than 30 years of work has also documented natural selection occurring faster than Darwin had imagined.

Because long-term monitoring of natural populations is rare due to funding and logistical challenges, the Grants' work is important because of the continuous record they and their colleagues have produced, by studying more than 25 generations and 20,000 birds. This work was popularized in Jonathan Weiner's Pulitzer Prize–winning work *The Beak of the Finch* (1995) and continues today. Much of the Grants's research informs the discussion of finches in this chapter.

Common name: Medium ground finch

Scientific name: *Geospiza fortis*

Comparisons: The medium ground finch is intermediate in size between the small and large ground finches. The medium ground finch is almost indistinguishable from the large ground finch, yet the medium ground finch may have a smaller, more pointed beak.[4] The medium ground finch is also more likely to occur in flocks.

See it: Arid zones on most islands (except the northern islands of Darwin, Wolf, and Genovesa)

Residence status: Endemic

[2] In 1945, Lack had argued that differences in bill size among Galápagos finches were due to genetic drift and were a species-recognition system rather than adaptations for different diets. Two years later, as a result of theoretical insights rather than field observations or new measurements of finch beaks, Lack admitted that his "views have now completely changed."

[3] Peter Grant discussed Galápagos finches with Lack in 1971, and began studying them two years later. Thirty-five years later, the Grants dedicated *How and Why Species Multiply* (2008) to Lack, and described themselves as "the bearers of a torch he passed on."

[4] There is significant variation in beak size among populations of finches on different islands. For example, the largest individuals of *G. fortis* on Santa Cruz have larger beaks than the smallest of the large ground finches (*G. magnirostris*) on Rábida.

Conservation status: Least Concern[5]

What you will see and what it means: Being intermediate in size between two sympatric species (and potential rivals), the medium ground finch is considered to be more generalist in its food preferences. Specifically, its medium-sized beak may allow this finch to eat a greater range of seeds (Color Image 7).

Are beak sizes really that important?

The medium ground finch may have meant more to contemporary evolution than any other organism in Galápagos.[6] Among the most-cited papers in evolution is one titled "Intense natural selection in a population of Darwin's finches (Geospizinae) in the Galápagos," published in 1981 in *Science* by Peter Boag and Peter Grant. For three years (1975–1978), Boag and Grant studied the reproduction and survival of finches on the small, rocky islet of Daphne Major. Boag and Grant captured and color-banded over 1,500 individual medium ground finches and collected data on seed availability each year. In 1977, a severe drought on Daphne Major dramatically reduced the abundance of seeds (especially small seeds), which in turn was correlated with 85% mortality in the finch population. At the beginning of this drought, medium ground finches ate the small, soft seeds that were abundant. However, the drought prevented the regeneration of the island's plants. As a result, the supply of small, soft seeds dwindled, and soon nothing was left for the finches to eat except the large, tough seeds of *Tribulus cistoides* (also called "goat's head," or "puncture weed") that were now comparatively abundant. Finch survival was not random: individuals that survived were those capable of eating the larger, tougher seeds. This fact was empirically demonstrated by measurements of beak sizes both before and after the drought: 642 birds measured before the drought had an average bill depth of 9.42 millimeters; 85 birds that survived the drought had an average bill depth of 9.96 millimeters. This difference (~6%) is statistically significant (Figure 5.1).

Beaks are the finches' principal tool for gathering food. Was this change in average beak size evidence of natural selection in action? Did the environment exert a selective pressure (a drought, and the resulting hardy seeds) that disproportionately favored certain phenotypes (larger, stronger beaks)? If so, one could logically predict that wetter years would yield *smaller* seeds that would favor *smaller*-beaked finches. This is exactly what happened: The dramatic El Niño event

[5] We use International Union for Conservation of Nature (IUCN) designations for all "Conservation Status" data. For some organisms, insufficient data exists for an official designation. The IUCN categories (and associated meanings) are: *Extinct* (the last individual has died); *Extinct in the Wild* (the only surviving individuals are in captivity or some unnatural location); *Critically Endangered* (this organism faces a very high risk of extinction); *Endangered* (faces a high risk of extinction); *Vulnerable* (faces a risk of extinction); *Near Threatened* (is close to qualifying for a threatened category); *Least Concern* (does not qualify for any threatened category, and is believed to be widespread and/or abundant); *Data Deficient* (there is insufficient information to assess this group's status); or *Not Evaluated*. These categories were developed in 2001 and are still used by IUCN.

[6] In homage to what the medium ground finch has meant to biology, the genome of this bird was the first of 100 vertebrate genomes to be sequenced by the BGI-Genome 10K collaboration. The genetic sequence of the medium ground finch became publicly available in August 2012.

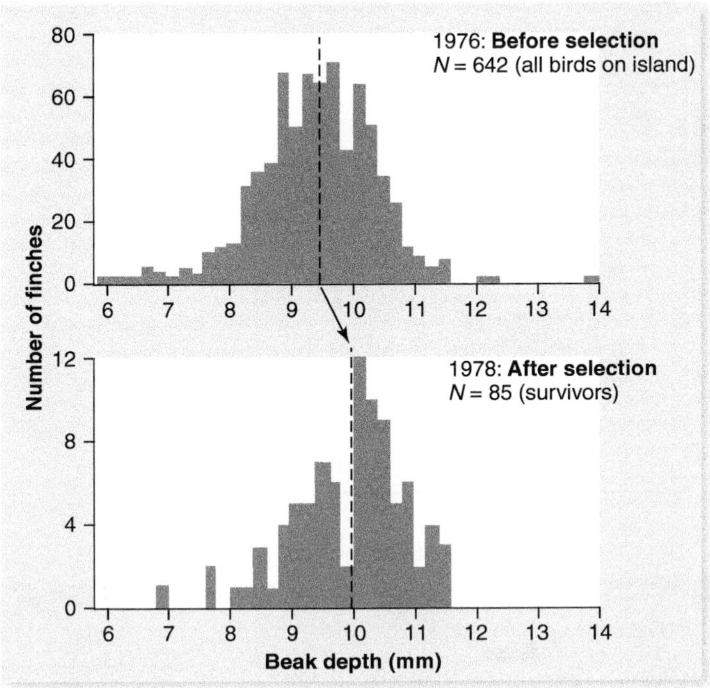

FIGURE 5.1 When a severe drought in 1977 reduced the abundance of seeds, the survival of medium ground finches was not random. Disproportionately more large-beaked birds survived than did their smaller-beaked counterparts. Dotted lines indicate average beak sizes before selection (9.42 mm) and after selection (9.96 mm). (From Boag and Grant, 1981)

of 1982 and 1983 altered the vegetation on Daphne Major to one dominated by grasses and herbs, which have smaller, softer seeds than those produced by *Tribulus cistoides*. This change favored small-beaked birds, and average beak sizes decreased.

Is beak size heritable?

For the puzzle to be complete, beak size must be heritable, a fact that has been successfully demonstrated. Several studies of Galápagos finches have focused on determining the heritability of a trait, whereby *heritability* refers to the proportion of phenotypic variance that is due to genetics. For example, Peter Boag has analyzed finch beak characteristics (width, depth, length) in hundreds of wild-caught parent-offspring combinations, to establish how much of each character is inherited from a finch's parents. His conclusion, that "about 76% of the phenotypic variation in external morphology of *G. fortis* is heritable," has been supported by follow-up work with these and other finches (Figure 5.2).

Finch beaks illustrate evolution in action

These studies of the medium ground finch are included in numerous biology texts to illustrate "evolution in action." The case can easily be made when you consider the main tenets of the theory for evolution by natural selection (Chapter 1):

- Organisms produce more offspring than the environment can sustain.
  - In the short term, resources may not appear to limit production of offspring, especially when a species is first colonizing a new area.

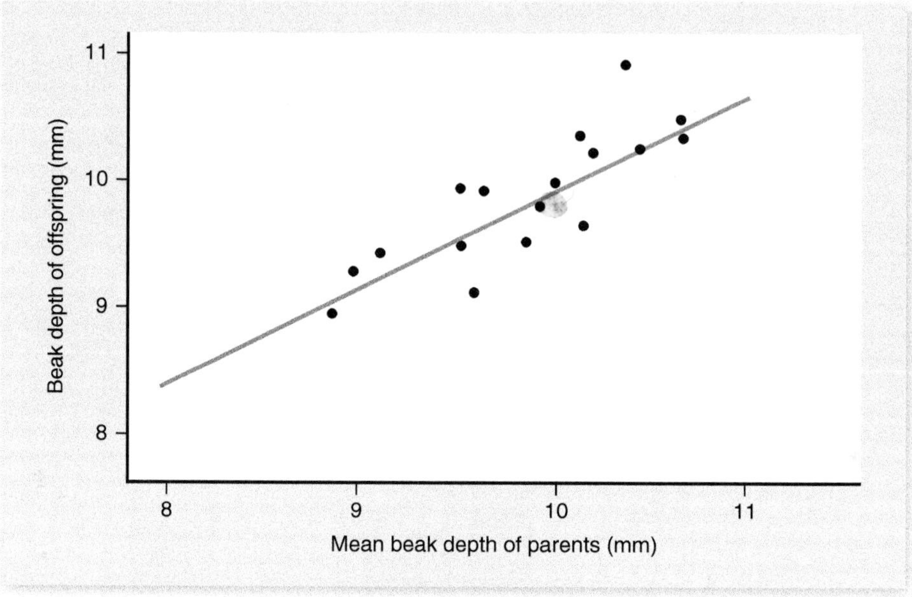

FIGURE 5.2
For natural selection to act on a given trait, some component of that trait (in this case, beak size in medium ground finches) must be inherited. These data show that much of a finch's beak characteristics are inherited from its parents.

However, during the drought of 1977, there was simply not enough food for all the finches to survive on Daphne Major.

- Organisms in a population exhibit individual variability.
  - Many features of ground finches—including the size, depth, and length of their beaks—are variable. In addition, some of this variability is associated with increased survival, and therefore with increased reproductive success.
- Some of the variability among organisms is heritable.
  - In the medium ground finch, for example, beak characteristics exhibit high heritability (Figure 5.2).

Thus, these finches deserve their role as oft-cited exemplars of natural selection, for the evolutionary changes in finches' beaks are an adaptation to food supplies. The selective processes previously described are likely the same mechanisms that led to speciation from an initial population of colonizers, millions of years ago. As Boag and Grant noted when concluding their 1981 paper, "Occasional strong selection of heritable characters in a variable environment may be one of the keys to explaining the apparently rapid adaptive radiation of the Geospizinae in the Galápagos."

Common name: Large ground finch

Scientific name: *Geospiza magnirostris*

Comparisons: *G. fortis*, the medium ground finch, is smaller and has a narrower beak.

See it: Arid zones of most islands (except Floreana, Española, and San Cristóbal)

Residence status: Endemic

Conservation status: Least Concern

What you will see and what it means: The large ground finch is characterized not only by overall size—by far the largest of the finches, the adult male can exceed 16 centimeters—but by its large, robust beak, which is an adaptation for eating bigger and harder seeds (e.g., *Tribulus cistoides*).

Ground finches and character displacement

When two similar species co-occur, as is often the case with the medium and large ground finches, competition can be intense. Given the potentially high costs of competition (e.g., time and energy expenditure, physical stress from fighting or fleeing), it makes sense that selection may favor physical or behavioral modifications that reduce competition. Sympatric species—that is, species that occur together—that reduce conflict by diverging in some trait exhibit *character displacement.* A simple test for character displacement involves analyzing two species that co-occur in some areas, yet occur independently of each other as well. For example, medium and small ground finches co-occur on Santa Cruz, but were originally *allopatric* (i.e., occurring alone) on Daphne Major (the medium ground finch) and Los Hermanos (the small ground finch) islands. Where the finch populations were isolated, they exhibited a range of beak sizes, with an average beak depth of about 10 millimeters; where the birds were sympatric, they appeared to diverge in beak size, with average beak depths of about 12 millimeters (medium ground finch) and about 8 millimeters (small ground finch; Figure 5.3).

Character displacement in action

Although several similar examples of character displacement exist, Galápagos finches provide a peek at the actual *process* of character displacement. In 1982, a population of large ground finches was established on Daphne Major with the arrival of two females and three males. The principal food of the large ground finch is the seed of *Tribulus cistoides,* the same "goat's head" seed that the medium ground finch relies on during the dry season. Thus, the two finch species were potential competitors, and selection was predicted to result in divergence in beak sizes. After two decades, in which the population of large ground finches increased, a severe drought occurred from 2003 to 2004. Mortality on Daphne Major was high for both species; notably, the surviving medium ground finches had smaller beaks and consumed smaller seeds. The medium ground finch had undergone character displacement, thereby reducing competition with a similar species: beak sizes did not change until the large ground finch population was sufficient to exert competitive pressure on the medium ground finches. The drought of 2003/2004 produced the *opposite*

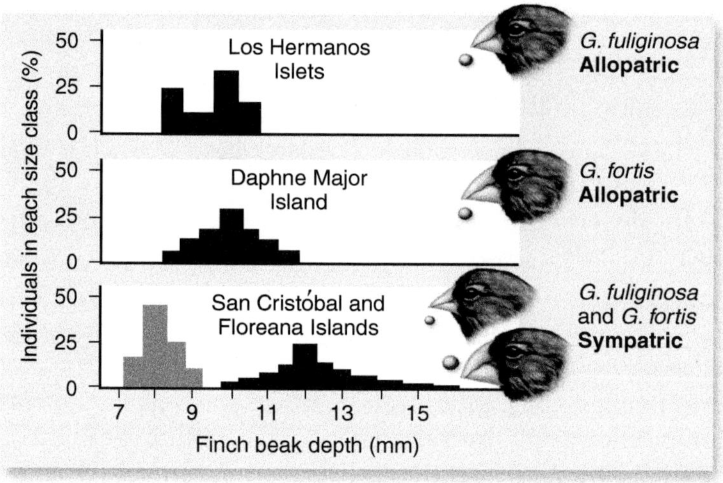

FIGURE 5.3 Character displacement is an evolutionary phenomenon in which competing organisms reduce competition by diverging with respect to some component of their niches. For example, when medium and small ground finches co-occur (i.e., are sympatric), they diverge in beak depth, with beak depths of medium ground finches averaging 12 millimeters, and those of small ground finches averaging 8 millimeters. Where they do not co-occur (i.e., are allopatric), there is considerable overlap in beak depth, with an average of 10 millimeters for both species.

effect on finch beak sizes, whereby average beak depth of medium ground finches *increased* after the drought. As Peter and Rosemary Grant noted in their description of this event, "[t]his is the strongest evolutionary change seen in the 33 years of the study."

Common name: Small ground finch

Scientific name: *Geospiza fuliginosa*

Comparisons: The medium ground finch may have a slightly longer beak, and the sharp-billed ground finch has a more pointed beak.

See it: Arid zones on most islands (except Darwin, Wolf, and Genovesa); also seen higher, in *Scalesia* forest zones on some islands

Residence status: Endemic

Conservation status: Least Concern[7]

What you will see and what it means: With its smaller beak, the small ground finch can exploit a different ecological niche, eating the small seeds of *Sesuvium edmonstonei* (Galápagos carpetweed) and *Tiquilia fusca* (grey matplant).

[7] Population status is based on IUCN Red Book data from 2008/2009, and does not reflect current concerns about introduced parasites such as *Philornis downsi* (see Chapter 11).

How can we understand variation in clutch sizes?

If you see nests of the small ground finch, you may note that those in the highlands tend to lay smaller clutches of two or three eggs, whereas lowland birds lay three to five eggs, with four being the most common clutch size at lower elevations. Why do highland birds lay fewer eggs per clutch?

Clutch size, or number of offspring produced per reproductive cycle, is an important life-history trait that has received much attention in evolutionary studies. In general, large clutch size is associated with increased mortality (especially immediately post-hatching), decreased parental care, and smaller relative size at emergence. Also, clutch sizes vary as a function of food supply, whereby larger families are possible when resources are plentiful and reliable. In the small ground finch (a species that breeds in wet, productive highlands as well as seasonal lowlands), you might predict greater clutch size in the highland populations. Yet in some cases, large clutches of dependent young can make a family more conspicuous to predators, thus imperiling all the offspring.

To understand more about clutch size variation in these birds, Sonia Kleindorfer spent several months visiting finch nests in the highlands and lowlands on Santa Cruz. She documented significant variability in life history traits between these two populations: highland birds averaged 2.5 eggs per clutch during every sampling period (and never had more than three eggs in a clutch), whereas lowland birds had clutch sizes between zero and five eggs. Moreover, in the highlands, finches often re-nested after a predation event, whereas no re-nesting was observed in the lowlands. Finally, highland nests suffered ten times greater predation—by introduced cats, introduced rats, and endemic short-eared owls—than did the nests of lowland small ground finches (52% versus 5%, respectively). Kleindorfer attributes the variability in lowland clutch sizes to rainfall and food availability, whereby more rain = more food = more eggs in the nest. In the highlands, however, where food is always plentiful and predation is intense, finches may lay fewer eggs to reduce their probability of detection. In summary, the environmental constraints imposed on clutch size appear to vary—from food availability at low altitudes to predation pressure at higher altitudes.

Common name: Sharp-beaked ground finch

Scientific name: *Geospiza difficilis*

Comparisons: The small ground finch has a blunter beak; also, the two species co-occur only in the highlands of Santiago, Pinta, and Fernandina. Populations of *G. difficilis* vary significantly on different islands; for example, individuals on Genovesa weigh 12 grams, whereas those on Santiago weigh 27 grams.

See it: On the small, northern islands of Wolf, Darwin, and Genovesa, as well as in *Zanthoxylum* forests in the highlands of Santiago, Pinta, and Fernandina (and possibly the volcanoes of Isabela). *G. difficilis* lived on other islands until humans destroyed their habitats (Chapter 11).

Residence status: Endemic

Conservation status: Habitat destruction has eliminated the sharp-beaked ground finch from Santa Cruz, San Cristóbal, Floreana, and possibly Isabela. Overall, however, the sharp-beaked ground finch population is considered by the IUCN to be of Least Concern.

What you will see and what it means: On low islands such as Genovesa, the sharp-beaked ground finch eats pollen, nectar, and small seeds, whereas those on Pinta, Santiago, and Fernandina have blunter, stronger beaks and eat arthropods, mollusks, fruits, and seeds. The sharp-beaked ground finch is notorious for its resourcefulness, feeding on various items that range from the lice and blood of Nazca boobies to the ectoparasites of iguanas. On Darwin and Wolf, a subspecies of this finch is so bloodthirsty it is called the "vampire finch" (these finches also break open and eat booby eggs). Watch for sharp-beaked ground finches loitering around iguana and booby colonies.

What Is a Species?

It is surprising that, given the frequency and facility with which the term is used, there is no universally accepted definition of "species." Historically, and in many contemporary uses, a species is defined as a group of actually or potentially interbreeding populations of individuals. This is the *biological species concept,* and its applications are vast. However, there are limitations, as documented in the cases of asexual organisms and hybridizing populations. In addition, proponents of the *phylogenetic species concept* argue that the reproductive definition is evolutionarily irrelevant; instead, a species should be the smallest taxonomic unit, which is an intact group of organisms that share common ancestry. Yet the point at which one population is considered two distinct species is itself somewhat arbitrary, and is based on perceived similarities—be they genetic, morphological, or behavioral. The situation is further confused by numerous other definitions of species.

In the typical model of speciation, a population is somehow subdivided into two or more allopatric (i.e., physically isolated) populations. These populations experience different environments and different mutations, and thus diverge in their evolutionary trajectories. Over time, the populations are sufficiently distinct to be considered separate species. This distinction may be due to some sort of *reproductive isolating mechanism,* which can be either a *pre-mating isolating mechanism* or a *post-mating isolating mechanism.* Pre-mating isolating mechanisms are often behavioral differences (such as bird songs) or habitat preferences that bar individuals from courtship or copulation. Post-mating isolating mechanisms often result from genetic incompatibility, whereby offspring may result but they are not viable.

Species concepts and the sharp-beaked ground finch

For those who enjoy wrestling with species concepts, the sharp-beaked ground finch is a dream come true. Galápagos finches in general illustrate an adaptive radiation, with various stages in the process of speciation. The sharp-beaked ground finch occurs in a number of established "varieties," warranting its evaluation for novel species. In addition, populations of this finch are morphologically and ecologically (in terms of habitat preferences) more distinct than are populations of any other species of finch. Further confounding their status, some populations of sharp-beaked ground finches are almost indistinguishable from the cactus finch (the Wolf and Darwin populations) or the small ground finch (the Genovesa population). In fact, genetic analyses support the distinctiveness of the Genovesa population, placing it in a group with the other ground finches and the cactus finch, and separate from those of populations of other sharp-beaked ground finches (Figure 5.4). To complicate things even more, the Genovesa population is morphologically similar to other sharp-beaked ground finch populations, except for its overall size, which is similar to that of the small ground finch. Even the authors of a recent report on this finch's status—Peter and Rosemary Grant and Kenneth Petren—are hesitant to make species recommendations. Instead, they note that "[f]rom a phylogenetic perspective the best taxonomic treatment of the Genovesa population is to consider it a distinct species. . . . Under the biological species concept, however, assignment [as a separate species] is not well supported." They conclude, "[r]ecognizing these difficulties we follow the traditional practice of considering the Genovesa population to be a well-differentiated population of the *G. difficilis* species."

Common name: Cactus ground finch, or common cactus finch

Scientific name: *Geospiza scandens*

Comparisons: With its characteristic long, down-turned beak, the common cactus finch is unlikely to be confused with sympatric finches (namely, the small, medium, and large ground finches).

See it: Arid zones of most islands (except Darwin, Wolf, Genovesa, and Española)

Residence status: Endemic

Conservation status: Least Concern

What you will see and what it means: Again, it's all about the beak! The feeding ecology of the common cactus finch highlights how the beak is adapted more for extreme conditions than for everyday feeding habits. While the cactus finch's beak appears to be specialized for eating cactus, these finches rely exclusively on the fruit of *Opuntia* only when other types of food are in short supply.

FIGURE 5.4
Although morphologically similar, molecular evidence suggests that populations of sharp-beaked ground finches may represent at least two separate species of *Geospiza*. In particular, the Genovesa population seems to belong to a group that includes the other ground finches and the cactus finch, but is separate from the other populations of sharp-beaked ground finches.

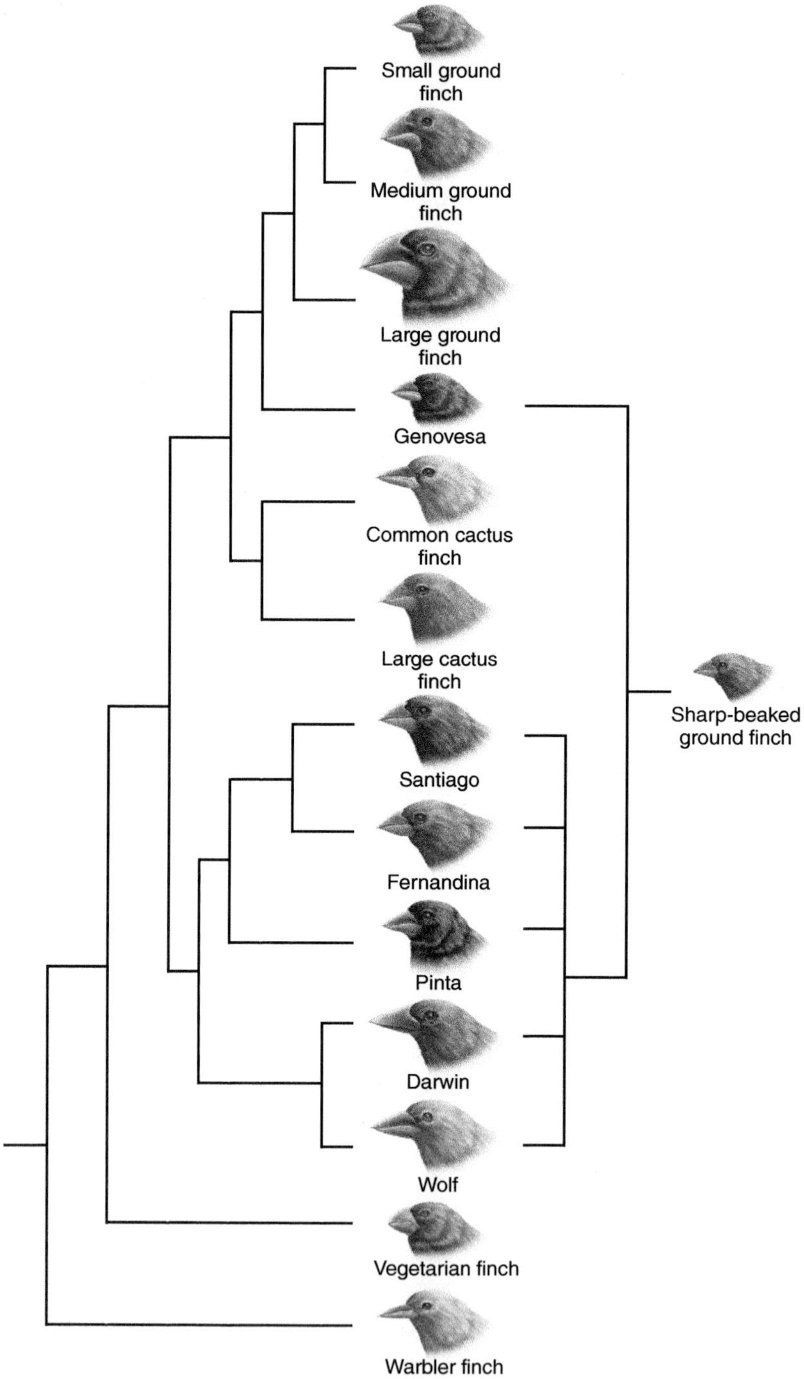

Do the finches hybridize?

Given their similar sizes and ecological niches, you might predict that selection would increase the differences between the common cactus finch and the medium ground finch; by diverging in certain characteristics—such as beak size and shape—these finches could exploit different food sources and avoid some of the costs of competition. (This type of character displacement was addressed earlier in this chapter in the discussion of the large ground finch.) Yet on Daphne Major during the 1990s, the cactus finch and medium ground finch converged on several measurements for beak shape; in particular, the beak of the cactus finch became blunter and more like that of the medium ground finch. Was this due to hybridization between the two species?

While reproductive isolating mechanisms (discussed above) clearly exist between the two species (otherwise they could not remain separate species and still co-occur), the barrier to hybridization is not perfect. Cactus finches and medium ground finches occasionally mate in the wild, and viable offspring—themselves capable of reproduction—result from these pairings. Successful hybridization raises a suite of questions about mate choices, speciation, and coevolution. For example, how extensive is hybridization between finch species? What factors govern the decision to mate with a heterospecific individual rather than one of the same species? Do the hybrid offspring mate with other hybrids, or do they *backcross*—that is, do they mate with an individual of their mother's or father's phenotype? And finally, is the increasing similarity between cactus finches and medium ground finches on Daphne Major due to hybridization—and thus, genetic homogenization—between the two species?

When cactus finches hybridize

In general, finches discriminate between potential mates based on morphology (e.g., beak characteristics) and behavior (e.g., song). Despite there being no differences in reproductive fitness of hybrids, hybridization is rare: in 20 years of observations on Daphne Major, few medium ground finches (<1% of pairs) mated with cactus finches. Of the hybridization events that were observed, finches did not appear to select heterospecific mates because of a lack of available same-species mates. So, with other options available, why would a finch mate with an individual of another species?

Hybridization may result from a failure to imprint on the song of the "correct" species. In many birds, including these finches, females select mates that sing like the females' fathers; males sing songs essentially identical to those they learned from their fathers. For example, in 56 father-son pairs of cactus finches, songs were faithfully transmitted from father to son, without evidence of misimprinting. In another study, more than 95% of the 482 female finches mated only with males who sang songs of males of their own species, indicating that females use songs to identify their own species.[8] However, imprinting mistakes do occur. On Daphne Major, for example, a few cactus finch males, singing medium ground finch songs, mated with medium ground finch females. Thus, hybridization occurs, and misimprinting on the songs of heterospecific males may be much of the reason.[9]

Introgression between cactus finches and medium ground finches

Even a rare (but recurring) event like hybridization can alter genomes. *Introgression* describes gene flow between species that hybridize and then *backcross*—specifically, when a first-generation hybrid mates with an individual of one of the parent species rather than another hybrid individual. During hybridization of cactus finches and medium ground finches, backcrosses may occur disproportionately between hybrid offspring and cactus finches, rather than between hybrids and medium ground finches. As a result, medium ground finch DNA is likely to become part of the cactus finch lineage. Evidence for this trend is genetic and morphological: several genetic markers that were unique to the medium ground finch in 1983 were frequent in the cactus finch in 2000; and during the same period, the cactus finch became more like the ground finch in beak size and shape. Time will tell the outcome of this trajectory—will the finches diverge again at some major climatic event? Will introgression ultimately lead to "despeciation" as mating barriers break down? Whether we live long enough to know these answers, one thing is certain: Finch populations in Galápagos continue to evolve, and the long-term evolutionary products are difficult—if not impossible—to predict.

Common name: Large cactus finch

Scientific name: *Geospiza conirostris*

Comparisons: The large ground finch, which is slightly larger when fully mature, only co-occurs with the large cactus finch on Genovesa.

See it: Arid zones of Genovesa and Española

Residence status: Endemic

Conservation status: Least Concern

---

[8] Many researchers have wondered if an individual finch's song is a genetic trait passed from father to son, or is a cultural trait that a son learns from its father. These hypotheses were tested by comparing the songs of sons with those of their paternal and maternal grandfathers. If the song is similar to that of both grandfathers, it was probably transmitted genetically, since a son inherits genes from both grandfathers. However, if the song is similar only to that of the paternal (and not the maternal) grandfather, it was probably learned (i.e., transmitted culturally), since the paternal grandfather gave the song to the son's father, who would have given the song to his son. The results were clear: A son's song resembles that of its paternal grandfather, indicating that the song is transmitted culturally. This cultural transmission of songs affects gene flow between species, helps females recognize their own species, and accounts for why medium ground finches seldom mate with cactus finches.

[9] David Lack paid relatively little attention to finches' songs because he had no way to record the songs.

Is the large cactus finch monogamous?

The life history of the large cactus finch is well established: females, the limiting sex, choose males based on plumage characteristics (preferring darker males, who are older and more experienced). Females then incubate the eggs, and both parents feed the nestlings. Biparental care is probably essential in this species, because when one parent dies or leaves during the nestling stage, the young are always abandoned and die.

This two-parent imperative likely imposes constraints on promiscuity, with penalties including outright abandonment—thus reinforcing monogamy. At the very least, a premium should be placed on *social monogamy,* whereby there is the appearance of monogamy, but with instances of nonpaternity due to female promiscuity. In some bird populations, rates of nonpaternity (evaluated by genetic markers) can be as high as 40%. In the large cactus finch, the estimate is a mere 5%—that is, 5% of the offspring in a population are raised by a male that is not their biological father.

What motivates a female cactus finch to cheat on her mate?

In the 1980s, Peter and Rosemary Grant hypothesized that a female is most likely to seek or accept extra-pair copulations if she is paired with a low-quality male. In the case of the large cactus finch, as with many other Galápagos finches, a low-quality male is one that is predominantly brown rather than black. Lighter plumage is a characteristic of younger, inexperienced males, and females may see quality in a darker male that has proven his ability to survive for several years. Because DNA testing for paternity was not yet common, the Grants evaluated female motivations instead. Observations of females that desert their mates for another male were revealing: in all mate-switching scenarios, females chose an older, darker, more experienced male for her new partner. Her choice makes sense in light of the fact that females increase their fitness by mating with these more desirable males; darker males fledge significantly more young than do their lighter competitors. Thus, in spite of the high potential cost of desertion, there can be tenable benefits to cheating.

Common name: Vegetarian finch

Scientific name: *Platyspiza crassirostris*

Comparisons: With its large size (16 cm when mature) and characteristic coloration (light underparts, dark upper, dark streaking on the female's chest), the vegetarian finch is unlikely to be confused with any sympatric finches.

See it: Transition zones of Isabela, Santa Cruz, Fernandina, Santiago, San Cristóbal, Floreana, Marchena, and Pinta

Residence status: Endemic

Conservation status: Although the vegetarian finch is considered by the IUCN to be of Least Concern, it may be heading toward extinction on Floreana; a recent survey found only one vegetarian finch there.

What you will see and what it means: As the only member of the genus *Platyspiza,* the vegetarian finch is thought to be among the earliest species of Galápagos finches to diverge from the ancestral group. This finch feeds almost exclusively on plants—leaves and stems as well as flowers, fruits and seeds—and often uses its beak to remove dead bark from young Galápagos croton (*Croton scouleri*) plants to feast on the sugary bark.

Common name: Medium tree finch

Scientific name: *Camarhynchus pauper*

Comparisons: The small tree finch has a blunter beak than the medium tree finch.

See it: Only in the highlands of Floreana

Residence status: Endemic

Conservation status: The medium tree finch is classified as Critically Endangered on the IUCN Red List, a classification often assigned to organisms with restricted ranges. This designation is due to a combination of factors, primarily habitat destruction from agricultural practices on Floreana. This finch is also, like most other Galápagos finches, vulnerable to the effects of introduced pests such as avian pox.

What you will see and what it means: Relatively few travelers venture to Floreana's highlands, and if you encounter a medium tree finch you will be doubly fortunate. A visual positive identification is difficult-to-impossible, given that this finch co-occurs with the large and small tree finches and is only slightly larger than the small species. Look for the pointed, down-turned beak in a bird about 12 centimeters tall.

Philornis downsi *and the fate of the medium tree finch*

The blood-sucking parasitic fly *Philornis downsi* has been on the islands since the 1960s, and was first detected in finch nests in 1997. In the past decade, several reports have warned of *P. downsi*'s potential for destruction. Female flies lay their eggs in finch nests, after which the larvae emerge and begin eating the tissues of nestlings. Larvae can consume over half a chick's blood by volume, leading to near-total mortality in some populations. *P. downsi* poses a special problem for the tree finches, as the parasite appears to thrive in the moist highlands along with the small, medium, and large tree finches. A 2004–2006 study of tree finches in the highlands of Floreana revealed that 100% of nests sampled were infested with fly larvae and pupae. The cost of the infestation is clear: *P. downsi* was the main cause of nestling mortality in the tree finches, with particular consequence for the medium tree finch. Specifically, 41% of medium tree finch nestlings died from the effects of *P. downsi*. Given the limited range of this finch, this heightened vulnerability to a novel parasite poses a critical problem for conservation biologists.

Common name: Small tree finch

Scientific name: *Camarhynchus parvulus*

Comparisons: The warbler finch, which is sympatric, similarly colored, and also small (about 10 cm), has a thinner beak than the small tree finch.

See it: In the *Scalesia* and transition zones, as well as some arid zones, of most islands (except Española, Marchena, Genovesa, Darwin, and Wolf; possibly extinct on Rábida as well)

Residence status: Endemic

Conservation status: Least Concern

What you will see and what it means: Although you will see various finches throughout Galápagos, one thing you may *not* see is the nest of the small tree finch. In fact, the ability to construct a concealed nest is part of the sex appeal of the male finch, a proclivity demonstrated by recent work on mate choice in this species.

Mate-choice mathematics in the small tree finch

Intersexual selection, whereby selective pressure involves the mating preferences of one sex for another, has been extensively studied in birds. Typically, females choose males based on phenotypic traits such as plumage coloration, feather length, call frequency or duration, or features of the male's territory. In the small tree finch, a preference for older males (identified by black plumage) has also been documented, an observation that begs the question: why are older males so appealing?

In her recent studies of small tree finches on Santa Cruz, Sonia Kleindorfer hypothesized that females benefit by mating with older males because older males build better, more concealed nests that are consequently less vulnerable to predation. Her observations were designed to test four predictions associated with this hypothesis:

1. Concealed nests are less vulnerable to predation than their less-concealed counterparts.
2. Older males build nests that are more concealed.
3. Females are more likely to choose males that build the most-concealed nests.
4. Fledging success (a measure of offspring survival) is higher for the offspring of older males.

Four years of monitoring the successes and failures of 176 tree finch nests revealed that older males build more-concealed nests, concealed nests experience less predation than more-visible nests, and females prefer older (darker) males. Ten males studied for long-term trends had less fledging success in 2000 (2 of 10 nests), when they were young, than they did in 2004 (7 of 10 nests), when they had aged, and had become darker and—presumably—better nest-builders. Collectively, these data suggest that choosing older males gives female small tree finches an indirect-fitness benefit, by enabling them to raise their offspring in a safer nest.

Solitary versus group nesting in small tree finches

During a walk through the highlands around Los Gemelos on Santa Cruz, you may see small tree finches nesting in mixed-species aggregations that include large tree finches, small ground finches, woodpecker finches, and warbler finches. These nesting patterns are intriguing, for communal nesting—either with individuals of the same or different species—carries established costs (e.g., competition for limited resources) and benefits (e.g., protection from predation) that may also affect tree finch mate choices. While conducting the field study discussed previously, Kleindorfer and her colleagues hypothesized that females would choose male small tree finches from mixed-species nesting aggregations, and these females would enjoy enhanced indirect fitness via increased offspring security. Females *did* disproportionately select males in mixed groupings, in contrast with solitary males (68% versus 35%, respectively). In addition, chicks in solitary nests experienced greater predation than did chicks in mixed nesting associations. Thus, there is a defined benefit to nesting in mixed groups.

Philornis downsi *and the costs of group nesting*

There are costs to group nesting as well. For example, in the case of the small tree finch, the number of *P. downsi* parasites per nest is greater for small tree finches in mixed nesting groups—65% of nests in groups were parasitized, as opposed to 35% of solitary nests. Clearly, *P. downsi* exacts a grim fitness cost from its hosts. However, this parasite has only been a threat for a few years, and nesting in mixed-species aggregations is a behavior that evolved long before the introduction of *P. downsi*. One could predict greater representation of solitary nesters in the near future, as those who opt to nest in groups experience greater mortality from *P. downsi*.

Common name: Large tree finch

Scientific name: *Camarhynchus psittacula*

Comparisons: The equally large vegetarian finch, which co-occurs with the large tree finch, lacks the large tree finch's curved beak, which gives the large tree finch a parrot-like appearance.

See it: In humid, evergreen forests in the highlands of Fernandina, Isabela, Pinta, Marchena, Santiago, Santa Cruz, Santa Fé, and Floreana

Residence status: Endemic

Conservation status: Least Concern

What you will see and what it means: Watch for the large tree finch's characteristic feeding behavior, whereby it uses its curved beak to peel back the bark of trees and eat arthropods living below.

How can we understand the existence of small, medium, and large finch varieties?

Again, beaks distinguish the finch species. In ground finches and tree finches, we identify the species—small, medium, or large—by bill size and shape. It would be difficult to deny the importance of variability in, and selection on, beak size and shape in the ongoing evolution of Darwin's finches. But questions remain, particularly with respect to proximate explanations for beak morphology—for example, the molecular origin of beak variation. What is the genetic basis of beak size and shape? Can we track similarities among, and differences between, the finches—from genes to the fully formed individual?

The emerging field of evolutionary developmental biology (*evo-devo*)—which unites the disciplines of molecular biology, developmental biology, and evolution—has revealed surprising similarities within groups of living organisms. For example, a regulator protein involved in human skull development, *bone morphogenetic protein 4,* or *BMP4,* also plays a pivotal role in finch beak development. *BMP4* activates genes that promote bone mineralization. Recently, developmental biologist Cliff Tabin and his colleagues collected eggs from six finch species—ground finches and cactus finches—that vary markedly in beak morphology. Each finch species has a unique *BMP4* "signature," with the timing and extent of *BMP4* expression reflecting specific beak characteristics. Finches with larger, thicker beaks show earlier and more pronounced *BMP4* activity during embryonic development. As confirmation of this connection, adding *BMP4* to developing chick tissues produced birds having thicker and wider beaks; conversely, injecting embryos with a protein that blocks *BMP4* produced chicks with smaller beaks. Another protein, *calmodulin* (or calcium-modulated protein, *CaM*), is a calcium-binding protein that also contributes to beak development. Higher levels of *CaM* are correlated with the longer beaks of cactus finches.

Apparently, fundamental differences between these finch species is not due to dramatic genetic alterations, but rather result from small changes in the timing and extent of expression of these two proteins. Presumably these slight alterations in gene expression afforded finch progenitors fitness advantages, thus creating the present-day distribution of birds defined by their beaks. Evolutionary developmental biology adds another exciting chapter to our understanding of evolution, and it is hardly surprising that Galápagos finches are playing a key role once again.

Common name: Woodpecker finch

Scientific name: *Cactospiza pallidus*

Comparisons: With its large size (18 cm), pale coloration, and elongated, downturned beak, the woodpecker finch is among the most distinctive of the finches.

See it: Transition and *Scalesia* zones on Fernandina, Santiago, Isabela, Pinzón, Santa Cruz, and San Cristóbal

Residence status: Endemic

Conservation status: Least Concern

What you will see and what it means: Like that of the other tree finches, the beak of the woodpecker finch is an adaptation for feeding on arthropods lodged under bark or in tree holes and crevices. However, the woodpecker finch has attracted additional attention for its use of twigs and cactus spines to retrieve otherwise inaccessible food items—particularly during food-limited seasons. You may, if you are lucky, see a woodpecker finch try to retrieve a food item from a tree hole; if unsuccessful, the finch may snap a thin twig or cactus spine and carry this tool back to the hole (Color Image 8). After using the tool to stab or lever the prey into the reach of its beak, the bird may tuck the implement under its foot for future use, and then use its beak to finish the task.

Tool use in the woodpecker finch

Although examples of premeditated tool use are notable in many animals, and have been much studied in the higher apes (e.g, humans, chimpanzees), little is known about the development of this behavior in birds. As a result, tool use by the woodpecker finch has motivated studies on the evolution of this behavior. Neither young nor adult finches appear to learn how to use tools from watching tool-using birds. Rather, tool use seems to result from an innate predisposition to learn during a sensitive post-hatch stage of development. Woodpecker finches learn tool use techniques through trial and error, and out of necessity.

Common name: Mangrove finch

Scientific name: *Cactospiza heliobates*

Comparisons: The woodpecker finch, which is considerably larger, is similar in color and beak shape to the mangrove finch.

See it: In two mangrove swamps on the northwest coast of Isabela

Residence status: Endemic

Conservation status: With fewer than 50 potentially breeding pairs, the mangrove finch is the most endangered of Galápagos finches, and is classified as

Critically Endangered on the IUCN Red List. Suggested reasons for the decline of this finch are numerous, and include habitat destruction, introduced predators (e.g., rats and cats), and the introduced parasite *P. downsi* (discussed previously).[10]

What you will see and what it means: While you are unlikely to encounter one of these birds, the mangrove finch is notable for its relatively extreme habitat specificity, as well as its ability to use tools in feeding. In fact, this finch's specialized niche only exacerbates its conservation status: mangrove finches feed on red mangrove buds, but they cannot nest in red mangrove. Instead, they nest in the outermost branches of black mangrove trees. Conservationists attempting to relocate the remaining finches to a safer habitat (e.g., one less threatened by rats and cats) face the challenge of meeting the mangrove finch's particular habitat requirements.

Common name: Green warbler finch

Scientific name: *Certhidea olivacea*

Comparisons: While visually indistinguishable from the grey warbler finch, the green warbler finch can also be confused with the small tree finch, which is similar in size and overall color, but lacks the fine beak that makes this finch so "warbler-like." Warbler finches are among the oldest of Galápagos finches. Their beaks are not like those of their seed-eating relatives of South America or the Caribbean.

See it: *Scalesia* zones on Santiago, Rábida, Pinzón, Isabela, Fernandina, and Santa Cruz

Residence status: Endemic

Conservation status: Least Concern

What you will see and what it means: See below (in discussion of *C. fusca,* the grey warbler finch)

Common name: Grey warbler finch

Scientific name: *Certhidea fusca*

Comparisons: Like the green warbler finch, the grey warbler finch can also be confused with the small tree finch.

See it: Arid zones on Pinta, Marchena, Darwin, Wolf, Genovesa, Santa Fé, San Cristóbal, Española, and Floreana

Residence status: Endemic (recently recognized as evolutionarily distinct from the green warbler finch)

---

[10] In response to concerns about the effects of *P. downsi* on several birds in Galápagos (including the rare mangrove finch, medium tree finch, and Floreana mockingbird), GNPS and CDF hosted a *Philornis* Workshop from January 31 through February 3, 2012, in Puerto Ayora, Santa Cruz. Participants included scientists from eight countries, including Ecuador, the United States, Trinidad, and Argentina. Workshop participants established several research and conservation priorities, among them a plan to protect the critically endangered mangrove finch.

Conservation status: Least Concern

What you will see and what it means: Unlike other insectivorous finches on the islands, warbler finches can capture prey during flight. But how will you know which warbler finch you are watching? Your best guide is their distribution: if you are on one of the larger, central islands (e.g., Santiago, Rábida, Santa Cruz), the finch is probably *C. olivacea,* the green warbler finch. If you are visiting one of the smaller, outer islands (e.g., Genovesa, Española), it is probably *C. fusca,* the grey warbler finch.

Cryptic speciation in warbler finches

Collectively, warbler finches are found on every major island in Galápagos, making them among the most widespread of the finches. Although formerly considered a single species (*Certhidea olivacea*), two distinct species are now recognized—*C. olivacea,* or the green warbler finch, and *C. fusca,* the grey warbler finch. The two species are morphologically similar and have similar calls. However, a 1999 analysis of genetic markers in hundreds of individuals, representing most of the islands' finch species, revealed something unexpected: the warbler finches could be easily divided into two groups based on genetic differences. In fact, using four different algorithms to construct finch phylogenies, no reconstruction of their evolutionary history even places the two *Certhidea* as sister species! Instead, every phylogeny constructed from DNA evidence shows *C. olivacea* as basal to the other finches. In other words, the green warbler finches are thought to represent the first lineage to split from the ancestral finches on the islands (Figure 5.5).

Ongoing research is focused on uncovering more of the history of *Certhidea* speciation. One tantalizing clue is that the two warbler finches sing different songs, a distinction that might have served as a pre-mating isolating mechanism—that is, grey warbler finch females might not be attracted to the song of the green warbler finch, and vice versa, thus maintaining separate lineages. For example, song is a barrier to gene exchange between finches in the ground finch group (see above). However, during playback experiments conducted in the field, some of the green warbler finches (on Santa Cruz) responded to recordings of grey warbler finches; likewise, some grey warbler finches (on Genovesa) responded to green warbler finch calls. Thus, it seems unlikely that separate species have been maintained by song preferences.

The distribution of warbler finches is also a puzzle. It is generally acknowledged that finches fly between islands, and there are many examples of selection reducing competition between similar species on the same island. So, why don't the two species of warbler finches co-occur? Genetic distance (a measure of DNA differences between organisms) does not seem to be correlated with geographic distance. However, there *is* a correlation between habitat type and genetic distance, suggesting that dispersing birds breed in areas similar to those places where they were raised. For example, the islands of Darwin and Española are on far ends of the archipelago, yet have similar habitats, and both house grey warbler finches (Figure 5.6).

Without recent DNA analyses, we would not have known that the two species of warbler finches were, in fact, distinct. They are morphologically and behaviorally similar and appear to be a single species spread throughout the islands. Darwin and many of his antecedents—e.g., David Lack, who argued for a ground finch–like ancestor for the islands' finches—never knew about cryptic speciation in the warbler finches. As Kenneth Petren and colleagues stated in 1999, "[t]he *Certhidea* results suggest that, in some instances, morphology may be a poor guide to the genetic distinctness of populations."

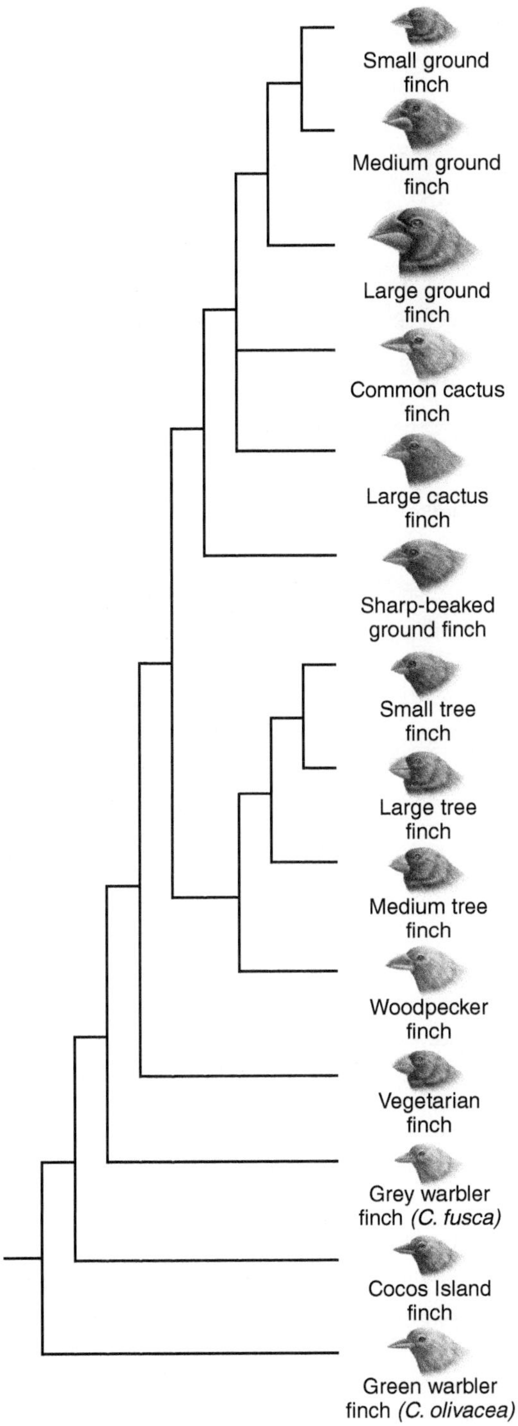

FIGURE 5.5
An evolutionary tree of Galápagos finches, based on genetic (i.e., microsatellite length variation) analysis, reveals some surprising relationships. Namely, the two *Certhidea* species, formerly classified as one warbler finch, may not even be sister taxa. The green warbler finch was the first lineage to split from the ancestral finches on the islands. (From Petren, et al., 1999)

BIRDS OF PREY

Common name: Galápagos hawk

Scientific name: *Buteo galapagoensis*

Comparisons: The only other hawk that occurs in the islands, the osprey, has bright white underparts that distinguish it in flight.

See it: In virtually all types of habitats, on Santiago, Española, Isabela, Fernandina, Pinta, Marchena, Pinzón, Rábida, and Santa Fé

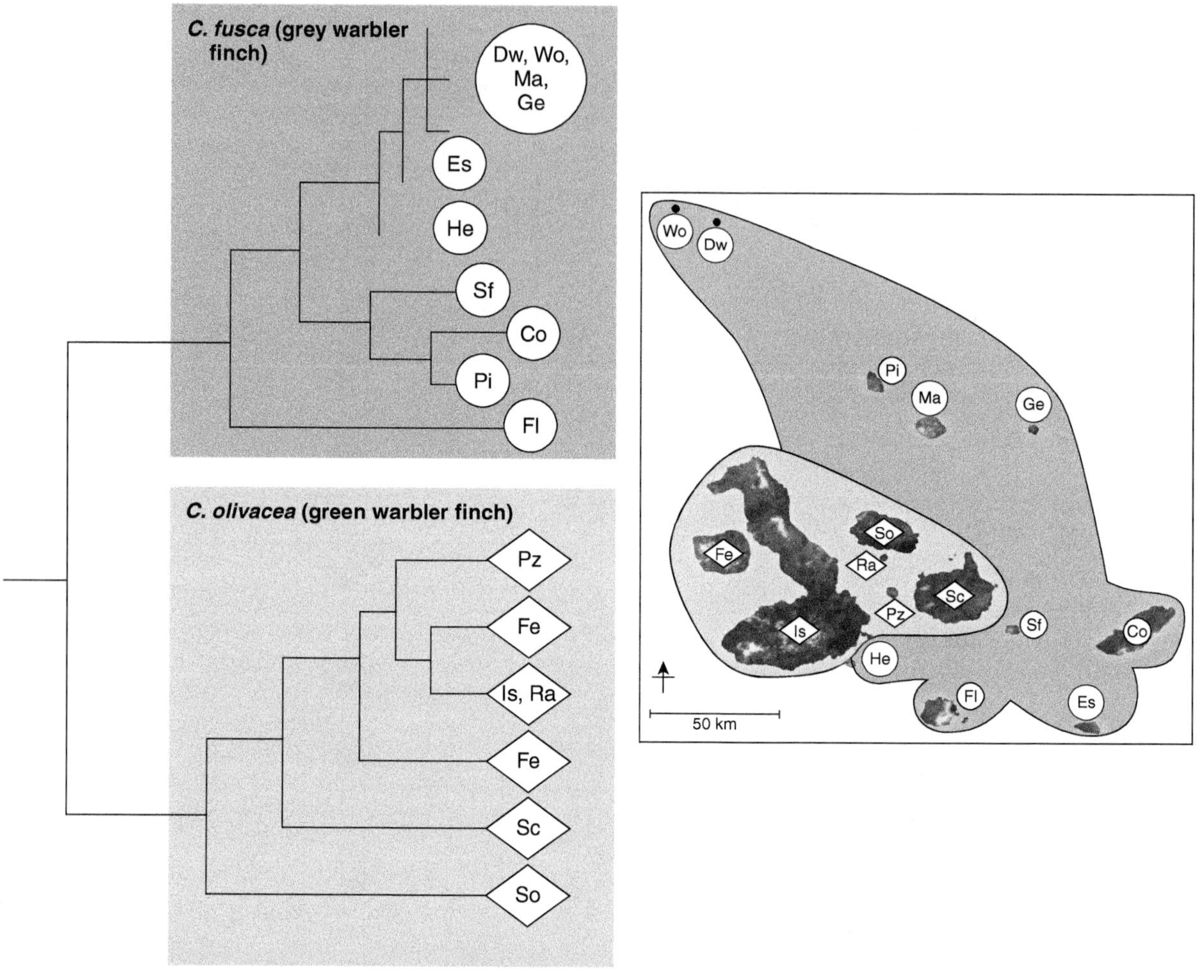

FIGURE 5.6 An analysis of habitat preferences may shed some light on how two similar species such as *Certhidea olivacea* and *C. fusca* can co-occur. A correlation between habitat type (especially the maximum elevation on an island) and the type of *Certhidea* present suggests that dispersing birds breed in areas with habitats similar to those where they were born. Islands are abbreviated as follows: Darwin (Dw); Wolf (Wo); Marchena (Ma); Genovesa (Ge); Española (Es); Hermanos (He); Santa Fé (Sf); San Cristóbal (Co); Pinta (Pi); Floreana (Fl); Pinzón (Pz); Fernandina (Fe); Isabela (Is); Rábida (Ra); Santa Cruz (Sc); Santiago (So). (From Tonnis, et al., 2005)

Residence status: Endemic

Conservation status: Although once common throughout the archipelago, the Galápagos hawk has experienced a population decline (largely due to persecution by humans on inhabited islands) and is now classified as Vulnerable on the IUCN Red List.

What you will see and what it means: The Galápagos hawk gives many visitors a great opportunity to observe predation in action. For example, a hawk may swoop into a colony of nesting boobies and take one of the chicks when its parents are away. Or you may encounter a hawk eating carrion in the middle of a trail (Figure 1.2). The hawk's temerity around humans, combined with its top-predator status (Color Image 9), ensure that an encounter will be memorable. However, in evolutionary circles the hawk is most notable for its mating system of *cooperative polyandry* (i.e., a single female mates with two or more males, who may cooperate in raising the young).

Cooperative polyandry in Galápagos hawks

The mating system of the Galápagos hawk is unusual, in that one female typically mates with two or three males (up to eight males have been observed with one female). *Polyandry* is rare in birds, where social—or behavioral—monogamy is most common, and *polygyny*—one male mating with two or more females—is not unusual. In Galápagos hawks, polyandry is doubly interesting, for the males appear to cooperate in rearing a female's young. Typically, one chick is fledged, yet it is cared for by two or more males. Genetic analysis of paternity has not identified any clear bias in caregiving by the biological father versus the other males at a nest. It also appears that all males copulate with the resident female and have an equal probability of paternity. And in five of six two-chick broods that were analyzed, the nestmates showed mixed paternity—that is, they shared a mother, but had two different fathers. Lastly, all males within a breeding group are unrelated, the result of leaving a natal nest to search for females with territories. But why would a male actively participate in increasing another male's fitness? Why would a male hawk choose to mate with a female that is already mated? What conditions lead to cooperative polyandry?

Some questions about the Galápagos hawk's mating system have been approached by comparing populations of hawks on different islands in the archipelago. On Santiago, where most territories have polyandrous groups, nests average between two and three males per group. On Española, however, all territories sampled are inhabited by monogamous pairs. These differences are presumably not due to sex-ratio limitations, for on Santiago, where most territories are polyandrous, there is a surplus of unpaired females. However, a shortage of resources, such as suitable territories or sufficient food, could lead to cooperative breeding. As yet, there is no accepted explanation for the behavioral differences between hawks on Española and Santiago. Española, at 6,056 hectares, is much smaller than Santiago (58,465 ha), yet so is Pinta (5,940 ha), where polyandry is common. An investigation into the connection between hawks and introduced goats may be revealing: on Española, goats have been gone for decades and hawks are monogamous; on Santiago, goats were recently eradicated (see Chapter 11), and the polygamous hawks had been enjoying goat meat for a long time. Could goat meat be the resource distinction that sets these two islands apart? Was the hawk population "artificially" elevated due to the easy prey of introduced goats (and possibly introduced rodents as well)? If so, what behavioral changes might occur on Santiago, with the recent elimination of goats? These questions await further research; in the meantime, we can only speculate about the connection between goats, hawks, and mating systems.

In summary, if a male defends the territory and raises the young within a group, his chance of fathering one of the successful offspring increases, provided he continues to copulate with the resident female. This is particularly tenable in a long-lived bird such as the Galápagos hawk—a single male may spend over 10 years rearing chicks for one female. He is bound to get lucky and father offspring at some point. While monogamy can give males their best chance at individual fitness gains, polyandry may often be a decent second-best option.

Common name: Short-eared owl

Scientific name: *Asio flammeus galapagoensis*

Comparisons: The only other owl on the islands, the Galápagos barn owl (*Tyto alba punctatissima*), is nocturnal and possesses a distinct heart-shaped face.

See it: On Genovesa, Santa Cruz, San Cristóbal, Floreana, and Isabela

Residence status: *Asio flammeus galapagoensis* is an endemic subspecies. Other *Asio flammeus* occur throughout Asia, Europe, and the Americas.

Conservation status: Least Concern

What you will see and what it means: Given that visitors are not allowed on most noninhabited islands at night, you may be surprised to encounter an owl in broad daylight. On Genovesa, it is not unusual to see short-eared owls perched on the rocks in plain sight. These owls hunt in the morning and late afternoon, preferring to feed on small birds (such as petrels) as well as introduced rats.[11] They are well camouflaged against the rocks, so watch your step! (Color Plate 10)

MOCKINGBIRDS

While the finches of Galápagos have become allied with Darwin's name, it was the mockingbirds that had the strongest impact on the development of Darwin's theory of natural selection (Chapter 2). The four species of mockingbirds on the islands occur almost everywhere, eat almost anything, and are, even by Galápagos standards, very intrepid. Don't be surprised if a mockingbird lands on your hat or tugs at your shoelace. But do resist the urge to pet or feed it.

The Origin of Galápagos Mockingbirds

During HMS *Beagle*'s visit to the archipelago, Charles Darwin gathered mockingbirds ("mocking thrush") from four islands. As he sailed home, Darwin noted that these birds differed from island to island. And, though these specimens

[11] On islands with competitors such as Galápagos hawks and barn owls, short-eared owls are nocturnal.

resembled similar birds on the mainland, they were also distinct. The "thoroughly aroused" Darwin began to formulate his ideas on how this geographical variation arose, and the rest is history—that is, until genetic studies in the last decade clarified the actual relationships among island mockingbirds, and identified their most likely mainland sister species.

In the years since Darwin wrote of his visit to Galápagos, the conventional wisdom about mockingbirds has essentially remained the same: the suggested ancestor and closest relative to island mockingbirds is the long-tailed mockingbird (*Mimus longicaudatus*) of Ecuador and Peru. The islands were colonized by an ancestral population from the mainland, and the current radiation is the result of that single colonization event—that is, the four island species of mockingbirds constitute a *monophyletic* grouping, with shared ancestry. These four species have been designated a separate genus (*Nesomimus*), suggesting that together they are evolutionarily distinct from any mainland group of mockingbirds.

David Anderson and several colleagues (including Rosemary and Peter Grant and Peter Boag; see above) used DNA sequence data—from mitochondrial DNA, or *mtDNA*—to evaluate the conventional wisdom about mockingbird evolution in the archipelago. The investigators analyzed short, single-gene DNA sequences from 60 birds, including representatives from each of the four suggested species on Galápagos, as well as mainland representatives from several different mockingbird species. Their results support a single colonization event, and common ancestry for the Galápagos mockingbirds. However, genetic similarities do not mirror the current classification scheme. Namely, the Galápagos species are not sufficiently distinct to warrant their own genus, and should instead be classified with the mainland *Mimus* group of mockingbirds.

Also, the genetic evidence does not support classification into four species (Figure 5.7). Rather, there appear to be three groups of *Mimus*-like birds on the islands: one group that consists solely of the currently recognized *Mimus trifasciatus,* or the Floreana (Charles) mockingbird; one group that contains birds from Genovesa, San Cristóbal, and Española and includes populations that are currently considered three separate species (*M. parvulus, M. melanotis,* and *M. macdonaldi*); and a third group, allied with the Charles mockingbird, but including populations currently recognized as the Galápagos mockingbird (*M. parvulus,* from Isabela, Santa Cruz, and Santiago). Further work by Paquita Hoeck and colleagues, using nuclear rather than mitochondrial DNA, largely confirms these findings.

Despite its geographic proximity, the long-tailed mockingbird of mainland South America is not the island mockingbirds' closest mainland relative. Instead, the Galápagos populations are genetically more aligned with mockingbirds of Central America and the Caribbean. Specifically, mitochondrial DNA evidence points to the Bahama mockingbird, *M. gundlachii,* as the closest living relative to Galápagos mockingbirds. However, this deduction is based on a relatively small (<3,000 base pairs) comparison of DNA, and may be revised

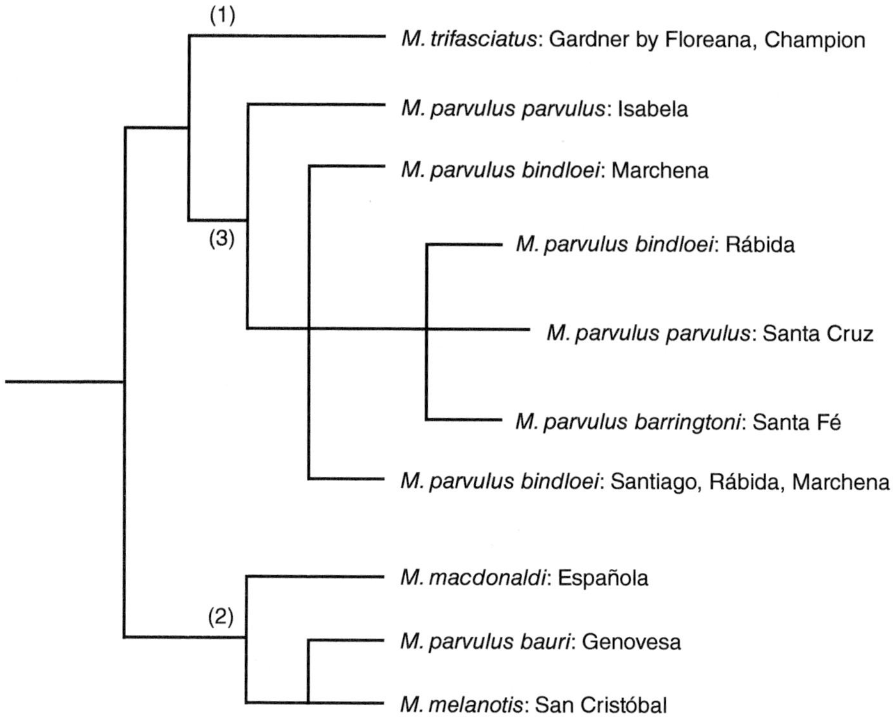

FIGURE 5.7 Molecular evidence disputes the conventional scheme of grouping mockingbirds. Rather than the four species listed in most texts (including this one), the archipelago's mockingbirds appear to belong to one of three groups: (1) the currently recognized *Mimus trifasciatus,* or the Floreana (Charles) mockingbird; (2) a group of birds from Genovesa, San Cristóbal, and Española that includes populations currently considered three separate species (*M. parvulus, M. melanotis,* and *M. macdonaldi*); and (3) populations currently recognized as the Galápagos mockingbird (*M. parvulus,* from Isabela, Santa Cruz, and Santiago). (From Arbogast, et al., 2006)

with further analysis. A Caribbean connection is intriguing, however, as similar affinities have been suggested for the finches (see discussion above), flamingos, and snakes (genus *Antillophis*).

Lastly, the genetic difference between island populations of Galápagos mockingbirds, combined with the estimated mutation rates for mitochondrial DNA, suggest that the adaptive radiation of these birds has occurred in the last 1.6 to 5.5 million years. Many of the currently above-water islands in the archipelago are within this age range, so the evidence converges on a relatively simple history: the islands were colonized by a population of birds from the mainland or Caribbean, possibly the Bahama mockingbird or close relative, and these birds underwent an adaptive radiation in the past few million years.

Common name: Galápagos mockingbird[12]

Scientific name: *Mimus parvulus*

Comparisons: The Galápagos mockingbird does not co-occur with the other mockingbirds on the archipelago, making identification relatively straightforward.

See it: On Santa Cruz (Figure 5.8), South Plaza, Santiago, Santa Fé, Isabela, Fernandina, Pinta, Marchena, Genovesa, Darwin, and Wolf

Residence status: Endemic

Conservation status: Least Concern

What you will see and what it means: You are likely to see many of these birds. Galápagos mockingbirds are, like Galápagos hawks (discussed above), cooperative breeders, with several adults occupying a territory and caring for the young. However, unlike the hawk, some of these adults are nonbreeders; these "helpers at the nest" do not copulate with the resident female, and there is no probability of paternity.

Helpers at the nest and the problem of fitness

Given that reproductive output (fitness) is the currency by which evolutionary success is measured, helpers at the nest—a phenomenon seen in many birds—pose a biological problem: how could a system evolve that selects for *nonbreeding* members of the population? When does altruistic behavior make evolutionary sense?

Helping behaviors, particularly those that involve foregone or delayed reproduction, have been addressed with several hypotheses centered on the evolution of (apparently) selfless behavior:

- Helpers may be relatively low-quality offspring that maximize their fitness *indirectly*, by concentrating on the reproductive success of their parents or siblings.

FIGURE 5.8 Galápagos mockingbirds, such as this one from a Santa Cruz study population (note the bands on its right leg), are fearless and common.

[12] While this four-species classification of mockingbirds may be phylogenetically inaccurate, we have used it here because it is the terminology you will encounter during most tours of the islands.

- Helpers may be banking on reciprocity, whereby when they have their own offspring (*direct fitness*), they will receive help at the nest.
- Helpers may actually increase their own eventual fitness by helping, especially if they acquire useful skills in the process or increase their chances of inheriting desirable breeding territory.
- Helpers may be making "the best of a bad lot." When resources are scarce, foregoing reproduction—at least temporarily—to focus on the fitness of kin, may be better than nothing.

Multi-year observations of mockingbirds on Genovesa indicate that these helpers are typically male, and most often are offspring from one of the resident pair's previous broods. Females typically leave home after fledging, so a related nonbreeding adult female is unlikely to be seen at the nest. Males, however, do not disperse and are more likely to reap the indirect-fitness benefits of helping; specifically, nests with helpers have greater fledging success than nests without helpers. Certain extrinsic and intrinsic factors make helping a logical strategy: On Genovesa, all available nesting habitat is quickly saturated, leaving some reproductively mature birds without a place to raise chicks. Because these mockingbirds are relatively long-lived, territory turnover-rates are low.

A study of mockingbird behavior on four islands in Galápagos investigated the hypothesis that cooperative breeding is enforced by a shortage of preferred habitat. Genovesa, Champion, and Española are smaller islands with limited habitat available for breeding territories, and helping behavior is common here. San Cristóbal is larger, with more available habitat, and helping behavior has not (as of a 1989 report) been documented. In summary, helping behavior seems to be a context-dependent adaptation for increasing a bird's indirect fitness, while he awaits the opportunity to focus on direct fitness.

Common name: San Cristóbal (or Chatham) mockingbird

Scientific name: *Mimus melanotus*

See it: In several habitats on San Cristóbal

Residence status: Endemic

Conservation status: Endangered

What you will see and what it means: The San Cristóbal mockingbird is similar to other archipelago mockingbirds in its varied food choices, temerity, and general morphology. However, this population is distinct in that these birds do not exhibit cooperative breeding, a feature that may result from the wide range of available breeding habitat on the island (see discussion in Galápagos mockingbird, above).

Common name: Floreana (or Charles) mockingbird

Scientific name: *Mimus trifasciatus*

See it: Small populations remain on the satellite islets of Champion and Gardner-by-Floreana (so called to distinguish it from Gardner-by-Española)

Residence status: Endemic

Conservation status: Critically Endangered

What you will see and what it means: You are extremely unlikely to see Floreana mockingbirds during a visit to Galápagos. Tourists are not allowed on Champion or Gardner, and these birds have been gone from Floreana for over 100 years. The fragility of these populations is especially troubling in light of the DNA evidence (discussed above) confirming their identity as a distinct species of mockingbirds (Color Image 11).

Genetic drift and the problem with small populations

Some populations are designated Vulnerable or Critically Endangered merely because of their small size, rather than the knowledge of a tangible threat. These designations are due to the threat genetic drift can pose to small, isolated populations (Chapter 1). Genetic drift is the loss of alleles that can result from random fluctuations in reproduction—when, by chance, not all individuals reproduce in a given generation. Drift can lead to a reduction in overall genetic variability and may underlie a population's inability to respond to environmental changes.

The Floreana mockingbirds disappeared from the main island in the mid-1800s. Their demise is tentatively attributed to the introduction of rats and goats to the island; rats eat nearly anything, including bird eggs, and goats compete with mockingbirds for the prickly-pear cactus. Hunting (by humans), as well as predation by introduced cats and dogs, may also be culpable.

The mockingbirds of Champion islet were the focus of a 12-year study of the demographics of small populations. Champion is less than 10 hectares in area, and has been isolated from other mockingbird populations for over 100 years. Mockingbirds on the islands are—by bird standards—relatively sedentary, and seldom fly over water, a behavioral trait that severely restricts gene flow. The resident population of Floreana mockingbirds on Champion hovers at around 20 breeding-age individuals and is presumably vulnerable to genetic drift and the effects of environmental perturbations. In fact, a comparison of morphological features between island populations revealed that the birds on Champion were less diverse—especially with respect to bill length—than was a population on Genovesa.

A follow-up study analyzed genetic diversity in mockingbirds on the archipelago: 19 populations of birds (nearly 900 individuals), from 15 islands, were analyzed for DNA variation. All measures of genetic diversity—e.g., heterozygosity (or genes with variable alleles) and the total number of alleles—were significantly correlated with island size: on larger islands, populations show increased genetic diversity. Champion, by far the smallest island housing mockingbirds, had the least diverse population. Also, a comparison of the diversity of historic specimens (collected by the California Academy of Sciences expeditions; Chapter 2) versus contemporary populations indicates that the Champion birds have lost 39% of their heterozygosity in the last century. Given the importance of heterozygosity in providing the template of variability upon which selection acts, these mockingbirds are near extinction.

Common name: Hood mockingbird

Scientific name: *Mimus macdonaldi*

See it: Española

Residence status: Endemic

Conservation status: Vulnerable

What you will see and what it means: These birds are curious and fearless, so your chances of an encounter are good. When you see a Hood mockingbird, note what it is eating. The Hood mockingbird is spectacularly omnivorous, feeding on standard fare such as arthropods and fruit, as well as more esoteric selections such as sea lion placenta, the external parasites (e.g., ticks) and sloughed skin of iguanas, and the blood of wounded sea lions and lava-scraped tourists.

OTHER LAND BIRDS

Common name: Galápagos dove

Scientific name: *Zenaida galapagoensis*

Comparisons: With its red feet and blue eye-ring, the Galápagos dove is unlikely to be mistaken for any co-occurring bird. However, the rock dove (or common pigeon), which is similar in shape but larger than the Galápagos dove, has been introduced to several of the islands.

See it: Arid zones of all the major islands

Residence status: Endemic

Conservation status: Least Concern

What you will see and what it means: You probably *won't* see a Galápagos dove fly. This bird shuns flight if possible, a feature documented as far back as 1835, when Darwin observed a boy on San Cristóbal with a stick, and noted "as the doves came to drink he killed as many as he wanted and in half an hour collected them together and carried them to the house." Galápagos doves are the only native species of the pigeon family in the archipelago (Color Image 12).

Common name: Galápagos rail

Scientific name: *Laterallus spilonotus*

Comparisons: With its characteristic shape, blue/brown color, and bright red eyes, the rail is unlikely to be mistaken for any other bird.

See it: In the highlands of Santa Cruz, San Cristóbal, and Isabela

Residence status: Endemic

Conservation status: Like many island birds, the rail is reluctant to fly and is innately curious. These characteristics make the rail especially susceptible to predation by introduced cats and dogs, and its numbers have declined in heavily impacted areas. The Galápagos rail is currently listed as Vulnerable by the IUCN.

Common name: Galápagos flycatcher

Scientific name: *Myriarchus magnirostris*

Comparisons: The female vermilion flycatcher and the immature yellow warbler are both similar in color (gray, brown, and olive feathers) and size, yet the bill of the Galápagos flycatcher is larger and more robust.

See it: Although the Galápagos flycatcher is found throughout the archipelago in a variety of habitats, it is most common in dense arid and transition zone vegetation. It often nests in cavities of dead or dying cacti, which it lines with cotton and feathers.

Residence status: Endemic

Conservation status: Least Concern

Common name: Vermilion flycatcher

Scientific name: *Pyrocephalus rubinus*

Comparisons: While the bright-red adult male of the species is unmistakable, the immature and female birds are similar in coloration to the Galápagos flycatcher. A smaller species, *P. dubius,* has been proposed for the San Cristóbal form of vermilion flycatcher, but this bird has not been seen for many years and is presumed extinct because of habitat destruction in the highlands.

See it: Arid zones and highlands throughout the islands

Residence status: Native[13]

Conservation status: Least Concern

Boobies

The Galápagos Islands are home to three iconic species of gannet-like seabirds: the red-footed booby, the blue-footed booby, and the Nazca booby. They are similar, but are distinguished by their color patterns (especially in the face and

[13] A "native" organism is one that occurs naturally in the area, but is not endemic.

feet) and their feeding strategies. Even though they may nest in close proximity, the boobies feed in different depths and at different distances from shore.

Common name: Blue-footed booby

Scientific name: *Sula nebouxii*

Comparisons: The adult blue-footed boobies, with their bright blue feet, are nearly impossible to misidentify. The young are large and adorned with a mass of white down, and look more like frigatebird chicks than the young of any other coastal bird. However, blue-footed boobies nest on the ground in a clearing distinguished by its 360°-ring of white guano, whereas frigatebirds nest in shrubs. It can be difficult, however, to distinguish between blue-footed and Nazca boobies at a distance, especially in flight. When viewed from below, the blue-footed booby has darker wings; the Nazca booby's wings are white, with a dark brown trim, and it has a distinctive dark mask around the base of its beak.

See it: Coastal areas on most islands; large nesting colonies can be seen on Española, Daphne Major, and North Seymour. Unlike Nazca boobies and red-footed boobies, blue-footed boobies are seldom seen far from shore.

Residence status: *Sula nebouxii excisa* is an endemic supspecies; other *S. nebouxii* occur throughout the Pacific coast, from California to Northern Chile.

Conservation status: Least Concern

What you will see and what it means: You are not likely to be disappointed by these photogenic birds. They are physically striking, remarkably blasé about human visitors, and comical in many of their behaviors. Their life histories are in full view, and unfold in spectacular surroundings. You are likely to see some or all of the following events: courtship dancing and "singing" (really honking and whistling), egg laying, incubation, hatching, parental care, symbolic nest-building, aggression between nestlings, group hunting at sea, attacks by frigatebirds, and predation (e.g., by Galápagos hawks).

Breeding boobies typify the laissez-faire nature of island fauna that so struck initial travelers to the islands. In fact, visitors often report having to step over nests, around dancing pairs, and away from probing beaks. The very name "booby" is derivative of the Spanish "bobo," for "stupid," or "clown," possibly in a nod to their unwary demeanor in the face of danger, or possibly from their awkward stride when on land. Among the other curious features of the boobies are their subtle *sexual dimorphism,* their stereotypical mating dance, their remarkable hunting and diving skills, and of course, those blue feet.

How do males and females differ?

Unlike frigatebirds, in which mature males can be distinguished by their bright red, inflatable throats, male boobies differ only subtly from the females. Male blue-footed boobies are slightly smaller than females, whistle during courtship (rather than honk), and appear to have smaller pupils. In fact, males have more yellow pigment in the iris, and females have dark spots that ring the pupil, making it appear larger. It is not known what function, if any, these spots perform in the eyes of the females.

Why blue feet?

Many organisms use color as a signaling device to warn potential predators of danger (as in the dramatic color patterns of poison-arrow frogs in the tropics), signal reproductive status (e.g., the bright red throats of sexually mature male frigatebirds; Color Image 17), or convey information about the health of an individual (bright orange guppy males are more fertile than their pallid competitors). Blue-footed booby males and females have feet that vary in brightness and color saturation (from light green to deep blue), and these feet play a prominent role in their courtship dance. Birds will honk (female) and whistle (male) to each other, then raise and lower their feet in slow motion; up against the white backdrop of the booby's abdomen, foot color is especially evident (Color Image 13). Both males and females seem to value brighter, greenish feet in a mate, rather than a duller, deeper blue. But why? What does foot color say about the quality of a potential mate, if anything?

If foot color is an honest signal of a bird's quality, it should indicate something about an individual's current condition (is he healthy? well-fed? free from parasites?) or overall genetic quality (is she free from inherited disease?). Observations of blue-footed boobies reveal that (1) females in better condition—and presumably able to be more selective—mate with males with greenish feet, and (2) feet that are deeper, duller blue characterize males that are food-deprived. Field studies, in which the feet of experimental males were painted to mimic the feet of hungry males, confirmed the importance of color as a signal: females that were initially paired with these manipulated males showed decreased interest in courtship and copulation. In a bird that depends on two parents for incubating and feeding the young, it is adaptive to reject a mate that cannot care for himself.

Further work by Alberto Velando established a connection between a carotenoid-rich diet, foot color, and immune-system function. Carotenoid pigments, such as beta-carotene, are antioxidants and immunostimulants; the use of these pigments in foot color suggests that they have been diverted from performing their detoxification and immune functions. When Velando and his colleagues controlled food and carotenoid intake in their boobies, they learned that (1) the feet of birds deprived of food for 48 hours (that are thus nutrient-deprived) become duller in this short period, a pattern that can be reversed with food supplements; (2) birds deprived of dietary carotenoids have altered foot color as well, a pattern that can likewise be reversed with supplements; and (3) bright-blue-footed males injected with a known antigen

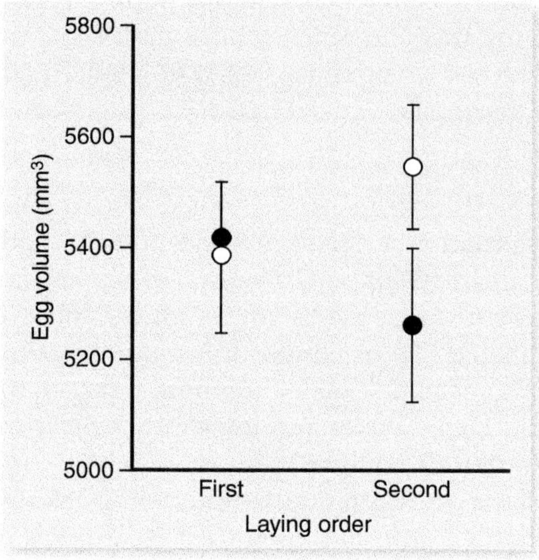

FIGURE 5.9 The feet of blue-footed boobies advertise the bird's overall health and potential fitness. In a foot-color manipulation, experimental males were modified to have duller feet. Female boobies responded by laying significantly smaller second eggs when paired to seemingly lower-quality males. Experimental animals (i.e., those with duller feet) are indicated by closed circles, and control animals (i.e., those with brighter feet) are indicated by open circles. (From Velando, et al., 2006)

(i.e., phytohemagglutinin) had a stronger immune response (measured as the production of T-lymphocytes) than males with dull blue feet.[14] In combination, these results indicate that foot color is an honest signal of male health, and that female boobies choose mates accordingly.

These experiments have also revealed a female's ability to modify her behavior as her mate's perceived quality is diminished: when a male partner's feet were altered—by painting or food deprivation—females decreased their reproductive investment. Namely, female boobies responded by laying significantly smaller second eggs when they seemed to be paired to lower-quality males (Figure 5.9). In this way, a mother may be reducing her overall commitment to offspring that are likely to be lower quality themselves.

Is foot color inherited?

Of course, for any of this information to be evolutionarily relevant, some component of the variation in blue feet has to be heritable, as does the booby's ability to discriminate based on foot color. Again, Alberto Velando's work

[14] Aging also dulls foot color in blue-footed boobies, thus explaining why—in contrast to the finch examples discussed previously—female blue-footed boobies are less attracted to older males. In addition to being less able to care for themselves, older male blue-footed boobies are also more likely to harbor damaged DNA in their sperm (and are therefore more likely to transmit genetic mutations to offspring than are younger males).

sheds some light on nature-versus-nurture in the expression of foot color. In an egg-switching experiment in which hatchlings were raised by a foster father, the weight of offspring at 15 days after hatching was influenced by both the biological and foster father; foster fathers with brighter, greener feet raised healthier chicks, and biological fathers with brighter feet had a similar effect on the chicks they did not raise. Velando and colleagues concluded that "the correlation between male colour and chick condition was mostly due to parental effects (23% of variance), although a small proportion was due to genetic effects (8% of variance, including maternal effects)." When females pair with colorful males, their offspring enjoy improved health, but these effects are only partly due to genetic variability. Thus, foot coloration may align with the "good-parent hypothesis" for mate choice, in that partners are selected for their provisioning ability rather than their genetic contributions.

If you can observe courtship in blue-footed boobies, recall how much valuable information is conveyed during this comical display. Those feet, which are so striking and memorable to tourists, are pretty important to the boobies themselves.

Common name: Nazca booby

Scientific name: *Sula granti* (previously classified as *Sula dactylatra*, the masked booby, which is widespread throughout the Pacific)

Comparisons: The blue-footed booby, which is smaller, lacks the dark mask of the Nazca booby, and has blue feet.

See it: Seaside cliffs on most islands; nesting colonies can be seen on Española, Genovesa, and San Cristóbal.

Residence status: Endemic

Conservation status: Least Concern

What you will see and what it means: Unless you *try* to avoid them, you will encounter Nazca boobies. While their colonies are not as large or visible as those of blue-footed boobies, the birds themselves are large, striking, and omnipresent in the islands. Astute observers may note the following unusual features of Nazca booby parenting: (1) Nazca boobies do not incubate their eggs with the abdominal brood patch characteristic of other birds; instead, they wrap their feet around the eggs to keep them warm; and (2) Nazca boobies lay two eggs, but raise only one chick. Nazca boobies are named for the Nazca tectonic plate on which the archipelago rests (Chapter 3).

Foot-mediated incubation in Nazca boobies

Nazca boobies create simple nests by making a small scrape in the ground, typically near the edge of a cliff overlooking the ocean. Male and female parents incubate their offspring by surrounding the eggs with their feet, thus transferring heat from a dense network of capillaries in the webbing. This functions in the same way as the abdominal pouches—themselves highly vascularized—used by most other birds. The foot webbing of Nazca boobies is more vascularized in incubating individuals than in those that are not incubating, and the feet of incubating boobies reach higher temperatures than the ground, the air, or even the booby's own abdomen. Thus, the feet of Nazca boobies act as surrogate brood pouches in maintaining the warm temperatures needed for chick development.

Achieving fitness through infanticide

Many egg-laying organisms produce more offspring than they can raise during a normal breeding season. There are several potential explanations for this excess: for example, with intense predation, laying many eggs may be necessary to ensure the survival of a few; or some potential offspring may be destined to become food for their siblings. *Siblicide,* in which an individual kills its sibling, is documented in many birds. Siblicide occurs often in blue-footed boobies and appears to depend on fluctuating resources. When resources are abundant, parents can provide sufficient food and protection for two or three chicks. But when resources are scarce, one chick—typically the first hatched—may kill its sibling. In a study of the blue-footed boobies of Isla Isabel (in the Mexican Pacific), Hugh Drummond tested the hypothesis that the oldest chick will kill its younger sibling when the weight of the oldest falls 20 to 25% below its potential. When Drummond applied neck bands to chicks to prevent normal feeding and artificially induce food deprivation, the older chicks deprived of food increased their aggressive behaviors toward siblings over 300% (in a control group, aggression—measured as frequency of agonistic pecking—did not change in the same amount of time). Thus, siblicide in blue-footed boobies appears to be a response to dire circumstances. However, when there is enough food to go around, aggression is minimized and two chicks are fledged.

Obligate siblicide in the Nazca booby

Nazca boobies exhibit *obligate siblicide,* in which the first-hatched chick *always* kills its younger sibling (Figure 5.10). The eggs are laid 4–5 days apart and, soon after the second chick emerges, the first chick attacks and evicts it from the nest. Outside of the nest, the weak, immobile chick dies from starvation, predation, or exposure. The attending parent does not stop the aggressive behavior or its outcome.

Given that egg laying is energetically costly, as is the incubation of a second egg, this behavior is puzzling. Why lay two eggs, just to have one of the chicks kill the other? Several hypotheses have been proposed:

- The "insurance-egg hypothesis" posits that second (or "marginal") offspring have equivalent potential as first (or "core" offspring). Thus, should disaster befall the first egg or hatchling, the marginal hatchling could take its place.
- The idea that marginal offspring provide a fresh source of food to predatory offspring is called the "icebox benefit hypothesis." The implication of the icebox hypothesis is that parents who provision their first offspring with consumable siblings enjoy fitness benefits that other parents do not.

FIGURE 5.10 In typical fashion, this Nazca booby (on Punta Espinosa, Española) has incubated two eggs. The chick on the left will kill its sibling immediately after it emerges from the egg.

- According to the "adoption benefit hypothesis," parents can achieve greater fitness through marginal offspring that are raised by neighbors. This idea rests on the observations that marginal offspring are typically alive and still viable when they are evicted from the nest, and accidental adoptions do occur in many bird species.
- If sibling competition results in the best offspring surviving, then parents will benefit by allowing this culling to take place. If siblicide arose to "let the best chick win," then this "progeny-choice benefit hypothesis" makes sense.

A study of nearly 3,000 Nazca booby families on Española tested these hypotheses. Not a single two-chick family fledged two chicks. In most cases—1,877 out of 2,799 families (67%)—two eggs hatched, but only the first chick survived. The adoption benefit of marginal offspring occurred for fewer than 0.1% of families, in which a second chick was successfully fledged by an adoptive family. Similarly, the data fail to support the idea that the second chick eliminated a weaker, sicker older sibling (progeny choice), nor were any family members observed eating ejected chicks (icebox hypothesis). However, in 326 (11.6%) of the broods, the second chick replaced a sibling that either failed to hatch, or who died prior to fledging. Thus, evidence supports the insurance-egg hypothesis: in Nazca boobies, obligate siblicide confers a fitness advantage to parents in the form of insurance against the loss of the first chick.

A cycle of abuse in Nazca boobies

Life can also be difficult for surviving Nazca booby chicks. When parents leave the nest to forage for food, the sole chick is often subjected to aggressive overtures from neighboring adult boobies. Recent work by David Anderson and colleagues reveals a correlation between an individual's history of abuse as a chick, and its tendency to be an abuser as an adult. The stress hormone corticosterone is higher in abused chicks, and the authors hypothesize that this may prime the boobies to be aggressive adults.

Common name: Red-footed booby

Scientific name: *Sula sula websteri*

Comparisons: Red-footed boobies are the smallest of the three species of boobies in Galápagos. Approximately 95% of red-footed boobies have brown feathers, and the remaining 5% have white feathers. The Nazca booby has a yellow—rather than blue-gray—bill.

See it: At sea throughout the archipelago, and nesting on Genovesa, San Cristóbal, North Seymour, Wolf, and Darwin

Residence status: *Sula sula websteri* is an endemic subspecies; other *S. sula* occur throughout the tropical oceans.

Conservation status: Least Concern

What you will see and what it means: Although red-footed boobies are the most common boobies,[15] you are least likely to encounter them in the archipelago. Visitors to the northern island of Genovesa typically see a few nesting in low-lying branches (Color Image 14), but most travelers are limited to an encounter at sea. Look for the small, torpedo-shaped body dive-bombing the ocean for fish and squid. You won't see them on islands where Galápagos hawks nest.

Why do red-footed boobies avoid Galápagos hawks?

Although blue-footed boobies and Nazca boobies nest throughout the islands, red-footed boobies nest on only five of the many islands of Galápagos. This restricted distribution is unlikely due to habitat availability, for many other islands have suitable nesting materials and are adjacent to the warm, deepwater fishing areas these boobies require. An island-by-island analysis of Galápagos hawk and booby distribution suggests that red-footed boobies and hawks do not co-occur. Hawks require a landmass of sufficient size to sustain their hunting demands; this eliminates several islands from consideration. For their part, red-footed booby chicks are especially vulnerable to predation, because their parents must leave them for long periods to forage at sea. In contrast, the blue-footed boobies, which do co-occur with hawks, forage close to land and can better protect their offspring from attacks. The occurrence of red-footed boobies on Genovesa, however, is puzzling, as this island is large enough to sustain hawks. The red-footed boobies on Genovesa may owe their safety to a fortuitous absence of lava lizards. These small brown lizards are a staple of the diets of Galápagos hawks and may, by their absence, influence the distribution of red-footed boobies.

[15] There are approximately 20,000 breeding pairs of blue-footed boobies, 50,000 breeding pairs of Nazca boobies, and 250,000 breeding pairs of red-footed boobies in Galápagos.

OTHER SEABIRDS

Common name: Waved albatross

Scientific name: *Phoebastria irrorata*

See it: Along the cliffs of Española, from March to January

Residence status: Native

Conservation status: The waved albatross was recently listed as Critically Endangered by the IUCN.

What you will see and what it means: If you visit Española during the breeding season (essentially year-round, with a pause in January through March), you will see waved albatrosses; you cannot miss them. These birds are huge, with a wingspan that can exceed 2 meters. To see them in flight is spectacular, but on land they have a hilarious gait and whimsical mating ritual. If you are there during the height of courtship, enjoy the show: 2 large, bottom-heavy birds circling each other in a goofy waddle, clacking bills, opening their mouths wide, and pointing their bills to the sky and honking (Figure 5.11). This behavior can go on for several minutes, and appears to not only serve in mate choice, but also in maintaining the pair bond for years of child-rearing. The waved albatross is named for the wavy feathers on its chest.

Why is the waved albatross endangered?

By having such a restricted breeding population (on Española in Galápagos and a much smaller colony on Isla de la Plata near the mainland), the waved albatross is vulnerable to environmental changes and human influence. In addition, surveys conducted throughout the last decade concluded that the albatross population in Galápagos is shrinking. While some suggested causes for their decline have been largely dismissed (e.g., the ingestion of plastic does not seem to impact these albatrosses as much as others), the waved albatross is likely affected by Peruvian fishing operations, over 1,000 kilometers away.

While brooding chicks, albatrosses are relatively secure, with parents foraging within 100 kilometers of shore, in the relative safety of the Galápagos Marine Reserve. Española is far from human habitation, protected by the marine reserve, and devoid of known threats to these birds. However, when chicks are old enough to be left alone, their parents make lengthy feeding voyages (up to 3,000 km), favoring the small fish and squid that are plentiful in the tropical upwelling off the Peruvian coast (chicks can receive more than a liter of food per feeding). Unfortunately, albatrosses' fishing areas overlap with Peruvian gillnet and long-line fishing operations, and this overlap has resulted in numerous albatross deaths from *bycatch* (i.e., the unintended harvest by fishing boats) and personal consumption. In 2003, the Peruvian group Pro-Delphinus began recording information on fishery bycatch. In a 3-year period, fishing operators collected 43 dead waved albatrosses, aged

FIGURE 5.11 The comical mating rituals of the waved albatross (shown here at Punta Espinosa, Española) play a role in both courtship and pair bonding, thereby helping to ensure biparental care for the young.

1 to 33 years old, bearing leg bands from Española. Of these, 24 died as bycatch and the others from unknown causes. These deaths are important, for if one of the breeding pair of albatrosses does not return to the nest, its mate abandons the egg or chick to feed itself.

In a similar study, the waved albatrosses recovered by fishing operations were disproportionately (82%) male, which could lead to sex-ratio differences in the colony and a reduction in the number of breeding pairs. Some investigators have hypothesized that these differences in fishing-related mortality are due to the male's ability to out-compete females (who are smaller) for prime sources of food—such as the baitfish used in the nets and long lines of fishing boats. If this is true, the male wins one round (i.e., access to an easy source of food), only to lose another (i.e., an increased likelihood of being snared). Humans and albatrosses appear to be competing over a precious resource—the productive fishing grounds near the Peruvian coasts. Unless measures are taken to protect foraging albatrosses, their numbers will probably continue to decline due to the combined effects of mortality and skewed sex-ratios.

Monogamy and the waved albatross

Albatrosses are often hailed as the epitome of monogamy. Young albatrosses are at sea for about five years before they return to Española to seek a mate. By most appearances, the pair-bonds stick together for years, replaying their courtship routines and raising chicks together. However, genetic analyses have revealed surprises in supposedly monogamous birds, especially in the songbirds, who often have young from *extra-pair fertilizations*—that is, a male may be caring for offspring that he did not sire. Yet the seabirds, especially the tube-nosed seabirds—such as the albatrosses, storm petrels, and shearwaters—tend to have few offspring from "cheating" events. In short, most seabirds are both *socially monogamous* and *genetically monogamous*. However, studies of albatrosses in Galápagos challenge this trend. For example, an analysis of 16 albatross families on Española indicated that four (25%) of the chicks were not the genetic offspring of the male at the nest. A follow-up study of 154 families identified extra-pair paternity in 26 (16.9%) of the chicks. This percentage of extra-pair fertilizations is the highest ever documented for a seabird, and is similar to the frequencies encountered in many songbirds. Like many songbirds, the waved albatross appears to be more socially monogamous than it is genetically monogamous. In a long-lived bird such as the albatross, cheating is difficult to explain: a male who suspects his mate has cheated would be evolutionarily rewarded (in increased fitness) by abandoning the nest in search of another receptive female. In other words, the cost of cheating should be high. Ongoing research is aimed at identifying why the waved albatross is so exceptional in this regard.

Common name: Galápagos penguin

Scientific name: *Spheniscus mendiculus*

See it: In the waters and coastal rocks surrounding Fernandina, Isabela, Floreana, Bartolomé, and Santiago

Residence status: Endemic

Conservation status: The Galápagos penguin is classified as Endangered on the IUCN Red List; the small population of this penguin is especially vulnerable to climatic (e.g., El Niño) and human impacts (e.g., introduced pathogens).

What you will see and what it means: Galápagos penguins usually feed in small flocks (2 to 10 individuals) within 200 meters of shore. If you see penguins in Galápagos, take a moment to appreciate the fact that these are penguins *at the equator* and that they constitute the northernmost population of penguins on the globe. Images of penguins perched on lava (Figure 5.12), with cactus in the background, are incongruous with the many species of penguins inhabiting the southern polar region. Also, watch how they move: they are not great swimmers, they cannot fly, and their stride is clumsy. Thus, you may wonder—what are penguins doing in Galápagos?

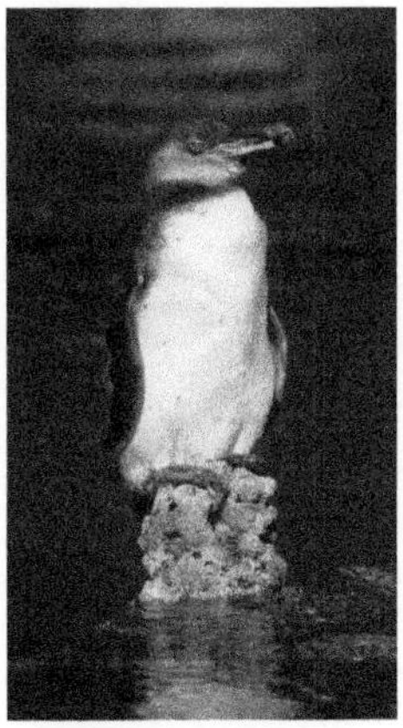

FIGURE 5.12
A Galápagos penguin perches on lava emerging from Elizabeth Bay.

The biogeography of Galápagos penguins

Biogeography is the study of the distributions of organisms. Biogeography addresses evolutionary explanations for why some organisms occur in certain areas, but not others, as well as how these organisms came to occupy a given location. We can use biogeography to solve the riddle of equatorial penguins, a riddle magnified by the facts that the range of the Galápagos penguins is so restricted—even within the archipelago—and that the nearest penguins are the Humboldt penguins (*Spheniscus humboldti*) on the Peruvian coast. Understanding penguin distribution is particularly important to conservation biologists, who are justifiably concerned about the Galápagos penguin; with a global population of fewer than 1,000 breeding pairs, there are fewer of these penguins than any other penguin in the world.

A recent study combining morphological, behavioral, and genetic analysis suggested that all penguins originated in the Antarctic. Molecular dating methods place this origin at about 70 million years ago, on the Gondwanaland landmass. At this time, the combined continents were cooler because more of the land was located in the polar regions. As the continents broke apart, penguins radiated to habitats with suitably cold, nutrient-rich waters.

The cold, nutrient-rich waters of the western Galápagos help explain the occurrence of penguins in the archipelago. The cool-water currents that converge on the islands (e.g., the Humboldt and Cromwell Currents; see Chapter 3) keep water temperatures ideal for nutrient production, in turn supporting an enormous food web for island inhabitants—from phytoplankton to zooplankton to fish and the penguins that feed on them. In a good year, the waters are productive, penguins thrive by feeding on schools of fish in cool coastal waters, and a pair of penguins can raise two or three separate clutches of nestlings. However, if surface-water temperatures rise above 24°C, near-shore productivity plummets, penguins make increasingly longer trips to forage for food, and no offspring are even attempted. These constraints make it unlikely penguins will ever colonize remote islands, and they also explain why penguin colonies are not located on islands of the archipelago in warmer waters (e.g., Española and Darwin).

The beleaguered Galápagos penguins

Given their small numbers, their isolation from other penguins, and the vagaries of life in Galápagos, it is remarkable that these animals exist at all. A combination of human-caused disturbances threaten the continued existence of these birds:

- Although El Niño events occur naturally, their frequency is increasing, a pattern thought to be associated with human-caused climate change. This is troubling, for El Niño effects on Galápagos penguins have been dramatic: In a 1972 event, 199 of 200 nests surveyed failed to produce chicks (in comparison, during a cool-water La Niña year [1971], 62 nests produced 82 chicks). During El Niño years, surface-water temperatures are high (>24°C), oceanic productivity is low, there are fewer schooling fish (i.e., penguin food), and penguins are too weak to reproduce. Consequently, the population of Galápagos penguins declined almost 80% after the 1982–1983 El Niño. Without several years between El Niño events, penguin populations cannot rebound; increasing frequencies of El Niño events associated with climate change may be a more serious impact of global warming than gradual changes in ocean temperatures. More frequent El Niño events, combined with the following threats, could be disastrous for Galápagos penguins.

- Penguins can be harmed by fishing—both directly (caught as bycatch) or indirectly (through cascading effects on their food source). A *trophic cascade* is an ecological phenomenon in which entire food webs can be dramatically altered by the addition or removal of one species, typically a top predator. Fishing near reefs in Galápagos may lead to a trophic cascade; as sea urchin predators are removed, sea urchin populations increase. More sea urchins consume more algae, thus depleting an important food source for zooplankton and the fish that consume them. As the planktivorous fish decline, penguins go hungry and begin making longer foraging trips. Essentially, fishing has the potential to create scenarios similar to those encountered during El Niño years, although the mechanisms are different (removal of a predator versus the overeating of algae). The reality of trophic cascades necessitates careful management of the Galápagos marine reserve, paying special attention to the harvest of key predatory fish.
- Feral dogs and cats are the descendents of domestic animals that have returned to the wild. Considering that the Galápagos flightless birds—such as penguins and cormorants—did not co-evolve with dogs and cats, these birds are easy prey for the new predators. A 1983 report noted that 33% of the dog feces sampled contained penguin remains. Because this finding was prior to the eradication of feral dogs from Isabela, and Fernandina has no dogs, dogs may no longer be an imminent threat to penguins. However, cats are a different story. According to a 2011 review, feral cats on islands are responsible for about 14% of the amniote (mammal, bird, and reptile) extinctions around the world. Cats co-occur with (and prey on) penguins on Isabela, but the overall impact of cats on penguins is unknown. Tellingly, a 2010 survey detected antibodies for *Toxoplasma gondii* in penguin blood. *T. gondii* is a protozoan parasite that cannot exist without its definitive host, the cat, yet *T. gondii* relies on various intermediate hosts to move from cat to cat. *T. gondii* may not be a threat to penguins, but the exposure of penguins to *T. gondii* suggests

an intimate association between feral cats and Galápagos penguins, an association that cannot be good for the penguins.

- Avian malaria is a disease caused by another protozoan parasite, *Plasmodium,* and transmitted by mosquitoes. The recent introduction of the mosquito *Culex quinquefasciatus* to Galápagos raised concerns about the potential arrival of this mosquito-borne disease. And a 2003–2005 survey detected *Plasmodium* in 5% of almost 400 penguins sampled, on five different islands. So, *Plasmodium* has made it to Galápagos; what this means for penguins and other endemic avifauna remains to be seen.

There are undoubtedly other potential threats to Galápagos penguins. For example, an unintended oil spill (e.g., as occurred in 2001 with *Jessica;* see Figure 2.13 and Chapter 11) could be devastating. And the synergistic effects of the preceding concerns will presumably magnify the burden on this small isolated population. In the words of ecologist Dee Boersma, "[l]ife is not likely to get easier for penguins."

Common name: Galápagos petrel

Scientific name: *Pterodroma phaeopygia*

Comparisons: Petrels are small, dark seabirds that feed by dabbling at the water's surface. The Galápagos petrel can be distinguished by the black marks on its white forehead, its large, white rump, and its slightly forked tail.

See it: Breeds on Santa Cruz, Floreana, Santiago, San Cristóbal, Genovesa, and Isabela

Residence status: Endemic

Conservation status: Critically Endangered (due to the effects of introduced mammals, especially dogs, cats, and rats)

Common name: Galápagos shearwater

Scientific name: *Puffinus subalaris*

Comparisons: The shearwater, which can occasionally be seen in mixed-species flocks of seabirds, can be distinguished by its contrast coloration—dark brown upper and white underparts. Until recently, the Galápagos shearwater was considered a subspecies of Audubon's shearwater (*P. lherminieri*).

See it: Breeds on Santa Cruz, Española, Champion, and Wolf

Residence status: Endemic

Conservation status: Not Evaluated

Common name: Red-billed tropicbird

Scientific name: *Phaethon aethereus*

Comparisons: This striking bird is easily distinguished by its red beak and long white tail feathers, which are called streamers (Color Image 15).

See it: Throughout the archipelago, but best viewed along the exposed cliffs of Genovesa and South Plaza

Residence status: Native

Conservation status: Least Concern

What you will see and what it means: Red-billed tropicbirds are fast and graceful in the air, but clumsy on land. Their tiny feet are positioned so far back on their body that walking is impossible; instead, the birds lie on their belly and push themselves along cliffs with their feet.

Common name: Brown pelican

Scientific name: *Pelecanus occidentalis urinator*

Comparisons: With its large, heavy body, and long, sturdy beak, the brown pelican is unmistakable.

See it: Throughout the islands

Residence status: The Galápagos brown pelican is considered a subspecies of the widespread brown pelican.

Conservation status: Least Concern

What you will see and what it means: You are bound to see the brown pelican. They can be seen clumsily plunge-diving for fish, alone or in mixed-species flocks. You will also see pelicans hovering at fish markets, awaiting scraps from the day's catch (Figure 5.13).

Common name: Magnificent frigatebird

Scientific name: *Fregata magnificens*

Comparisons: The great frigatebird is almost indistinguishable from the magnificent frigatebird. The magnificent frigatebird has a purplish sheen to its dorsal feathers, while those of the great frigatebird are greenish. Chicks of great frigatebirds have a red eye ring, white chest, and white throat; magnificent frigatebirds have a blue eye ring, white chest, and dark throat.

See it: Throughout the islands, and breeding at Genovesa, North Seymour, and San Cristóbal

Residence status: Endemic subspecies

Conservation status: The magnificent frigatebird is classified as Least Concern on the IUCN red list. However, a 2011 analysis of magnificent-frigatebird populations suggests the Galápagos population has been isolated from the others for hundreds of thousands of years, and is sufficiently distinct to warrant its own species designation. If the magnificent frigatebird of Galápagos is a separate species, its conservation status—with only about 1,000 breeding pairs—may require revision.

What you will see and what it means: See discussion below (with great frigatebird)

FIGURE 5.13 Brown pelicans are frequent and demanding visitors at the fish market on Charles Darwin Avenue in Puerto Ayora, Santa Cruz.

Common name: Great frigatebird

Scientific name: *Fregata minor ridgwayi*

Comparisons: See above (under magnificent frigatebird)

See it: Throughout the islands, and breeding at North Seymour, Española, and Genovesa

Residence status: Native

Conservation status: Least Concern

What you will see and what it means: You cannot miss these avian pirates. Because frigatebirds nest in low bushes, often at about the level of a human's eyes, you may be able to observe any or all aspects of their life histories—courtship, nest building, incubation, and feeding of hatchlings and fledglings. The life histories of frigatebirds are impressive: not only are they long-lived (several have been documented at older than 40 years), they are committed parents—even by bird standards—who care for their single offspring for a year or more (Color Image 16).

The mystery of sexual dimorphism in frigatebirds

Frigatebirds are striking for their obvious sexual dimorphism—reproductive males, with their bright-red, greatly enlarged throat-pouch[16] (Color Image 17), are dramatically different from female frigatebirds. What can these differences between the sexes tell us about mating behavior in frigatebirds?

Conventional wisdom states that sexual dimorphism should be minimal in species that give biparental care, but the sexes may show extreme differences when one parent does the majority of the provisioning. This prediction rests on the logic that selection for mates is most intense when one sex provides little more than sperm or eggs; that is, if all you are getting from a mate is genetic material, those genes better be good. In birds especially, male ornamentation is positively associated with intense female choice and female-biased parental care. In confirmation of these relationships are Galápagos animals with strong biparental care and minimal distinctions between the sexes (e.g., the boobies and waved albatross), as well as those with uniparental care and pronounced sex differences (e.g., sea lions).

Given that female mammals are physiologically bound to care for their offspring (via internal incubation and lactation), it is not surprising that male mammals tend to provide little—if any—care for the young. However, male mammals must expend a lot of energy in competing for access to females (as do male sea lions), or bearing costly ornaments to increase their chances of copulating with a choosy female. Conversely, because male and female birds are equally capable of incubating and feeding their young, biparental care—and associated monogamy—are typical of many birds.

So, how can we understand the sexual dimorphism that is so evident in frigatebirds? In addition to their unmistakable balloon-like red throat pouch and dorsal patches of iridescent plumage, male frigatebirds expend energy in behavioral displays, from which the female makes her choice. But then, the pair embarks on a relatively egalitarian domesticity—nest building, incubation, feeding, and protecting offspring. Frigatebirds appear to be monogamous: a genetic study of 92 nesting families revealed only 1% nonpaternity (i.e., the male at the nest was usually the chick's biological father). The frigatebirds' biparental care, in a species in which the male and female differ so much, challenges conventional wisdom. This life history trait is evolutionarily derived: a phylogeny of 200 birds, including the five species of frigatebirds, indicates that biparental care is the ancestral condition, shared by all included taxa. However, only the frigatebirds have such pronounced sexual dimorphism (Figure 5.14). Why did the frigatebirds pursue a different evolutionary trajectory? Why do male frigatebirds invest so much in costly displays?

Operational sex-ratios in the great frigatebird

Answers to the preceding questions may lie in the frigatebird's *operational sex-ratio,* which is the ratio of males to females in the breeding population. A study

[16] It takes a male frigatebird approximately 10 minutes to inflate its pouch.

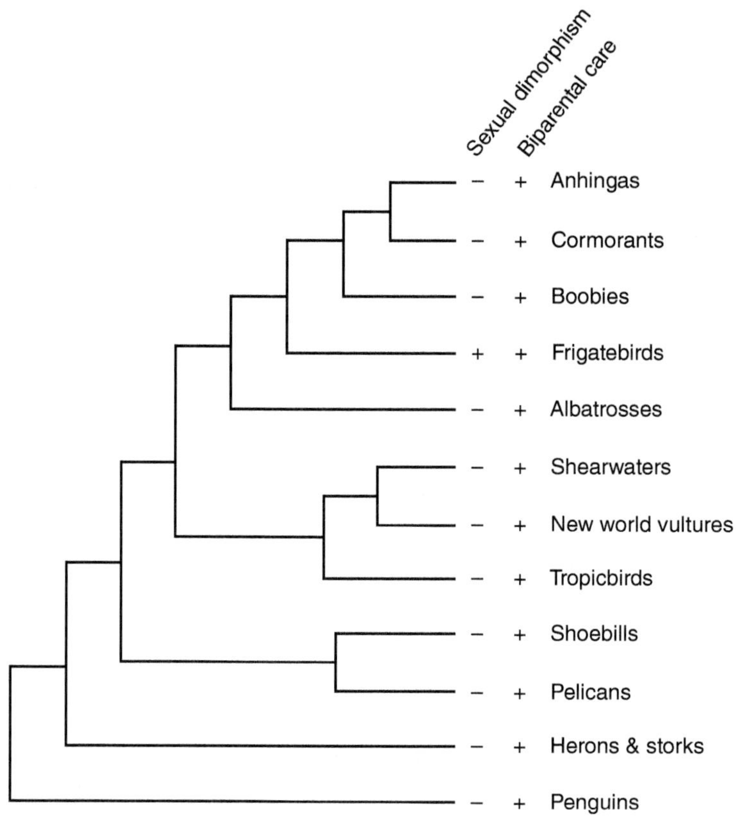

FIGURE 5.14 Frigatebirds are notable among seabirds for pronounced sexual dimorphism, while also exhibiting monogamy and biparental care. This phenomenon may be due to skewed sex-ratios in many frigatebird populations. (From Dearborn, et al., 2001)

of sex ratios in great frigatebirds of Hawaii's Tern Island revealed a strongly male-biased population, with five or six displaying males for every reproductive female. With so many options, a female could afford to be choosy, and selection may have benefitted males that advertised their qualities (e.g., via bright red throats, or courtship displays).

But why is there such a markedly male-biased sex ratio? In birds, sex is determined genetically, via sex chromosomes, and should result in a 1:1 ratio of males and females. Where are all the female frigatebirds? Were they selectively culled (aborted) prior to egg laying, as occurs in pigeons? Do they experience greater mortality, as seen in the waved albatrosses of Española (discussed previously)? Do females form same-sex groups away from the breeding colony, as do Galápagos hawk females (also discussed previously)? In addition to these possible explanations for altered sex-ratios, how can we understand skewed sex-ratios from an evolutionary perspective?

By now, it should make sense that parents will invest more in offspring that are most likely to increase the parents' inclusive fitness, and this reality prevails in Galápagos. In some cases, the sex of the young has fitness consequences for the parent, whereby a male or a female is more likely to survive to reproductive age, or one sex is more likely to out-do the other in making grand-offspring. Given that males can produce seemingly limitless sperm, and females are restricted in their ability to produce eggs, male offspring should—all things being equal—give parents the best shot at becoming grandparents. But all things are rarely equal, and when resources are limited or when parents are otherwise stressed, it makes sense to produce more of the sex that is less demanding.

In frigatebirds, the less demanding sex may be the male, simply due to his smaller size (female frigates are over 20% larger than males). If a smaller male chick is easier to provision, shouldn't selection favor some sort of sex-ratio allocation in favor of males, especially when resources are scarce? A test of this hypothesis focused on the stress that foraging seabirds (e.g., frigatebirds) encounter when sea temperatures increase and prey abundance decreases. Female frigatebird offspring declined with seasonal increases in water temperature during a breeding season (in 1998). These results (1) suggest that young females are costly to their parents, and (2) support the idea that there may be more male frigates because they are easier to raise. However, a follow-up study (in 2003) did not replicate these findings. The skewed operational sex-ratios in frigatebirds may explain the birds' unusual sexual dimorphism, but the sex ratios themselves remain a mystery.

Aerial piracy by frigatebirds

Look to the sky for evidence of *kleptoparasitism* by frigatebirds. Kleptoparasitism, which is the stealing of food, has earned the frigatebirds the moniker "pirates of the sky," and the designation is apt. However, stolen food is not the primary component of the frigatebird diet—frigates will swoop down for sea turtle hatchlings on the sand, jumping fish just above water, and small fish and jellies ("jellyfish") just below the water's surface. Frigatebirds also exploit the effort of other birds, such as flightless cormorants and penguins that drive prey fish to the water's surface (see discussion of flightless cormorants, below).

Any chance to observe a frigatebird securing food is bound to be inspiring; they are spectacular in flight. Also, because their feathers are not waterproof, frigatebirds must retrieve prey from the sea gingerly to avoid becoming waterlogged. However, a glimpse at piracy in the air is unforgettable. A frigatebird will pursue other birds returning to the nest with captured prey, often physically harassing the target birds until they release their catch. In Galápagos, the most common victims are the blue-footed boobies, but frigatebirds will also harass other boobies, red-billed tropicbirds, gulls, and pelicans. On rare occasions, a frigatebird will induce vomiting in its target by pecking it about the neck repeatedly. Should you happen upon a feeding flock of boobies, look for frigatebirds circling the perimeter and dabbling at the water's surface. You might get a peak at real-life piracy on the open sea.

FIGURE 5.15
The flightless cormorant (shown here on Fernandina) has several adaptations that have allowed it to forego flight for diving—vestigial wings, a dense body, and large, muscular legs and feet. The flightless cormorant in Galápagos is the only species of cormorant that has lost the ability to fly.

Common name: Flightless cormorant

Scientific name: *Phalacrocorax harrisi*

See it: On Fernandina (at Punta Espinosa) and Isabela (northern and western coasts). Flightless cormorants seldom stray inland more than a few meters.

Residence status: Endemic

Conservation status: Vulnerable

What you will see and what it means: While their rudimentary wings (Figure 5.15) and inability to fly may be their most distinctive features, flightless cormorants also have interesting feeding and mating habits. Females, who can breed three times per year, usually lay two to three eggs. The largest chick gets the most food, and is usually the only chick to survive. Look for cormorants in large, multispecies feeding flocks near the coasts of Isabela and Fernandina. And watch for their stereotypical mating behaviors—a snake-necked dance, the bearing of gifts such as feathers and seaweed—and the signs of *facultative sequential polyandry.*

Sequential polyandry in flightless cormorants

In contrast to the cooperative polyandry observed in many Galápagos hawk communities (discussed previously), flightless cormorants may engage in *sequential polyandry,* whereby a female leaves her offspring with its father and makes another clutch with a new mate. The first mate, thus deserted, is left to raise the nestling on his own. This polyandrous mating system is *facultative,* in that females more likely desert and seek second mates when resources are plentiful enough to support two or more of her nests.

In many monogamous species, including most birds, pair bonds appear to be reinforced by the need for biparental care. However, should conditions arise where a single parent can go it alone, the deserting parent—who seeks a new mate and increased opportunities to reproduce—will be rewarded with increased reproductive success (i.e., fitness). In many species, especially in mammals (in which lactation commits females to a disproportionate amount of parental care), *polygyny* is common. A polygynous male who acquires multiple mates to raise his many young is bound to be more fit than a monogamous counterpart. But why would a female cormorant, with so much invested in egg production and laying, desert her own offspring?

The short answer is: because she can. The female increases her own fitness by coercing paternal care for more of her own offspring. But what's in it for the male? Why would a male enter into a polyandrous relationship? Simply, how can we understand this apparent role reversal in flightless cormorants?

This behavior may be a by-product of size differences: male cormorants are noticeably larger than females, allowing males to capture larger food items for the growing chicks. In the early stages of caring for the young, small foods are desirable, and both the male and female can forage for fish. In addition, one parent must protect the newly hatched chick from predators such as the Galápagos hawk, while the other parent forages for food. Thus, when chicks are young, biparental care is critical. However, larger chicks need larger fish, and are less dependent on parental protection. These conditions lead to a situation in which the larger male is in a bind: he is better suited to forage for his offspring, yet if he deserts they are likely to die from the weaker provisionings of their mother. In other words, the costs of desertion are just too high for the male.

The role of cormorants in multispecies feeding flocks

You may encounter multispecies feeding flocks, which are fascinating displays of several to hundreds of birds plunging into the waters after the same large school of fish. These flocks may include boobies, pelicans, and brown noddies—all closely monitored by a few opportunistic frigatebirds. What you may not see, unless you look closely, are penguins and cormorants down below, making the fishing much easier for the flying birds.

Penguins and cormorants are *pursuit divers,* meaning they swim underwater after identified prey. The boobies and pelicans are *plunge divers,* crashing into the water from above, using the momentum of the plunge to overwhelm prey. When prey are abundant, flocks of plunge divers may forage in concert, possibly profiting from the general confusion among fish faced with this bombardment. These plunge divers benefit when they follow pursuit divers, who often approach schools of fish from below, thereby delivering them closer to the surface and making them more visible from the air. Again, the cold waters of the archipelago create ideal circumstances for these feeding flocks: the high productivity associated with cold water leads to the existence of large near-shore schools of fish, as well as to the occurrence of the pursuit divers themselves. In fact, there is a negative correlation between surface temperatures and feeding flocks: where waters are cooler (e.g., around Fernandina and Isabela), feeding flocks are larger and more common. Also, the contribution of pursuit divers is evident: a study of feeding-flock duration near Fernandina showed that flocks that included cormorants and penguins last much longer—presumably because there continued to be a supply of available fish—than those without the pursuit divers.

By facilitating feeding for plunge-diving birds, flightless cormorants fill a role that is occupied by dolphins, tuna, and whales in other coastal ecosystems. Thus, this behavior can be seen as a type of *coevolutionary convergence,* whereby similar relationships have emerged in distinct lineages (cormorants, dolphins, and tuna do not share a recent common ancestor). Plunge divers appear to profit from the association, yet it is unclear if the pursuit divers are affected at all. Thus, this may be an example of *commensalism,* whereby one symbiont benefits and the other is neither positively nor negatively affected (Chapter 1).

The evolution of flightlessness in island birds

Because pursuit diving is energetically taxing, and is therefore less likely to occur in low-productivity waters, there must be a substantial reward to warrant its cost. Pursuit diving exerts an evolutionary cost as well: many pursuit-diving birds have invested so heavily in the physiological demands of diving that they have foregone ancestral traits. For example, pursuit diving is typical of many penguin species as well as cormorants in Galápagos, and these birds have all lost the ability to fly. In flightless cormorants, the inability to fly is associated with *vestigial* wings—that is, wings that served a function in ancestral birds, and continue to function in other lineages, but have become useless in these cormorants as selection on them was relaxed.

Along with vestigial wings and smaller wing-muscles, cormorants have other physical adaptations that predispose them to dive: their legs and feet are large and muscular, and their bodies are dense to assist with diving to depths exceeding 5 meters. Notably, the Galápagos cormorant is the only one of its 27-member genus to have lost the ability to fly. The unique history of the archipelago, whereby it lacks terrestrial predators, is presumably the reason flightlessness has evolved in these birds.

Common name: Lava gull

Scientific name: *Leucophaeus fuliginosus*

Comparisons: From a distance, the lava gull can be confused with the similarly sized swallow-tailed gull. However, the adult swallow-tail has red feet and a bright-red eye ring. The adult lava gull is completely gray, with white "brows" above and below the eyes.

See it: Lava gulls are among the rarest gulls in the world. The lava gull is widespread in Galápagos, but with a sparse distribution. It is rare to see more than one at a time. Lava gulls are scavengers, so they are often encountered in inhabited areas.

Residence status: Endemic

Conservation status: Vulnerable

Common name: Swallow-tailed gull

Scientific name: *Creagrus furcatus*

Comparisons: The swallow-tailed gull, with its red feet, red eye-ring, and white spot on its beak, is an unmistakable bird.

See it: Throughout the islands, and breeding in large numbers on Genovesa and South Plaza

Residence status: Endemic

Conservation status: Least Concern

What you will see and what it means: The swallow-tailed gull is the only gull that feeds at night; during the day it returns to its nest. Several adaptations make them successful on nocturnal forays. For example, the large eyes are typical of nocturnal animals, and the red eye-ring may improve night vision. Swallow-tail chicks are white, making them easier for the parents to locate at night, and the white spot on the adult's beak may help the chicks retrieve food from their parents (Color Plate 18). Swallow-tailed gulls feed on squid, which are naturally bioluminescent. Also, these gulls appear to use a form of echolocation that may help them hunt. If you are at sea at nighttime, listen for clicks overhead, accompanied by the flapping of large wings.

SHORE BIRDS

Common name: Lava heron

Scientific name: *Butorides sundevalli*

Comparisons: This small (<40 cm), stocky, dark-gray heron is easy to identify as an adult (Color Image 19). As a juvenile, it has brown and cream streaking resembling the striated heron (which has a yellow mark in front of each eye) and the juvenile yellow-crowned night heron (which is larger and has yellow legs).

See it: Lava coasts and mangrove lagoons throughout the archipelago

Residence status: Endemic

Conservation status: Least Concern

Common name: Yellow-crowned night heron

Scientific name: *Nyctanassa violacea*

Comparisons: The adult yellow-crowned night heron can be distinguished by its head plumage—black with white cheeks and yellow crown—and yellow legs.

See it: Coastal areas throughout the archipelago. Yellow-crowned night herons are usually solitary.

Residence status: Native (breeds throughout the islands)

Conservation status: Least Concern

Common name: White-cheeked pintail

Scientific name: *Anas bahamensis galapagensis*

Comparisons: This aptly named shore bird has bright white cheeks and throat. The rest of the body is brown, and the dark beak has a reddish mark at the base. The white-cheeked pintail is larger than the blue-wing teal—another dabbling duck, with which the pintail co-occurs.

See it: Ponds and lagoons throughout the archipelago

Residence status: Endemic subspecies

Conservation status: Least Concern

Common name: Greater flamingo

Scientific name: *Phoenicopterus ruber*

Comparisons: This large, long-legged, large-beaked, pink bird is unmistakable (Color Image 20).

See it: Coastal lagoons throughout the archipelago (commonly seen in Puerto Villamil on Isabela, and Las Bachas on Santa Cruz)

Residence status: Native

Conservation status: Least Concern

Common name: American oystercatcher

Scientific name: *Haematopus palliatus*

Comparisons: The American oystercatcher, with its long, bright-red beak and yellow eyes bordered in red (Color Image 21), is unlikely to be confused with any Galápagos shore birds.

See it: Coastal areas throughout the archipelago, often in pairs. Oystercatchers seldom stray inland; the animals they eat live only in the intertidal zone.

Residence status: Native

Conservation status: Least Concern

CHAPTER 6

Reptiles

These huge reptiles, surrounded by the black lava, the leafless shrubs, and large cacti, seemed to my fancy like some antediluvian animals.
—Charles Darwin, *Journal of Researches,* 1845

ALTHOUGH THERE ARE NOT MANY SPECIES OF REPTILES IN Galápagos, the archipelago is still, literally, crawling with the scaly creatures. And, if you count birds as reptiles—which many evolutionary biologists do—Galápagos is *ruled* by reptiles.

The Galápagos Islands are home to native tortoises, lizards, and snakes. There are also five endemic, and at least three introduced, species of geckos (lizard-like reptiles characterized by fringed toe-pads). Our focus is on the iconic giant tortoises and iguanas. However, you are likely to see a few green sea turtles and countless lava lizards, and they are also discussed here.

Colloquially, "tortoise" or "turtle" describes terrestrial turtles, whereas aquatic varieties are strictly "turtles." We do not discuss the four endemic species of snakes, for you are unlikely to see a snake for more than a few seconds (possibly as one flees a predatory bird). But first, a bit about the evolution of reptiles.

WHO ARE THE REPTILES?

The evolutionary relationships of reptiles are far from settled, but several molecular, morphological, and paleontological studies have provided new and exciting insights:

- Collectively, reptiles are in the *amniote* lineage of vertebrates, along with mammals, and excluding fish and amphibians. Amniotes encase their offspring in an amniotic egg.
- Within the amniote lineage, there are two distinct evolutionary groups: *synapsids* (e.g., mammals), which have a single opening in the skull behind each eye; and *diapsids* (e.g., birds and reptiles), which have two

openings in their skulls. These openings are due to shared ancestry and shared features of skeletal development. Questions remain about the evolution of the *anapsids,* a group of organisms that have no comparable opening in the skull (this group includes modern turtles).

- Birds are diapsids, along with crocodiles, lizards, and snakes. This group also includes the now-extinct dinosaurs.

So, the organisms we call reptiles include representative diapsids (lizards, crocodiles, and snakes) and anapsids (turtles).[1] Given that birds are also diapsids, you can see how amniote classification can be confusing. However, several features unite birds and reptiles: for example, birds and some reptiles (i.e., crocodilians) have four-chambered hearts; birds and reptiles both possess scales made of the protein beta-keratin (the tougher counterpart to the alpha-keratin in mammalian hair); and birds and reptiles have similar jawbones and single bones that attach the head to the neck (and allow for a greater range of head motion than mammals possess). These and many other lines of evidence establish the shared ancestry of birds and reptiles, to the exclusion of mammalian amniotes. However, the separate terminology (*bird* versus *reptile*) is ingrained in our understanding of animal diversity, so we continue to use the terms without much—if any—confusion.

GALÁPAGOS GIANT TORTOISES

Although hungry pirates and whalers killed the giant tortoises by the thousands for food (Chapter 2), in more recent years, conservation of the turtles has fueled multinational efforts and focused global attention on the islands.[2] As you will see, the giant tortoises of Galápagos are remarkable animals.

Darwin and the Giant Tortoises

During his visit to Galápagos in 1835, Charles Darwin met with Nicholas Oliver Lawson, a British sailor from Jamaica and the vice-governor of Floreana, who asserted that he could "pronounce with certainty from which island any tortoise had been brought from the shape of its shell." At the time, Darwin did not grasp the potential significance of Lawson's claim, thus possibly explaining why Darwin did not collect tortoise shells from all of the islands, and explaining why Lawson's comments were barely mentioned in the first edition of *Journal of Researches* (1839). However, when the second edition appeared in

[1] Mammals are synapsids (Chapter 7).

[2] Giant tortoises from Floreana, Pinta, Santa Fé, and Fernandina are extinct. Tortoises have never lived on Marchena, Genovesa, Darwin, and Wolf Islands.

1845, Darwin devoted much more attention to Lawson's claim by noting: "I have not as yet noticed by far the most remarkable feature in the natural history of this archipelago; it is, that the different islands to a considerable extent are inhabited by a different set of beings. My attention was first called to this fact by the Vice-Governor, Mr. Lawson."

While in Galápagos, Darwin believed that the giant tortoises were varieties of tortoises introduced from Madagascar and elsewhere in the Indian Ocean. (Today, they are considered subspecies of a single Galápagos species, *Chelonoidis nigra*.) Darwin reported seeing only dome-shaped tortoises (on Santiago and San Cristóbal) which, contrary to Lawson's claims, are difficult to distinguish. Darwin and his shipmates collected no tortoises on Floreana or Santiago, but did collect 48 tortoises from San Cristóbal, which they ate (and threw the shells overboard) before reaching Tahiti.

GIANT TORTOISES AND COMMON MISCONCEPTIONS ABOUT EVOLUTION

There are many misconceptions about how populations and species evolve, and one notorious error often includes Galápagos tortoises—namely, that adaptations arise to suit an established need. For example, you may hear someone say, "the saddleback shell evolved because these turtles needed to extend their necks to feed on the leaves of tall plants." Statements such as this are pervasive in the literature about the islands, and are problematic not because of the connection between turtle shell morphology and turtle feeding habits, but because of the implication that evolution occurs in a predetermined direction. Evolution does not occur to meet a need; rather, natural selection uses the variability inherent in populations to adapt organisms to changing environmental pressures.

Tortoise Shell Evolution

In the case of giant tortoises living on islands with tall cactuses, adaptations (such as shell morphology) that enabled turtles to reach plant leaves could give some individuals an advantage over others. This feeding advantage would have led those turtles with the more saddle-like shells to live longer and enjoy more reproductive success. In turn, saddleback tortoises (Figure 6.1) historically occurred on dry, low-elevation islands with tall *Opuntia* cactus (e.g., Pinta, Pinzón, and Española); domed tortoises (i.e., those whose shells lack a saddle-like notch above the neck) occurred on high-elevation, humid islands with extensive low-lying vegetation (e.g., Santa Cruz, Isabela).

This fit between a tortoise's shell and its habitat may reflect convergent evolution (Chapter 1). Carapace shape is not correlated with genetic similarities between the tortoise populations, suggesting that saddleback shells evolved more than once, as a result of similar selection pressures (i.e., dry habitats that could not support sufficient low-lying, easy-to-reach vegetation). Saddleback shells may

FIGURE 6.1 This saddleback tortoise, attempting to eat *Opuntia,* illustrates a potential advantage of the animal's deeply notched shell. Tortoises with the saddleback shell could raise their heads high and were common on dry, lowland islands without substantial undergrowth. Their dome-shelled counterparts (who could not raise their heads as high) occur on islands with highlands, and thus have year-round moisture and plentiful undergrowth for easy grazing.

also play a role in *sexual selection* (Chapter 1). Male saddleback tortoises are aggressive, and they gain increased access to females by out-competing rival males in "outstretched neck" displays (some can raise their heads 1.5 m). Female saddleback tortoises have a notched shell, but it is not as pronounced as that of males. The notched shells of the saddlebacks may have arisen initially as an adaptation for feeding, only to be co-opted by males for intrasexual competition.

Giant Tortoise Species and Subspecies

The Galápagos giant tortoise is a name applied to 14 or 15 distinct types, or *subspecies,* within the species *Chelonoidis nigra*. These subspecies are distinguished by their habitat and shell type. Of these subspecies, 10 still remain. A single island, Isabela, houses five distinct subspecies on five different volcanoes. Barren lava flows between the volcanoes have isolated these populations and prevented them from interbreeding. We focus our discussion on *Chelonoidis nigra* in general, and those tortoises you are most likely to encounter outside of captivity (i.e., *C. nigra porteri* on Santa Cruz and *C. nigra vandenburghi* on Volcán Alcedo, Isabela). Both of these populations have dome-shaped shells.

In 2012, scientists reported evidence of hybrid offspring of a Floreana population of *C. nigra*[3] long believed extinct. Interestingly, these 84 first-generation hybrid tortoises were found on Isabela's Volcán Wolf, separated from Floreana by over 300 kilometers of open sea. (The Floreana tortoises were presumably dumped at or near Isabela by whalers needing to lighten their cargo. Although

[3] Garrick, et al., 2012, in describing this discovery, use the specific epithet *Chelonoidis elephantopus* rather than referring to the Floreana tortoises as a subspecies of *C. nigra*. In late 2012, scientists also reported 17 tortoises having a Pinta ancestry at Volcán Wolf.

Galápagos tortoises don't swim, they can float, and prevailing currents would have carried the tortoises to Isabela.) Genetic analysis led investigators to estimate that at least 38 *C. nigra* parents were involved, and, given that some of the hybrids were less than 15 years old, that individuals of the Floreana group may still be alive on Isabela. If scientists cannot locate these individuals, future programs to conserve tortoises could attempt to reconstruct this lineage by selective breeding of the hybrid tortoises.

Common name: Santa Cruz Galápagos giant tortoise

Scientific name: *Chelonoidis nigra,* or *Geochelone nigra*

See it: Highlands of Santa Cruz, often around farmlands (e.g., Rancho Primicias)

Residence status: Endemic

Human impacts: As a result of repatriation (Chapters 2, 11), this giant tortoise has been upgraded from Endangered to Vulnerable.

What you will see and what it means: Most visitors to Galápagos encounter the turtles in captivity or, possibly, in the wild. If you visit the tortoise-breeding facilities on Isabela or Santa Cruz (at the Charles Darwin Research Station [CDRS]), you can see well-established captive-breeding programs. Turtles are bred here, marked upon hatching, and monitored for several years prior to their release (Chapter 11). You can see turtles in the highland farms on Santa Cruz (Figure 6.2) or, if you make the trek, around Isabela's Volcán Alcedo.

Giant tortoises are not shy, and they will share many features of their life histories with you. On Santa Cruz, look for turtles lumbering (at about 0.3 km per hour) along well-established tortoise trails. These paths are used for routine feeding as well as long-distance migrations, as turtles move from the lowlands to the highlands when the dry season approaches. Most movement occurs in early morning or late afternoon, and the middle of the day is spent eating grasses, fruits, and cactus. An adult tortoise consumes over 30 kilograms of food per day; as a result, most of its waking hours are spent chewing. Giant

FIGURE 6.2
A Galápagos giant tortoise from the highlands of Santa Cruz.

tortoises are often accompanied by flycatchers and finches, symbionts that will fly under a turtle's shell to remove annoying ticks. You may see a turtle raise its heavy shell to accommodate the birds' cleaning behavior. If you get too close to a turtle, it will withdraw into its shell, a process that requires the forceful expulsion of air (listen for the unmistakable "hiss").

Although tortoises mate throughout the year, you are most likely to catch them copulating in the rainy months in the highlands. Copulation can be a noisy, clumsy spectacle, with the large (>200 kg) male mounting the smaller (140–200 kg) female from behind. The male's ventral shell is concave, accommodating the female's rounded dorsal shell, and aligns their genital openings. Fertilization is internal, and afterwards the female travels to a dry, sandy area to lay 5 to 10 eggs. Digging a hole to incubate the eggs is a laborious process, requiring several days of the female's effort. After depositing the eggs, she plugs the nest hole with a mixture of sand and urine, and leaves her young to incubate.

Young that do not drown in the nest (from a too-wet rainy season) or get trapped in the nest (from an extreme dry season) emerge after incubating for several months. Hatchlings weigh about 50 grams and are 5–7 centimeters long. After 40 years, the turtles will be sexually mature and full-grown, having added up to 2 meters in length and more than 200 kilograms in weight (if male). No one knows how long giant tortoises can live because their lives have often exceeded the time elapsed since the mid-1800s, when people began to monitor them closely. However, one of the oldest-known Galápagos tortoises, Harriet, died in captivity in an Australian zoo in 2006; she was approximately 175 years old.

Giant Tortoises as Ecological Engineers

Although giant tortoises are remarkable and viewed by many as an evolutionary novelty, these lumbering reptiles were once found on all continents except Antarctica, and they have suffered only recently at the hands of humans. Today, giant tortoises exist only on the island of Aldabra in the Indian Ocean, and on a few islands in Galápagos. Recent studies on the ecological significance of these extant populations of tortoises suggest that they must have had huge impacts on the native ecology in their historic range. In fact, giant tortoises can be viewed as *ecological engineers,* whereby through their natural activities—feeding, mating, caring for young—these organisms alter entire landscapes.

The relationship between the prickly pear cactus (*Opuntia*) and Galápagos giant tortoises has fascinated scientists for over 100 years. Tortoises rely on the fruits and pads of this cactus during the dry season, and through herbivory, tortoises have exerted selective pressure on *Opuntia.* Islands with tortoises and, to a lesser extent, land iguanas, are home to the tall (arboreal) varieties of the cactus. Islands without tortoises and land iguanas are home to lower-lying varieties. This symbiosis between tortoises and *Opuntia* has far-reaching impacts. *Opuntia* is a key food source for many of the archipelago's inhabitants (e.g., land iguanas, finches, mockingbirds), often providing an especially critical resource during the dry season. As a result, tortoises' impacts on *Opuntia* may have broad ecosystem-level effects. In spite of the presumably negative impacts of grazing, *Opuntia* may actually benefit from the presence of giant tortoises. Indeed, tortoises may increase cactus distribution by (a) disrupting the soil and creating ideal conditions for germination; (b) consuming other vegetation, thereby reducing *Opuntia's* competition for sunlight and other resources; or (c) helping distribute *Opuntia* seeds by consuming the seeds and then defecating in a different location. In 2010, adult tortoises were released on Pinta to help restore the island's ecological balance (Chapter 11).

A recent study compared tortoise–cactus interactions at three different areas in or near CDRS in Puerto Ayora, Santa Cruz. Some study sites included cacti in pens that had housed tortoises for decades, others included cacti in pens without tortoises, and some sites were outside the station and had neither tortoises nor cacti in pens. Grazing by tortoises reduced the overall density of cacti, and there were few young cactus plants near mature plants. In areas free of grazing tortoises, cactus plants were much denser, and regeneration occurred closer to adult trees. Thus, tortoises alter the density and age–class structure of *Opuntia* in ways that could affect other organisms that rely on *Opuntia*. In addition, regeneration patterns of *Opuntia* showed little evidence of asexual reproduction (i.e., few cacti developed from buds on the fallen pads of mature cacti), suggesting that tortoises promote sexual reproduction in cacti. Sex, with the creation of novel sperm and eggs that combine to make unique individuals, increases genetic variability in a population. What this increased genetic variability will mean to the *Opuntia* population's genetic structure as tortoise repatriation programs continue (Chapter 11) is mere speculation at this point. For now, we can only await the results of long-term monitoring of ecosystems with and without repatriated tortoises.

SEA TURTLES

Although you might encounter the hawksbill, leatherback, and olive ridley sea turtles in Galápagos, you are most likely to see the green sea turtle, *Chelonia mydas*. Green sea turtles breed in the islands, coming ashore to lay eggs from December through June; however, they can be seen throughout the year feeding on algae in shallow bays. Your best bets for encountering a sea turtle are from a boat in Black Turtle Cove (Northern Santa Cruz) or in the snorkeling areas near Tintoreras by Isabela (La Calera), Tagus Cove (Isabela), Las Bachas (Santa Cruz), and Puerto Egas (Santiago).

Common name: Galápagos green turtle

Scientific name: *Chelonia mydas agassizi*

See it: Shallow bays throughout the archipelago (Color Image 22)

Residence status: The Galápagos green turtle is an endemic subspecies of the wide-ranging Pacific green turtle (*Chelonia mydas*).

Human impacts: Green turtles are classified as Endangered on the IUCN Red List, due to a decades-long trend of decreasing and locally extirpated populations.

What you will see and what it means: Most visitors to Galápagos see green turtles, but they may be tempted to misidentify them based on name alone. Most green turtles have brown, dark yellow, or even black shells, which can be confusing given their name; the "green" refers not necessarily to their shells, but instead to the greenish appearance of their fat and connective tissue.

Sea turtles represent a small branch of an otherwise terrestrial lineage of vertebrates (the amniote group Testudines, or turtles); a remnant of their evolutionary connection to the land is their need to lay eggs on shore. These turtles spend their lives at sea, feeding almost exclusively on algae, and then, after two decades of development, gravid females return to land to lay their eggs. Females lay their eggs at night. Although tourists are not allowed on most islands at night, you may see a few exhausted female turtles at water's edge in the morning; this is a typical sight on many beaches. You are more likely to see the ambling tracks of the female and the large depressions she makes in the sand, especially if you visit during the peak egg-laying months of December through February.

Life is not easy for a baby sea turtle

Laying eggs on shore is a perilous strategy. Although a 100-kilogram female is rarely threatened, her eggs are a desirable food source for introduced rats and pigs. After approximately 2 months of incubation, the hatchlings emerge and dig themselves out of the sand. If all goes well, they emerge at night and during high tide, and make a speedy trek to the relative safety of the water. However, should the baby turtles make it past hungry crabs and shore birds and into the water, they are still far from secure. In the water, the hatchlings are vulnerable to frigatebird attacks from the air, and shark attacks from below. Thus, it is not surprising that female sea turtles often lay several clutches of over 100 eggs per clutch, to ensure survival of at least a few offspring.

Long-distance migration by green sea turtles

In 2007, researchers reported that 12 tagged sea turtles leaving their nesting grounds in Galápagos traveled between 700 and 2,500 kilometers in 33 to 100 days, with some moving more than 50 kilometers in a day. In the best of scenarios, long-distance open-ocean migrations can be perilous. With the addition of offshore fishing pressures—sea turtles are often killed as accidental bycatch—and various oceanic contaminants (e.g., plastic refuse, manufactured chemicals), a successful migration is uncertain.

Temperature-dependent sex determination and the future of the green sea turtle

All turtles, like many other reptiles, exhibit an interesting sex-determination mechanism. Instead of having a genetic basis (like the human X/Y system), the gender of turtles is determined by the temperature at which an egg is incubated. In the case of green turtles, incubation temperatures above 30°C produce females. Mid-range temperatures, averaging 30°C, produce roughly equal percentages of males and females, and lower temperatures produce males. Temperature-dependent sex determination (TSD) leads to unresolved evolutionary questions, such as "What is the adaptive significance of TSD in turtles?"

TSD may be an artifact of past selection pressures, and it may persist in reptiles for no obvious benefit other than that it still "works." This explanation is difficult to test, and it can be applied to many otherwise-unexplained natural phenomena. However, another explanation is the "Charnov-Bull" hypothesis, named for the scientists who initially proposed it in 1977. Charnov and Bull hypothesized that TSD is adaptive for organisms in which males and females fare better at different temperatures. According to this hypothesis, female turtles enjoy fitness benefits at higher temperatures, and males have greater fitness at lower temperatures. In a sea of otherwise unsatisfactory explanations, Charnov and Bull's idea was acknowledged to be logically sound, but not satisfactorily tested until recently. In 2008, biologists at the University of Sydney in Australia separated incubation temperature from sex determination by manipulating eggs with sex hormones. This allowed the investigators to analyze how incubation temperature influences lifetime reproductive success in male and female Jacky dragon lizards. Their results met the predictions of the Charnov-Bull hypothesis: males incubated at typically male-producing temperatures had greater fitness than males incubated at typically female-producing temperatures. Likewise, females incubated at female-producing temperatures fared better than other females. Thus, in fluctuating environments, TSD may select for the sex most likely to succeed in the prevailing temperature.

But what does TSD mean for green turtles in a warming climate? While maternal behavior can influence offspring sex-ratio (e.g., by laying eggs in a shady spot to produce more males), the limits of the influence of female turtles are unknown. Recently, when investigators modeled the effects of various climate-change scenarios on beaches where green turtles nest along the northern Great Barrier Reef, extreme warming scenarios led to female-only clutches of turtles by 2070. However, given the dynamic nature of the climate of the Galápagos archipelago (with El Niño events occurring every 2 to 8 years), sea turtles nesting here should be adapted to some degree of climatic variability. Simply, there are too many variables, and not enough background knowledge about these turtles, to make confident predictions about what the future holds for sea turtles. On a warming planet, in an archipelago projected to change dramatically in the years to come, behavioral flexibility may mean the difference between the green turtle's continued persistence and extinction.

Species concepts and sea turtle conservation

You may encounter several names for the turtle you will see in Galápagos:

Chelonia mydas agassizi, the green sea turtle
Chelonia mydas agassizi, the black sea turtle
Chelonia agassizi, the green sea turtle
Chelonia agassizi, the black sea turtle
Chelonia mydas, the green sea turtle

It's confusing! While several studies have suggested that migratory behavior and morphology (e.g., a darker shell) distinguish the Galápagos-nesting turtles from *Chelonia* that live elsewhere, recent genetic analyses do not support this claim. For example, although *Chelonia* in the Atlantic and Mediterranean regions appear distinct from *Chelonia* from the Pacific and Indian Oceans, there is little reason to distinguish *C. mydas* from *C. mydas agassizi* (or *C. agassizi*). However, "demoting" a lineage from species status is tricky business, with potential implications for conservation. Is there nothing special about the Galápagos population of green turtles? In an article acknowledging that "the dire conservation status of the black turtle has been invoked as an argument for preserving species status," Stephen Karl and Brian Bowen claim that "the artificial geopolitical designations of the past are being perpetuated principally for management and regulatory reasons." Rather than dismiss these designations altogether, they propose the label *geopolitical species* for organisms such as the green sea turtle of Galápagos. This term is based not on scientific data, but rather on "geographical, political or cultural grounds."

Whether the green sea turtles you see in Galápagos are accorded species status or not, the archipelago is an important nesting area for *Chelonia.* Together, the Galápagos National Park and Marine Reserve preserve habitat that presumably resembles the conditions these turtles evolved to exploit.

LAVA LIZARDS

Seven to nine species of lava lizards, all in the genus *Microlophus,* occur in Galápagos. You will see them everywhere (except on Genovesa; Chapter 5). Lava lizards are small, are various shades of brown and gray, and are active throughout the day. The sexes are dimorphic; larger males have a distinct spinal crest. The most widespread, and most commonly encountered species, is *M. albemarlensis.*

Common name: Lava lizard

Scientific name: *Microlophus albemarlensis*

See it: Abundant all over (on rocks, sand, buildings, roads) Baltra, Fernandina, Isabela, North Seymour, Rábida, Santa Cruz, and South Plaza (Color Image 23)

Residence status: Endemic

Human impacts: *M. albemarlensis*' status is unclear as of this writing, but it may be listed as Vulnerable in the near future.

What you will see and what it means: You *will* see lava lizards, many of them. In some areas, lava lizards are relatively unfazed by humans and will allow you to watch them go about their business. Lava lizards are omnivorous; they will be stalking insects or munching on flowers as you take pictures. Not only are males larger than the females, but females are smoother and often possess a bright red throat-patch. If you visit several islands, observe how lizard coloration varies with respect to the substrate: lizards common to lava fields are darker, while those skittering in the sand are lighter. Like so many classic stories of how the environment (e.g., rock and sand color) gives some phenotypes (e.g., skin color) an advantage, the lava lizards that are well camouflaged from predators experience greater fitness (i.e., they reproduce more) than their more-visible counterparts.

Why are some lava lizards more skittish than others?

After a difficult time approaching lava lizards on Santa Cruz, you may be surprised how close you can get to them on South Plaza. What could explain this behavioral difference between lizard populations? Although Santa Cruz is home to rats, cats, and snakes, none of these predators occurs on South Plaza. Could predation pressure have selected for increased wariness by the lizards, or is there a more specific explanation? In fact, lizards on islands with cats are significantly more skittish than lizards on cat-free islands. Thus, the fairly recent introduction of cats appears to have selected for a behavioral feature in lizards—notably, their tendency to flee danger. If you become frustrated with your inability to get close to some lava lizards, recall that this behavior is an adaptation, possibly keeping these lizards from extinction.

How many times did lava lizards colonize Galápagos?

The geology of the archipelago—in which older and lower (and often submerged) islands are moving to the southeast (Chapter 3), and volcanic activity is creating newer islands in the west—has led to a proposed "conveyor belt" scenario to describe how some organisms may have colonized Galápagos (Figure 6.3). According to this proposal, colonization of the eastern islands was followed by westward dispersal, as older landmasses moved east and became submerged. That is, as islands sank, some organisms migrated to the higher western islands of the archipelago. This scenario fits with much of the biogeography of the islands, and explains how animals that evolved over 5 million years ago (e.g., pink land iguanas) can exist on islands that are younger than one million years old (e.g., Isabela). However, this model does not fully explain the dispersal of lava lizards.

The archipelago's lava lizards are likely the result of two separate colonization events, giving way to two distinct groupings: one, the eastern radiation, includes the two species endemic to Marchena and San Cristóbal Islands; the other, the western radiation, includes all the other lava lizards. The younger, eastern group became established on San Cristóbal less than 1.4 million years ago, while the older, western group began in Española. From Española, lava lizards moved north and west, resulting in an adaptive radiation that can easily be seen today.

Other species of lava lizards in Galápagos

The remaining *Microlophus* species are easily identifiable by their location:

- if you see a lava lizard on Española, it is *M. delanonis*
- on Floreana, *M. grayi*
- on Marchena, *M. habellii*
- on Pinta, *M. pacificus*
- on Pinzón, *M. duncanensis*
- on San Cristóbal, *M. bivittatus*

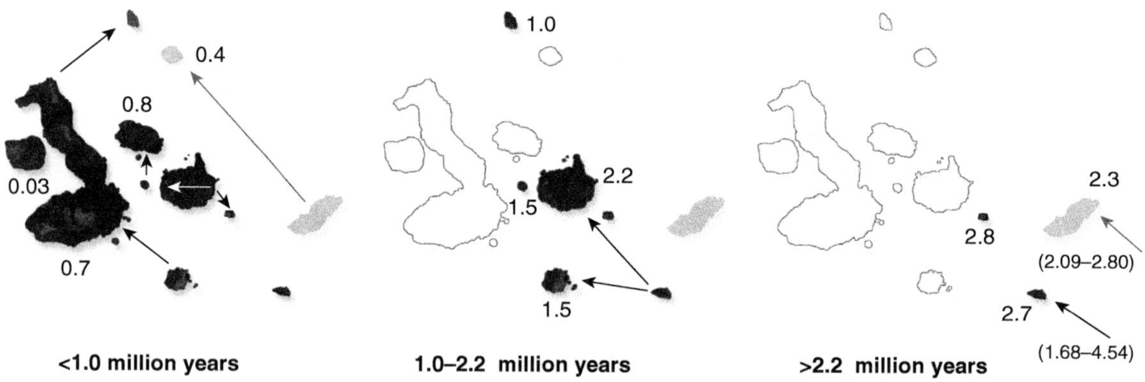

FIGURE 6.3 According to the "conveyor belt" scenario of colonization, arrival on eastern islands was followed by a westward dispersal, as older islands moved east and became submerged. In this figure, submerged islands are lightly shaded (e.g., Santa Cruz was underwater over 2 million years ago, but above water 1.0–2.2 million years ago). In addition, the dispersal of lava lizards in Galápagos appears to be the result of two separate colonization events—a younger, eastern radiation (in gray) that began on San Cristóbal, and an older, western radiation (identified by the darkest shading) that began on Española. The numbers by each island indicate approximately how long ago (millions of years) each island emerged. Arrows illustrate the hypothesized migration of lava lizards. (From Benavides, et al., 2009)

LAND IGUANAS

Hybrid Iguanas and Iguana Diversification

Marine and land iguanas diverged over 10 million years ago, a speciation event that must have occurred either on a now-submerged landmass in the archipelago or at a location removed from the islands altogether. The land iguanas then diverged in the past several million years, resulting in something akin to their present-day distributions. However, even with this ancient divergence, reproductive barriers are not fixed; in fact, you may observe a rare hybrid of marine and land iguanas. These hybrids are often dark, with a pattern of light bands on the back; their heads are shaped like those of the marine parent, yet they eat cactus like the land parent. While viable hybrids do occur, it is unlikely they represent a true barrier to species integrity. The absence of second-generation hybrids suggests some form of "hybrid sterility" in these individuals.

Common name: Land iguana, or common land iguana

Scientific name: *Conolophus subcristatus*

See it: Scrubby habitats next to loose soil or volcanic ash (for egg-laying) on many islands, particularly Baltra, Fernandina, North Seymour, Rábida, Santa Cruz, and South Plaza

Residence status: Endemic

Human impacts: The land iguana is listed as Vulnerable by the IUCN because of its restricted range (historically occurring on only five islands) and the threats from introduced species that eat the iguanas or compete with them for food. In fact, several populations of land iguanas have been damaged by human activities, including the Santiago population that was extirpated over 100 years ago. The Santa Cruz populations were severely damaged by feral dogs in the mid-1970s, and they have been the focus of an intense reintroduction program through the CDRS. Successful, and not-so-successful, eradication and reintroduction programs are discussed in Chapter 11.

In 2012, doctoral student Emily Wheeler reported an intriguing phenomenon in land iguanas; namely, that land iguanas living closer to human settlements and tourist sites harbor more antibiotic-resistant *E. coli* bacteria than do those living farther from humans. Wheeler and her colleagues propose that "reptiles living in closer proximity to humans potentially have higher exposure to bacteria of human origin." This work is preliminary, yet may ultimately reveal information about human impacts on the general health of this Galápagos endemic.

What you will see and what it means: Few animals appear as "prehistoric" as do the land iguanas, ambling through the shrubs with their gigantic claws, leathery skin, and long spinal crests. Lying in the sun, with their partially closed eyes and mouths set in a half-smile, they do not appear clever or alert (Color Image 24). But wait for a cactus pad to drop, and iguana pandemonium ensues (for one or two seconds), as several nearby iguanas scramble for this important source of water, nutrients, and calories. You might also catch some action during mating season (at the end of the year), when males pounce on females, often violently restraining them to copulate.

Land iguanas mature between 8 and 15 years of age. After mating, the females dig a burrow and lay from 2 to 20 eggs. The young iguanas hatch 3 to 4 months later and take several days to dig their way out of the nest. Land iguanas can live more than 40 years.

Did American Troops Kill All of Baltra's Land Iguanas?

Although most books have attributed the demise of land iguanas on Baltra to American troops stationed there during World War II (often describing the iguanas' disappearance as a "senseless slaughter"), the explanation is probably not that simple. Writer William Beebe noted in 1923—that is, almost 20 years *before* troops arrived—that all of Baltra's land iguanas were young and rare, and in 1940, Loren Woods made a similar observation. When American troops arrived in 1942, the Smithsonian reported that Baltra's land iguanas were "no longer very plentiful," prompting the commanding officer of the Baltra air base to prohibit "the killing of all animals and birds." Base personnel, who did not recall seeing any young iguanas at Baltra, denied killing the iguanas, and photos taken at Baltra often showed adult land iguanas, but no juveniles. However, goats, who compete for the same food as land iguanas in Galápagos, were abundant on Baltra during the U.S. occupation of the island. Indeed, they were the mascot of the American base at Baltra, and were so abundant on the island that *Time* magazine noted that "Galápagos goats idled nearby" when the base was closed in 1947. Thus, while it appears unlikely that people killed the Baltra iguanas directly, they may have done so indirectly by introducing invasive goats to the island at some point prior to the U.S. occupation. Thanks to recent repatriation efforts (Chapters 2, 11), Baltra is now home to a breeding population of land iguanas.

Mutualisms in Land Iguanas

If you snorkel in the archipelago, you are likely to see cleaner fish servicing client fish by pecking parasites (or dead scales) from the client's body. These symbioses are generally regarded as mutualistic, with the clients enjoying the benefits of parasite removal, and the cleaners enjoying an easy meal. These mutualisms are common underwater, and client fish are the typical participants (Chapter 8). However, the common land iguana presents an unusual example of a client that is a *lizard,* and the cleaner is a *bird*—specifically, one of a few species of Galápagos mockingbird (*Mimus*) and finch (*Geospiza*). Mockingbirds and finches clean iguanas by removing ticks, flies, and sloughed skin, in turn getting a nutritious meal.[4] Although mainland herbivores seem largely detached from the cleaner/client symbiosis, land iguanas, like tortoises, often raise their bodies and assume a receptive posture to accommodate cleaner-birds. This behavioral adaptation gives human visitors yet another interesting coevolutionary story to observe in Galápagos.

Common name: Santa Fé land iguana

Scientific name: *Conolophus pallidus*

See it: Exclusively throughout Santa Fé

Residence status: Endemic

Human impacts: The Santa Fé land iguana is listed as Vulnerable by the IUCN, presumably due to the small size and restricted range of the population—a single small island (24 km^2) susceptible to random environmental changes.

What you will see and what it means: If you visit Santa Fé, you will see several of these iguanas; the island is small and the iguanas are largely unobscured (although they are well camouflaged along the paths and against the vegetation). Look for these large, pale iguanas awaiting falling cactus pads. You will also see many of their burrows, which are convenient retreats into which they return at night to conserve body heat.

Common name: Galápagos pink land iguana

Scientific name: *Conolophus marthae* (proposed name for this newly identified species)

Contrast with: Given its limited distribution, the pink land iguana is unlikely to be confused with any other animal in the archipelago.

See it: Exclusive to Volcán Wolf, Isabela

Residence status: Endemic

Human impacts: Although its formal status has yet to be evaluated, there is cause for concern. With its limited range and small population size (101 at Volcán Wolf in 2009), along with competition from feral goats in northern Isabela, the pink land iguana may be Critically Endangered.

What you will see and what it means: You probably will not see these land iguanas; given the uncertainty about their conservation status, protection of this population—via limited contact with humans—is important.

If historical visitors saw these large, pink, black-banded iguanas, they failed to make note of it. In fact, there is no record of their discovery prior to 1986,

[4] Cleaner-bird symbioses occur elsewhere in the neotropics, and typically include herbivorous mammal clients (e.g., capybaras, deer, domestic ungulates such as cattle). In Galápagos, the iguanas—marine and land—and the giant tortoises appear to fill this niche.

and they were not presented as a novel species until 2009. In announcing this "new" species of land iguana, Gabriele Gentile[5] and colleagues wrote, "[d]espite the attention given to them, the Galápagos have not yet finished offering evolutionary novelties." And although these iguanas are new to human observers, they are hardly an *evolutionary* novelty. In fact, pink iguanas appear to be the most ancient of the archipelago's land iguana lineages, having diverged approximately 5.7 million years ago.

How can iguanas be older than the island they occupy?

Many of the archipelago's earlier islands are now underwater, having submerged in the past few million years (Chapter 3). Given that Volcán Wolf is less than 500,000 years old, and most of the currently above-water islands are less than 5 million years old, the pink iguana must have evolved on a now-submerged island. How it came to Isabela is a mystery.

Evolution of land iguanas

A 2008 study analyzing genetic variation in over 700 land iguanas supports grouping land iguanas not as the 3 divisions presented above, but as 5 distinct taxa: (1) the Santa Fé land iguanas; (2) Fernandina and southern Isabela iguanas; (3) the newly described pink "rosada" iguana (*C. marthae*) from Isabela; (4) Santa Cruz iguanas; and (5) South Plaza iguanas. (The translocated North Seymour population represents iguanas from two or more of these groupings, which may or may not have been an intentional part of the conservation plan.) If further work confirms these relationships, we will need to revise our current classification of land iguanas.

MARINE IGUANAS

Common name: Marine iguana

Scientific name: *Amblyrhynchus cristatus*

See it: Rocky coastlines adjacent to shallow reefs throughout the archipelago

Residence status: Endemic

Human impacts: The marine iguana is classified as Vulnerable by the IUCN; this designation is due to the small size of the population, as well as its vulnerability to climatic fluctuations (e.g., El Niño events) and human impacts (e.g., introduction of non-native species, oil spills).

What you will see and what it means: Few things scream "Galápagos!" like the iconic marine iguanas (Figure 6.4). They are endemic, omnipresent, and

[5] The proposed specific epithet (*marthae*) for this iguana is in homage to Martha, the stillborn baby girl (b. 2003) of Gabriele Gentile.

FIGURE 6.4 It is common to see marine iguanas surrounded by one of many potential symbionts. Mockingbirds, finches, and Sally lightfoot crabs remove dead skin and ectoparasites, and lava lizards (such as the one shown here) often sun themselves on the backs of marine iguanas. Marine iguanas, which are prey to few species, provide the lava lizards with a relatively safe—and warm—place to thermoregulate.

remarkable for so many features: they are the only seagoing iguanas in the world; they feed underwater, scraping algae off coral and volcanic rock (Color Image 25); they often "snalt," or expel a spray of salt water from their nostrils; and they are, like so many island denizens, remarkably fearless, and will walk over your shoes if you let them (don't let them; it's against park rules). After feeding in the cold ocean, marine iguanas return to land and bask in the sun to raise their body temperature.

There are seven "races," or subspecies, of marine iguanas, which vary in shape, size, and location. The largest iguanas are on Fernandina, the smallest are on Genovesa, and the most colorful males (during breeding season) are on Española. Marine iguanas lay two to three eggs, which hatch 2 to 4 months later. Marine iguanas can live up to 60 years.

Lekking in marine iguanas

Prior to and during the breeding season, many male marine iguanas establish territories for soliciting female copulations. This group of two to six territory-holding males is a *lek,* and the area they inhabit is chosen not for desirable resources, but strictly for reproductive purposes. In this way, lekking is different from another mating strategy, *resource defense polygyny,* whereby males defend a desirable resource and females usually choose these resourceful males as their mates. In a marine iguana lek, the central—usually the largest—male (the *hotshot* male) copulates the most. Meanwhile, *satellite* males roam the perimeter of other males' territories and can copulate by chasing females and overtaking them, often while the females are foraging away from the central male's territory. Also, *sneaker* males, who look like females and are undetected by territorial males, are nearby for stealth copulations when the primary male is away. Given that iguanas sun themselves *en masse* on lava rocks, in mixed-sex associations, it is not too difficult for a female-sized sneaker male to approach a female and cryptically initiate copulation from a pile of sunning iguanas.

Sneaky males and phenotype switching

Although sneakers and satellites can, in their respective roles, become fathers, the fitness prize goes to territorial (i.e., hotshot) males, a reward associated with the territorial male's age and size. A key component to this reproductive arrangement may be *phenotypic plasticity,* whereby males can switch between roles as conditions warrant. For example, an El Niño year that kills the largest males (see discussion below) opens opportunities for other males in the colony to reproduce. Therefore, it is in a nonterritorial male's best interests to be able to switch roles more-or-less on demand. How is this accomplished?

An experimental manipulation of 48 males (territorial, satellite, and sneaker) on Genovesa focused on the role of androgens (sex hormones such as testosterone) in the transition between reproductive types. Experimental males received injections of either testosterone-blocking substances (the territorials) or testosterone (the satellites and sneakers). Two days after these injections, the males were observed, plasma testosterone was evaluated with a blood sample, and iguana colonies were compared. The results indicate that testosterone induces phenotypic switching: injected territorial males had less testosterone, reduced courtship behavior, and fewer females attending their territory. Injected satellites tried to establish territories, often fighting other males to do so; and sneakers left their cluster of females to patrol territories like satellite males do. Notably, the changed behaviors of satellites and sneakers did not result in increased fitness; rather, satellites were wounded in testosterone-fueled fights, and sneakers left their supply of available females. Thus, testosterone is important, but not the whole story. Body size, colony composition, and circulating hormone levels may work in concert to create the complex mating dynamics of marine iguana colonies.

Stress and survival in marine iguanas

Few things appear to cull iguana populations like an El Niño event. In fact, severe El Niño seasons have reduced the population of marine iguanas by more than 80%. Some factors that affect survival are clear; for example, smaller iguanas require less food and are more likely to survive the famine conditions (i.e., decreased availability of algae) of an unusually warm year. However, questions remain: body size notwithstanding, why do some iguanas survive while others do not? Is survival random, or the result of selection on some aspect of iguana behavior or physiology?

Successful iguanas, some of which can survive months without food, can shrink during the famine that accompanies an El Niño; specifically, large (over 30 cm long, minus the tail) individuals that shrink the most survive longer than large iguanas that shrink less. (Shrinking only a centimeter increases survival by 10%, and greater amounts of shrinkage increase survival by as much as 35%.) The mechanisms underlying shrinkage are unknown, but given the large extent of the changes (sometimes up to 20% of total length) may involve absorption of bone.

Iguanas also vary in their release of the steroid hormone *corticosterone,* which influences reptile metabolism, immune function, and stress response. Corticosteroid levels are lower in iguanas that survive El Niño events. In one recent study, Martin Wikelski and Michael Romero evaluated stress responses in 98 large male iguanas by recording how individuals responded to a synthetic version of corticosterone, allowing the scientists to evaluate the iguanas' ability to suppress continued production of corticosterone under normal conditions. Then, they waited for an El Niño. After the mild El Niño conditions of 2002 and 2003, 23% of the iguanas died. Only one measurement differed significantly between the iguanas that survived and those that died: namely, surviving iguanas were better at suppressing the production of corticosterone, displaying a more acute negative feedback response to stress. In other words, surviving iguanas did not get as "stressed out" as the iguanas that died. In marine iguanas, being stressed can lead to a cascade of physiological responses such as protein metabolism or decreased immune function. Marine iguanas that do not respond as strongly to stress are more likely to survive El Niño events.

For this story to be complete, the exact nature of this variable response must be clarified. It is possible that, like many hormonal pathways, the number of hormone-specific receptors in the brain is what controls the strength of this response. If that is the case, the surviving iguanas could be those that have more corticosterone receptors in their brains. This trait could result from duplication mutations (whereby genes are copied in full) and then be inherited and positively selected.

Tourism and stressed iguanas

El Niño events are not the only source of stress for marine iguanas. Another recent study compared corticosterone levels in marine iguana populations frequented by tourists and those not accustomed to humans. Not only did iguanas from heavily visited sites have greater concentrations of corticosterone, but these stressed iguanas had a diminished immune response. For example, superficial wounds of stressed iguanas having higher levels of corticosterone took longer to heal, and stressed iguanas were more susceptible to bacterial infection. This physiological response to humans contradicts what is often claimed about Galápagos fauna: namely, that these organisms are not affected by tourist activity.

Coevolution in marine iguanas

Once marine iguanas become adults, they are largely free of predators. Aside from the dangers posed by introduced predators (especially cats and dogs; see discussion in Chapter 11), the natural exception is the Galápagos hawk, which eats juvenile iguanas and an occasional adult female. Recently, an interesting coevolutionary story emerged, whereby iguanas appear to heed the warning calls of mockingbirds, who call to alert other mockingbirds of a hawk's presence. When recorded calls of mockingbirds were played back in iguana colonies, 45% of the female and juvenile iguanas (of over 200 individuals observed) reacted with either increased vigilance or a flight response. When a marine iguana is sunning itself, it may not be facing in the right direction to see incoming hawks; thus, exploiting the alertness of a mockingbird (who is not constrained by this thermoregulatory behavior) could be the difference between life and death. This symbiosis between marine iguanas and mockingbirds is exciting not only because it is a rare example of an animal heeding the alarm call of a different species, but also because this is the first documented instance of a *nonvocal* species responding to this type of acoustic signal.

Early visitors to Galápagos were not kind to marine iguanas; for example, Darwin described them as "hideous-looking creatures, of a dirty black colour, stupid and sluggish," and James Colnett dismissed them as having "the most hideous appearance imaginable." However, marine iguanas continue to surprise us with developing stories of mating behavior, stress physiology, and coevolution. For such a highly visible, common, and fearless animal to still be so full of novelty is a testament to the scientific goldmine that is Galápagos.

CHAPTER 7

Mammals

There is no other quarter of the world where [reptiles] replace the herbivorous mammalia in so extraordinary a manner.
—Charles Darwin, *Journal of Researches,* 1845

ONE OF THE FIRST THINGS THAT VISITORS TO GALÁPAGOS realize is that, except for humans (and the animals that humans have introduced), mammals are not prominent in the various ecosystems of the islands. As noted in Chapter 3, the niches typically filled by mammals in other ecosystems are filled by different organisms in Galápagos. For example, although mammals are the dominant herbivores in many ecosystems, in Galápagos this distinction falls to the reptiles. And while in many ecosystems mammals are the dominant predators, in Galápagos it is the Galápagos hawk that is the top land predator. Yet mammals—marine and terrestrial—do shape Galápagos communities, and some of their stories are discussed in this chapter.

The islands' brief history with mammalian herbivores and predators is, tragically, at the root of many of the islands' problems today. For example, introduced goats quickly became the dominant herbivores on many islands, resulting in local extinctions and imperiling many of the native organisms (Chapter 11). Similarly, introduced dogs and cats have also been destructive, especially in ecosystems without any evolved resistance to mammalian predators. Today, humans are the dominant mammal in Galápagos; we have been altering the archipelago since the time the islands were visited by the first whalers and buccaneers (Chapter 2).

WHO ARE THE MAMMALS?

Mammals are amniotes, a group of animals that also includes birds and reptiles. Mammals, however, are *synapsid* amniotes, characterized by a single opening in the skull behind each eye. Much evidence indicates that all mammals share a single common ancestor; shared traits of the mammals include the presence of hair or fur and a collection of glands modified for different purposes (e.g., mammary glands for producing milk, scent glands for marking territories and self-identification, sweat glands). Of the various mammalian orders, or clades, in Galápagos you are most likely to encounter native animals from the Cetacea (whales and dolphins), Carnivora (pinnipeds such as fur sea lions and sea lions), Rodentia (rats), and Chiroptera (bats). Introduced mammals in Galápagos belong to the Carnivora (dogs and cats) and Artiodactyla (goats and pigs), as well as a few destructive rodents (introduced rats and mice). Although an oft-repeated claim about Galápagos is that there are no native land mammals, this is not entirely true; endemic rats and bats live in the archipelago.

Marine Mammals

Most endemic mammals you will see in Galápagos share an affinity for water, a fact that is hardly surprising given the archipelago's vast coastlines and long distance (1,000 km) from the mainland. You may see a whale's fluke or spout, or a pod of jumping common dolphins (Color Image 26), but you are most likely to be surrounded—on land or in water—by sea lions and fur sea lions (Figures 7.1 and 7.2). For many visitors, swimming with curious young sea lions is the highlight of a trip to the islands. Just keep your distance from the males; they can be aggressive and dangerous.

Seals versus sea lions

Pinnipeds include a family (Otariidae) of sea lions and a family (Phocidae) of seals. However, what is often called the Galápagos fur seal is not *really* a seal; rather, with its external ear-flaps and powerful front flippers, the fur "seal" in the archipelago is an Otariid, or sea lion, and is accurately referred to as the Galápagos fur sea lion.

Common name: Galápagos sea lion

Scientific name: *Zalophus wollebaeki*[1]

Comparisons: The Galápagos fur sea lion often co-occurs with the Galápagos sea lion on Fernandina, North Seymour, and Genovesa, and can be distinguished by the shape of the face (blunter in the fur sea lion), overall body size (smaller in the fur sea lion), and eyes (protruding from the face in the fur sea lion).

See it: Sandy beaches and rocky ledges throughout the islands; larger congregations of sea lions live on Española and Santa Fé (Figure 7.1).

[1] Until recently, the Galápagos sea lion was considered a subspecies of the California sea lion, *Zalophus californianus*.

Residence status: Endemic

Conservation status: With its small, fluctuating population size, along with a large decline (more than 50%) in the past few decades, the Galápagos sea lion is classified as Endangered on the IUCN Red List.

What you will see and what it means: You will encounter sea lions on any trip to the islands; if you do not venture from any of the towns, you will still see these large, barking, boisterous animals. In Puerto Ayora, mornings at the fish market are a cacophony of noisy sea lions, vying for scraps with pelicans. Napping sea lions may block your path through mangrove inlets of Puerto Villamil (Isabela) (figure 7.1) or clog the boardwalk through Puerto Baquerizo Moreno (San Cristóbal). You can even see these animals playing by the ferries that transport passengers between Santa Cruz and Baltra. Sea lions are the largest native or endemic terrestrial animal in Galápagos; full-grown males can weigh more than 250 kilograms.

Female sea lions reach sexual maturity in five years and can live to be 20 years old. Males, who do not guard a harem until several years after they

FIGURE 7.1 Napping Galápagos sea lions (shown here at Gardner Bay, Española) are a common sight throughout the archipelago.

reach sexual maturity, have shorter lifespans. Beside their larger size, males—unlike females—have a large "bump" on their forehead.

Although sea lions do not appear bothered by human activity, they thrive on their own protected beaches. If you visit a large community of sea lions—such as that at Gardner Bay of Española—you can see what their lives are like. You will see playful behaviors and a lot of napping on beautiful beaches, but you may also see bloody battles between large "bull" sea lions, a dead baby sea lion being eaten by hawks and mockingbirds (Fig. 1.2), or a young sea lion calling for its missing mother. With their preferred habitat—shallow bays and open, sandy beaches—and their fearlessness, sea lions reveal their entire life histories to human visitors.

Symbiotic sea lions

The coevolution of sea lions with several other species has shaped several communities in Galápagos. Some of these symbioses with sea lions are obvious; for example, sea lions prey on dozens of fish species, and Galápagos hawks and mockingbirds enjoy sea lion carcasses and placentas. In addition, a profound dependence on sea lions is reflected in the composition of coastal plant communities. Some of the volcanic islands of Galápagos are so nutrient-poor that many endemic plants depend on an imported supply of the critical nutrients nitrogen and phosphorus from sea lion feces.

The recent volcanic origin of Galápagos, its arid climate, and its isolation from the mainland combine to produce the distinctive flora of the archipelago. Most plants in Galápagos are salt-, wind-, and drought-tolerant, low-lying, and slow-growing varieties similar to those characterizing primary succession elsewhere (see Chapter 10). During the early stages of plant colonization, the availability of essential nutrients determines the composition of plant communities. It has not always been easy to determine the origin of these nutrients, but this assessment is now possible by analyzing isotopes—whereby elements such as nitrogen and carbon have different "elemental signatures" depending on where these nutrients originate. For example, $\partial^{15}N$, which is the ratio of nitrogen with 15 neutrons (^{15}N) to nitrogen with 14 neutrons (^{14}N), can be used to estimate whether the nitrogen in tissues—plant or animal—is primarily marine or primarily terrestrial. Specifically, a high $\partial^{15}N$ indicates marine, and not terrestrial, nitrogen input.

A recent investigation of sea lion and plant-community symbioses focused on the extent to which plant nutrients were marine-derived, using a sea lion vector. Investigators analyzed $\partial^{15}N$ in the tissues of plants on several islands in Galápagos, and noted whether these islands housed year-round sea lion populations. Although sea lions lived on all of the islands surveyed, their colonies were restricted to shorelines within 20 meters above sea level; these shorelines have soils replete with nitrogen-rich ammonia and nitrate. The isotopic signatures of low-elevation plants suggest that these plants thrive on essential nutrients supplied by sea lion feces and carcasses. In other words, the sea lions, by defecating and dying, bring many Galápagos plants what they need to grow on the harsh volcanic landscapes in the middle of the Pacific Ocean.

The story does not end with sea lions supplying nutrients to low-elevation plants. Rather, some common plants, like saltbush (*Cryptocarpus pyriformis*), grow tall enough to create a bit of shade that can protect sea lions from the intense equatorial sun. At sites such as Isla Lobos and elsewhere, you will see these large, dark, furry mammals vying for this shade under clumps of saltbush. Remember that this shade may have ultimately been provided by nutrients supplied by the sea lions themselves.

Status of sea lion species

For decades, the taxonomic status of the Galápagos sea lion, in relationship to the California sea lion, has been unclear. Even in the scientific literature, the Galápagos sea lion has recently been classified at both the species (*Zalophus wollebaeki*) and subspecies (*Zalophus californianus wollebaeki*) levels. However, a 2007 report outlined reasons for honoring the species designation: first, analysis of nuclear and mitochondrial DNA supports a split between the California sea lion populations and the Galápagos sea lion populations; second, molecular-clock estimates place their divergence at over two million years ago; third, the ranges of the two types of sea lions do not overlap; and finally, they exhibit different levels of genetic diversity, with California sea lion populations being more diverse than the Galápagos populations.

In summary, there appear to have been three distinct species within the genus *Zalophus*. Two of these species—that is, the California sea lion and the Galápagos sea lion—are discussed above. The third species, the Japanese sea lion (*Zalophus japonicus*), is thought to have recently become extinct. Sadly, the Galápagos sea lion has an uncertain future as well. Populations have declined in the past few decades (from over 40,000 animals in the 1970s to fewer than 16,000 in 2012), and climatic variability, combined with human impacts,[2] could be disastrous to these animals. Establishing that Galápagos sea lions merit species status may help outline conservation priorities.

Common name: Galápagos fur sea lion

Scientific name: *Arctocephalus galapagoensis*

Contrast with: The Galápagos sea lion is larger and has a pointier face (Figure 7.2)

See it: Fernandina, Isabela, and Santiago

[2] For example, in 2011, Juan José Alava and colleagues reported the presence of DDT and DDT metabolites (e.g., DDE) in Galápagos sea lions, noting an increase in levels of the contaminants in just a 3-year period (from 2005 to 2008). DDT can be detrimental to wildlife, and DDE disrupts male hormones at levels comparable to those that have been identified in Galápagos sea lions. DDT, which kills mosquitoes that transmit human malaria, is increasingly popular in tropical countries, spurred on by recommendations of the 34th G8 Summit in 2008.

FIGURE 7.2 Although Galápagos fur sea lions resemble Galápagos sea lions, the fur sea lions are smaller and have blunter faces.

Residence status: Endemic

Conservation status: The Galápagos fur sea lion is classified as Endangered on the IUCN Red List. This status is due to several factors, including its small, insular population (the result of harvests to near-extinction in the 1800s; Chapter 2).

What you will see and what it means: Fur sea lions are not as visible as sea lions, largely because fur sea lions prefer rocky ledges to open, sandy beaches. At first encounter, it may be difficult to distinguish fur sea lions from sea lions, especially without a side-by-side comparison. But if you can identify groups of fur sea lions, you may observe some fascinating features of their natural histories, such as mother-infant bonding behavior and sibling-sibling aggression. Curiously, you are most likely to see fur sea lions in the days around a full moon.

The fur sea lion/full-moon connection

Why would fur sea lions be more visible on days during a full moon? The answer may lie in the foraging patterns of the fur sea lions, and may reflect an avoidance of water, rather than a preference for the shore on bright nights. Fur sea lions feed at night, and presumably exploit the tendency for many prey species to migrate upward in the water column at night. For example, mollusks such as squid—a fur sea lion favorite—stay in deep, dark waters during the day, and move upward to feed in the relative safety of darkness. But these shallow waters may not be so safe on a moonlit night, thus curtailing much of this vertical migration. In turn, there is less for a fur sea lion to hunt during a full moon, so fur sea lions stay home. This means that fur sea lions are more likely lounging in their natal territories during the day, rather than off at sea awaiting a night of fruitless hunting. Also, by avoiding the water on bright nights, fur sea lions are less visible to sharks, a typical predator of pinnipeds (as you will note when you see shark-bite wounds and scars on these animals).

Maternal care in fur sea lions

Female fur sea lions copulate annually (often right after the birth of a baby), and can bear one pup each year. Typically, an adult female gives birth, and then spends the next year or two alternating between days ashore, lactating, and days at sea, foraging. If you see a colony of mothers and pups, you may be impressed that mothers and pups can reunite successfully after these foraging trips, despite the facts that fur sea lion pups move around a lot, and they all look alike. How do they find each other?

Although there is little evidence that visual cues are important in mother-pup recognition, sound and scent appear to mean everything. During the first few days of a pup's life, its mother is close by, sniffing her baby repeatedly and making long, low calls. In this way, the mother learns the baby's smell, and the baby learns the mother's sound. These signals are used for the next year or more, as young fur sea lions respond to their mother's call when she returns from foraging trips. Sound brings them close together, and they appear to confirm the connection by nose-to-nose sniffing and touching. These recognition behaviors are typical of both sea lions and fur sea lions, and they are a common sight where these animals congregate. Mother-infant fidelity is unwavering, a fact with obvious adaptive significance—why expend resources on a baby that is not yours?—but leads to some heartbreaking displays as well. For example, you may see a pup calling out for a lost mother, slowly starving to death while it wanders through a colony of fat, lactating females.

Parent-offspring conflict in fur sea lions

Although a lactating female is less likely to conceive than a non lactating female, it is not uncommon for a female to have a newborn and a still-lactating yearling or 2-year-old—both demanding her milk. In these cases, siblings compete directly for their mother's milk, and with its age and size, the older offspring will win without parental intervention. So, why would a female have two dependent offspring at one time? How can this reproductive schedule, documented in up to 23% of lactating females, make evolutionary sense? Do females intervene to protect the newborn pup?

Fur sea lions present an intriguing case of *parent-offspring conflict,* whereby the interests of the offspring (in this case, the older pup's demand for milk) conflict with the interests of a parent (the mother's increased fitness, achieved by nurturing a second pup). This scenario is similar in some ways to that of siblicide by the Nazca booby, whereby the second hatchling is killed by the first hatchling soon after birth (Chapter 5). Like the Nazca booby, the parent fur sea lion loses out on potential fitness, and wastes valuable energy, by producing a doomed pup. Unlike with the Nazca booby, however, the fur sea lion's younger offspring are not all killed. A closer look at which offspring survive, and which do not, suggests that the reproductive strategy of fur sea lions may be one of many adaptations to the fluctuating resources that characterize Galápagos.

A recent study of parent-offspring dynamics in Galápagos fur sea lions examined whether the "extra" offspring (born when its older sibling was still consuming maternal milk) represented the sort of insurance against loss that the second Nazca booby hatchlings provide. Alternatively, extra offspring could represent a mother hedging her bets on an especially productive year (as during a La Niña; Chapter 3). A multi-year study of hundreds of fur sea lions revealed that a mother incurs measurable costs from offspring produced while she is already lactating; these extra young have lower birthweights and reduced chances of survival. Extra offspring were less likely to survive if their siblings had yet to reach a "threshold" body weight (about 15 kg), at which the older pup can fish for its own food and can afford to go without as much maternal milk. In productive years, when food is plentiful, mothers aggressively defend their newborn pups against physical harassment from older siblings, either weaning the older pup or accommodating both pups. In lean years (e.g., during an El Niño), mothers are less prone to protect the newborn, and allow the older pup, either through bullying or competition for milk, to reduce her responsibility. Finally, extra pups experience greater mortality if the first pup is male rather than female. Male offspring are larger and more aggressive, and, as males have potentially limitless fitness, mothers can better afford to focus on an only child if that pup is a male.

In summary, the outlook is bad for a newborn fur sea lion, born during a lean year, with a skinny older brother. A newborn fur sea lion, born during a productive year, with no older siblings, or with a fat older sister, has a greater chance of surviving to reproductive age. Environment, life histories, birth order, and genetics all contribute to the survival of newborn fur sea lions. Evolutionarily, bearing two pups close together looks more like bet-hedging than insurance against loss. Bet-hedging also makes sense in light of the fur sea lion's prolonged gestation; because copulation and conception occur a year before giving birth, a female cannot assess near-future productivity in making her reproductive decisions. Should it be a good year, she increases her fitness with a second baby. Should it be a bad year, she can allow the first offspring to eliminate its competition.

Common name: Common dolphin

Scientific name: *Delphinus delphis*

Contrast with: Several other species of dolphins live in Galápagos, such as bottle-nosed, spinner, and striped dolphins. The common dolphin can be identified by its striking coloration, with bright white flanking against a black dorsal fin and cape. Occasionally, common dolphins travel in pods of over 100 individuals (Color Image 26).

See it: In open waters, frequently sighted from tour boats

Residence status: Native

Conservation status: Given that this species is widespread and abundant, the common dolphin is classified as Least Concern by the IUCN.

Endemic Mammals

Common name: Santiago rice rat, or Santiago Galápagos mouse

Scientific name: *Nesoryzomys swarthi*

See it: Santiago

Residence status: Endemic

Conservation status: Vulnerable

What you will see and what it means: Although you are unlikely to see the Santiago rice rat (Color Image 27), they are interesting by virtue of their survival. Four other species of endemic rodents have gone extinct since Darwin's visit to Galápagos in 1835.

Rats versus mice

The terms "rat" and "mice" are not scientific; that is, they do not refer accurately to any actual taxonomic divisions. Colloquially, "rats" are often larger, with hairless tails; "mice" are generally small, with larger ears relative to their body size. Given that "rice rats" are mouse-sized, it is hardly surprising that some texts refer to the Galápagos rice rats as mice.

Survival of the Santiago rice rat

Although it was presumed extinct, a viable population of the Santiago rice rat was found in 1977. While the rats' surprise existence is notable, it is doubly exciting because the Santiago rice rat coexists with the introduced black rats and house mice (see below). This sympatry between introduced rodents and native rodents is interesting, because black rats have been blamed for the extinction of three of the endemic rats in the islands—*Oryzomys galapagoensis, Nesoryzomys indefessus,* and *N. darwini*. How does the Santiago rice rat survive in the presence of the black rat?

When ecologists consider cases in which similar species co-occur, they often invoke a way of thinking called *niche theory.* According to niche theory, sympatric species with identical niches—for example, species that eat the same foods, at the same times, live in the same locations, and use the same resources to construct homes—cannot coexist for very long. Thus, similar species can only coexist if their requirements for survival are sufficiently different to minimize competition. Similar organisms have similar niches where they occur independent of each other, but different niches where they are sympatric; this means that some sort of *niche partitioning* has occurred. For example, niche partitioning occurs when the medium and large ground finches co-occur; namely, they modify their food preferences through a phenomenon described as *character displacement* (Chapter 5). Do rice rats and black rats likewise minimize competition by doing the same thing on Santiago?

To test for niche partitioning in sympatric species, investigators are essentially determining whether a species' *fundamental niche*—that is, the full range of the organism's habitats, foods, and behaviors—differs from a population's actual, or *realized niche,* which occurs in the presence of a competitor. For Galápagos rodents, does the realized niche of the Santiago rat differ from its fundamental niche in the presence of the introduced black rat?

A 2009 study of the food preferences of rice rats and black rats (in captivity) versus actual food selections (in the wild) showed that black rats prefer foods similar to their selected foods; the rice rats, however, consumed different foods in the wild than they did in captivity. This distinction may be due to niche partitioning, whereby the black rat out-competes the rice rat for these food items; the rice rat then maintains a stable population by relying on second-choice foods. In addition, rice rats eat *Opuntia* pads and fruits, especially during the dry season. Black rats do not eat *Opuntia,* and their populations typically crash during dry seasons. The combination of the black rat's population instability and rejection of *Opuntia* may explain how on Santiago, these two rodent species can co-occur.

Robert Dowler and colleagues at Angelo State University (Texas) have suggested that feral cats may be similarly culpable for many declines and extirpations of endemic rodents. In fact, feral cats occur on all islands in Galápagos where endemic rodents have gone extinct—Santa Cruz, San Cristóbal, and Baltra—but they do not occur on Santiago.

Common name: Galápagos rice rat

Scientific name: *Aegialomys galapagoensis* (formerly *Oryzomys bauri* or *Nesoryzomys bauri*)

See it: Santa Fé

Residence status: Endemic

Conservation status: The Galápagos rice rat is listed as Vulnerable on the IUCN Red List because of its restricted range. Although the Santa Fé population appears to be stable, the only other population of these rats was extirpated from San Cristóbal by introduced species.

Common name: Large Fernandina rice rat

Scientific name: *Nesoryzmys narboroughi*

Comparisons: The Fernandina rice rat is smaller.

See it: Fernandina

Residence status: Endemic

Conservation status: This rice rat is listed as Vulnerable by the IUCN because of its small, restricted population.

Common name: Fernandina Galápagos mouse, or Fernandina rice rat

Scientific name: *Nesoryzmys fernandinae*

Comparisons: *N. narboroughi* (the large Fernandina rice rat), which co-occurs with the Fernandina rice rat, is larger.

See it: Fernandina

Residence status: Endemic

Conservation status: This rat is listed as Vulnerable by the IUCN.

What you will see and what it means: You are unlikely to see this rodent, a prediction based on the rat's recent status. While originally described from skeletal remains in an owl pellet in 1980, as recently as 1996 the Fernandina rice rat was declared extinct by the IUCN. However, a survey of rodents in the archipelago disclosed the continued presence of this species, as well as that of the Santiago rice rat (above). With the rediscovery of the Fernandina rice rat, Fernandina is recognized as the only island in Galápagos housing sympatric species of native rodents.

Common name: Galápagos bat

Scientific name: *Lasiurus brachyotis*

See it: Santa Cruz and San Cristóbal, particularly in mangroves

Residence status: Endemic

Conservation status: Not Evaluated

What you will see and what it means: Although there are two species of bats in the archipelago, they are seldom seen. The Galápagos bat is endemic and more common than the native hoary bat.

Common name: Hoary bat

Scientific name: *Lasiurus cinereus*

See it: Santa Cruz, Isabela, San Cristóbal, Santiago, Floreana

Residence status: Native

Conservation status: The hoary bat, with its widespread distribution (throughout North and South America) and large numbers, is a species of Least Concern on the IUCN Red List. It is light brown, eats insects, and often roosts in mangrove trees or shrubs during the day. On islands where they coexist (e.g., Santa Cruz, San Cristóbal), the hoary bat forages high in the trees and air, whereas the Galápagos bat forages near the ground.

Introduced Rodents

In addition to the infamous black rat, the Norway rat and house mouse (*Mus musculus*) threaten the native fauna of Galápagos. The black rat has been in Galápagos for hundreds of years, while the Norway rat and house mouse are more recent introductions. In areas where the black rat and house mouse co-occur, the black rat appears to be dominant, suppressing the activities and number of the mice. The Norway rat seems to be confined to areas of human habitation. These rodents are mentioned briefly below, and discussed further in Chapter 11.

Common name: Black rat, or ship rat

Scientific name: *Rattus rattus*

Comparisons: The Norway rat, with which the black rat co-occurs on Santa Cruz and San Cristóbal, is larger, with hairy ears and a smaller tail.

See it: Widespread throughout the islands

Residence status: Introduced

Conservation status: Least Concern

Common name: Norway rat, or brown rat

Scientific name: *Rattus norvegicus*

Contrast with: The black rat is smaller, with hairless ears and a proportionately larger tail.

See it: Santa Cruz and San Cristóbal (and possibly Rábida)

Residence status: Introduced

Conservation status: Least Concern

Color Image 1 View from San Cristóbal's Frigatebird Hill, where Darwin landed in 1835

Color Image 2 Sombrero Chino, a classic shield volcano

Color Image 3 Bartolomé

Color Image 4 Floreana's Mirador de la Baronesa

Color Image 5 Tagus Cove (background) and Darwin Lake (foreground)

Color Image 6 South Plaza (foreground)

Color Image 7 Medium ground finch

Color Image 8 Woodpecker finch using a twig as a tool

Color Image 9 Galápagos hawk

Color Image 10 Short-eared owl on Genovesa

Color Image 11 Floreana mockingbird

Color Image 12 Galápagos dove

Color Image 19 Lava heron

Color Image 20 Flamingos in flight

Color Image 21 American oystercatcher

Color Image 22 King angelfish and green sea turtle in a cleaner-client mutualism

Color Image 23 Male lava lizard

Color Image 24 Santa Fé land iguana

Color Image 25 A marine iguana, eating algae underwater at Cape Douglas (Fernandina)

Color Image 26 A pod of common dolphins at Bolivar Channel

Color Image 27 Santiago rice rat

Color Image 28 Yellowtail surgeonfish

Color Image 29 Sergeant majors

Color Image 30 A yellowtail damselfish guards a patch of algae

Color Image 31 Mexican hogfish terminal male (large, top) and juvenile-stage female (small, bottom)

Color Image 32 The ubiquitous Panamic fanged blenny

Color Image 33 Giant hawkfish

Color Image 34 Coral hawkfish

Color Image 35 Pacific seahorse

Color Image 36 Jewel moray eel

Color Image 37 Golden cowrays

Color Image 40 Sally lightfoot crab

Color Image 41 Colonization by cactus

Color Image 38 Orange cup coral

Color Image 42 Saltbush

Color Image 39 Variable sea-star

Color Image 43 Lava cactus on Fernandina, with Volcán La Cumbre in the background

Color Image 44 *Opuntia* in bloom

Color Image 47 *Parkinsonia*

Color Image 48 Galápagos cotton flower

Color Image 45 Palo santo as seen from Mirador de la Baronesa (Floreana)

Color Image 49 Passion flower

Color Image 46 Muyuyo in bloom

Color Image 50 *Miconia*

CHAPTER 8

Fish

Below the water line [Roca Redonda] seemed one honeycomb of grottoes, affording labyrinthine lurking places for swarms of fairy fish. All were strange, many exceedingly beautiful, and would have well graced the costliest glass globes in which goldfish are kept for a show. Nothing was more striking than the complete novelty of many individuals of this multitude. Here hues were seen as yet unpainted, and figures which are unengraved.
—Herman Melville, *The Encantadas, or Enchanted Islands* (1854)

ACCOUNTS VARY, BUT AT LEAST 450 SPECIES OF FISH (AT LEAST 80 of which are endemic) live in Galápagos. You can see several of these fish without even getting wet—pufferfish and Galápagos sharks circle tour boats awaiting scraps of discarded food, flying fish soar above the water's surface, and young rays cruise through shallow bays. Yet, below the surface is where the real fish-watching happens. Scuba divers are treated to deep-water treasures such as fairy basslets and grape-eye seabass, but snorkeling can be astounding as well. Highlights of a snorkeling foray can include invertebrates (e.g., courting cuttlefish and scavenging sea stars), birds (swimming penguins and diving boobies), mammals (playful sea lions and curious dolphins), and reptiles (feeding iguanas and sea turtles), as well as thousands of fish. You will encounter famed tropical reef fish such as parrotfish, wrasses, and surgeonfishes, massive schools of snappers and silvery salemas, and improbably shaped (but aptly named) sunfish, porcupinefish, ratfish, and trumpetfish. In this chapter, we highlight some of the most common varieties of fish—native and endemic—to watch for, and focus on a few of their evolutionary and ecological stories.[1] Fish restricted to outlying islands (i.e., Darwin and Wolf), rarely observed fish, and deepwater fish are omitted from this discussion because you will probably not see these fish. Instead, this chapter discusses most of the fish you will see while snorkeling in Galápagos.

[1] Often, a single fish has several common names. When names are disputed, we have used terminology from Humann and DeLoach (2003). In most cases, biogeographic origins for the fish are from Edgar, et al. (2004).

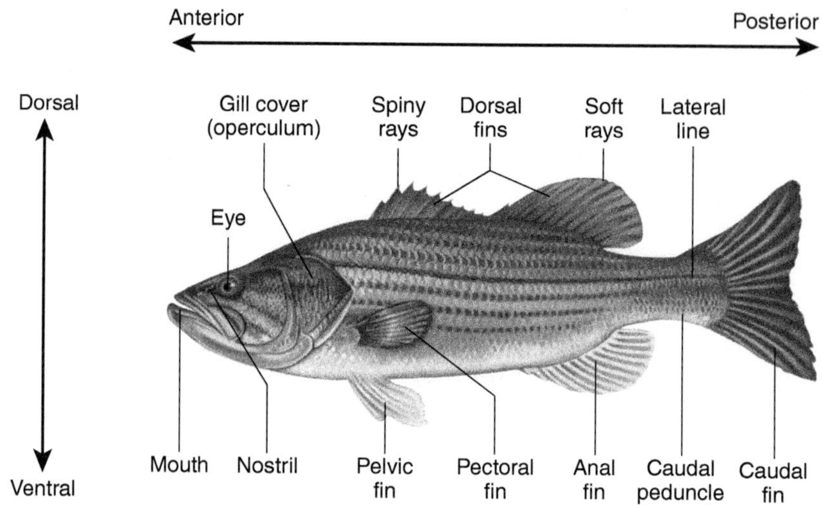

FIGURE 8.1 This simplified diagram illustrates the parts of a typical fish, as referenced in descriptions throughout this chapter.

WHO ARE THE FISH?

There is not a good definition of what makes something a fish. In general, fish are aquatic vertebrates with gills, appendages in the form of fins, and scales. However, there are fish without gills, fish without fins, and fish without scales. Further complicating the taxonomy of fish is the Sarcopterygii—a group of lobe-finned fish that includes lungfish and coelocanths, but which also gave rise to amphibians and amniotes (such as ourselves). Galápagos animals that we call fish belong to two groups—the Chondrichthyes (the cartilaginous sharks and rays; discussed later in the chapter) and the Actinopterygii (the ray-finned bony fish; discussed below).

The bony fish (Actinopterygii) are the dominant vertebrates on the planet, exceeding all other groupings in number of species and number of individuals. Most bony fish have a skeleton of bone, fin rays that support the fins, a swim bladder for buoyancy, and gills that are supported by bony arches and covered with an operculum (Figure 8.1). Scales of bony fish can be smooth, spiny, or absent.

WHERE DID GALÁPAGOS FISH ORIGINATE?

The distribution of fish in Galápagos reflects the islands' currents (Figure 3.8), proximity to continents, food availability, and water temperature. Assemblages of fish in the archipelago can be divided into five major biogeographic zones (Figure 8.2): (1) The Central-Southeastern zone includes the waters around eastern Isabela, Santiago, Santa Cruz, San Cristóbal, Española, and Floreana; (2) the Western zone includes southwestern Isabela and western Fernandina; (3) the Far Northern zone encompasses only Darwin and Wolf; (4) the Northern zone includes Genovesa, Pinta, and Marchena; and (5) the Elizabeth zone is restricted to Elizabeth Bay and Bolivar Channel, which are between Fernandina and Isabela.

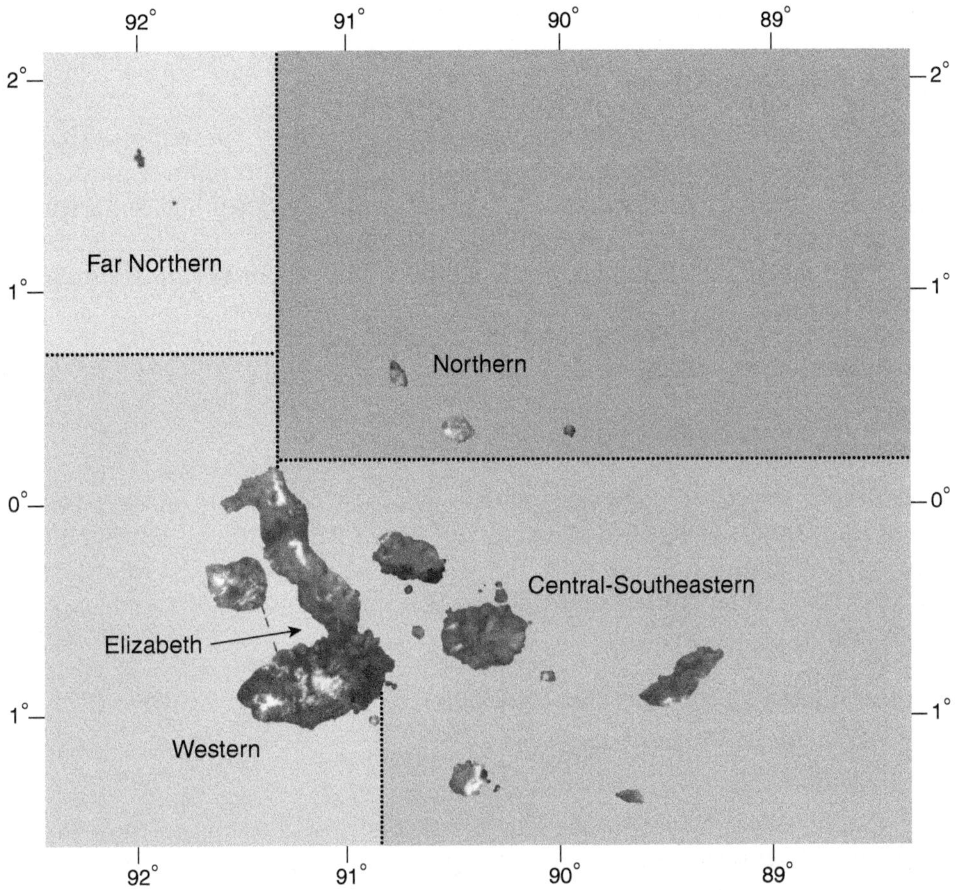

FIGURE 8.2 Galápagos fish can be loosely organized into communities by five major biogeographic zones: The Central-Southeastern zone; the Western zone; the Far Northern zone; the Northern zone; and the Elizabeth zone. (From Edgar et al., 2004)

Each of these biogeographic zones includes characteristic species of fish. The Far Northern zone has the highest diversity of fish and the largest number of fish recruited from the Western Pacific (the "Indo-Pacific" fish). However, there are relatively few fish from the southern South American coast (the "Peruvian" fish) in Galápagos. The Central-Southeastern and Northern zones include many widespread species, with the largest proportion of fish from the northern coast of South America (the "Panamic" fish). The Western zone is characterized by high levels of endemism and many Peruvian fish, but few Indo-Pacific fish. Lastly, the Elizabeth zone has the greatest proportion of endemism, possibly because the western islands are so isolated.

Only about 18% of fish in Galápagos are endemic; this is lower than the endemism rate for terrestrial species because marine species are less isolated than terrestrial species. Because fish can be dispersed by ocean currents, they are less likely to undergo speciation by adaptive radiation from a common ancestor than terrestrial organisms such as finches, tortoises, and *Scalesia*. The rates of endemism for echinoderms (e.g., sea stars and sea cucumbers) and mollusks (e.g., squid, snails) are similar to those of fish.

FAMILIES OF GALÁPAGOS FISH

Surgeonfish (Acanthuridae)

Surgeonfish belong to one of several groups of thin-bodied, colorful, oval-shaped fish found throughout the tropics. Surgeonfish have small protruding mouths and razor-sharp spines—like a surgeon's scalpel—at the base of their tail. These spines are often used for self-defense against other fish. Of the surgeonfish in Galápagos, you are most likely to see the goldrim, yellowfin, and yellowtail surgeonfishes.

Common name: Goldrim surgeonfish (also whitetail surgeonfish or whitecheek surgeonfish)

Scientific name: *Acanthurus nigricans*

Origin: Indo-Pacific

Description: The goldrim is a striking fish, with a dark body and a bright-yellow border running the length of the anal and dorsal fins. With its white tail and bright white mark under its eyes, the goldrim is unmistakable. Goldrim surgeonfish often clean green sea turtles (see "Cleaner-Client Associations in Galápagos Fish").

Common name: Yellowfin surgeonfish (or Cuvier's surgeonfish)

Scientific name: *Acanthurus xanthopterus*

Origin: Indo-Pacific

Description: This surgeonfish has a light-blue body, with a bright yellow pectoral fin and a yellow horizontal bar through the eye. Yellowfins are common and abundant, especially in warmer northern waters of the archipelago. The yellowfin surgeonfish is one of several marine fish that change color as they age; this change complicated the identification and classification of these fish for many years.

Common name: Yellowtail (or razor) surgeonfish

Scientific name: *Prionurus laticlavius*

Origin: Panamic

Description: The yellowtail has a gray-blue body, with a yellow tail and two distinctive black bars marking the head; one of these bars runs through the eye like a vertical patch. A row of three spines at the base of the tail is unmistakable. Yellowtails are common and abundant, often found in small schools, and relatively fearless (Color Image 28).

Moorish idols (Zanclidae)

A single representative of this family, the Moorish idol, was previously classified as a fellow Acanthurid (or surgeonfish). Moorish idols got their name from the Moors of Africa, who purportedly believed that the fish brought happiness.

Common name: Moorish idol

Scientific name: *Zanclus cornutus*

Origin: Indo-Pacific

Description: These triangle-shaped, white and yellow fish are unmistakable, largely because of their long dorsal fin with its trailing tip. Moorish idols have pointed noses and three broad, black, vertical bars. They often occur in pairs, drifting along the crevices of boulder-strewn reefs.

How El Niño Affects Fish of Galápagos

El Niños (Chapter 3) dramatically change the ocean. In addition to warmer waters overall, the annual upwelling of cool water is suspended, leading to exceptionally warm, nutrient-poor surface waters. Without essential nutrients, algal growth slows and the entire food web is affected. The 1982–1983 El Niño was particularly drastic. As discussed in previous chapters, many island vertebrates—marine iguanas, albatrosses, sea lions, and penguins—suffered from a lack of algae and fish from the ocean. The fish populations in the islands were not as dramatically affected—positively or negatively—as the land vertebrates. Fish, especially fish larvae, can disperse large distances in the ocean, as long as they find suitable food. However, many endemic fish have restricted ranges and could not adjust to the sudden and dramatic warming of the ocean and concomitant loss of food. One report revealed a decline in 17 species of fish, including nine endemics (e.g., the black-striped salema). However, warm currents from the western Pacific and Panama increased the frequency of 19 species of fish, none of which is endemic. In addition, the Indo-Pacific Moorish idol arrived in Galápagos for the first time during the 1982–1983 El Niño. Now there is a thriving population of Moorish idols in the archipelago. Although the 1982–1983 El Niño is often regarded as a natural disaster for many of the islands' inhabitants, for some fish, the unusually warm waters were a bonanza.

The endemic black-striped salema has a restricted range and was adversely affected by the 1982–1983 El Niño in Galápagos. In most years, these fish are abundant throughout the archipelago.

Angelfish (Pomacanthidae)

Angelfish, like surgeonfish, have thin, colorful bodies and small mouths, but they also have long dorsal and anal fins that can extend past the tip of the caudal fin. Only one angelfish—the king angelfish—occurs in Galápagos.

Common name: King angelfish

Scientific name: *Holacanthus passer*

Origin: Widespread

Description: The king angelfish has a large, dark blue body with metallic blue flecks on its head. With its trailing dorsal and anal fins, prominent white bar behind the pectoral fin, and bright yellow tail, this fish is easy to identify. Juveniles are even more colorful than adults. Angelfish, like many fish you will observe, engage in "cleaner" mutualisms with other species of fish and other types of sea creatures (Color Image 22).

Cleaner-Client Associations in Galápagos Fish

While snorkeling, you may come upon a large rock or coral mound upon which one or a few fish are hovering in a head-down position. Look closely, for if these head-down fish are accompanied by smaller fish, you have probably found a *cleaning station*. Cleaning stations are specific locations that "client" fish visit so that "cleaner" fish can remove the client's parasites. A common sight in Galápagos waters is a client parrotfish being cleaned by several juvenile rainbow wrasses. Many fish species act as clients at cleaning stations and many act as cleaners. Some, such as the barberfish and the king angelfish, take on both roles. Some cleaner fish are nutritionally dependent on client refuse, and they die if they cannot perform this service.

Cleaner-client symbioses are typically seen as *mutualistic*—that is, both the client and the cleaner benefit. The client benefits because its parasites are removed, and the cleaner gets an easy meal. You may observe cleaners that become greedy and attempt to eat the client's mucus coating or scales. When this happens, the client fish penalizes this behavior by aggressively chasing the cleaner away. Many cleaners have horizontal stripes that may serve as a signal to fish that need cleaning. In one experiment, researchers made several fish-mimics—some with stripes, some without—and evaluated the reactions of client fish. The striped mimics elicited the head-down, "clean me" behavior, while the fish mimics without stripes did not. If client fish are more likely to allow striped fish to approach, then one could predict selection for stripes in any fish that will benefit by proximity to the client. This prediction has been met by the sabretooth blenny (discussed below), a fish whose size and stripes resemble those of the Cortez rainbow wrasse. The sabretooth blenny, however, exploits the client's behavior toward stripes. The sabretooth often darts into cleaning stations and nips the client's scales or mucus.

Similar cleaning mutualisms typify many of the archipelago's fauna: king angelfish clean green sea turtles (Color Image 22); lava lizards perch on sleeping sea lions, consuming black flies that pester the sea lions; mockingbirds and finches eat the dead skin of land and marine iguanas; and finches eat ticks that annoy giant tortoises. You will see many of these interactions in Galápagos, and may even stumble upon a novel, undescribed mutualism.

Butterflyfish (Chaetodontidae)

Butterflyfish are similar (and related) to angelfish, but they are smaller and have pointier snouts and slightly concave foreheads. With their bright colors and flittering movements, these fish are reminiscent of butterflies.

Common name: Barberfish

Scientific name: *Johnrandallia nigrirostris*

Origin: Panamic

Description: You will see these alone or in great numbers, often in large schools. Barberfish are small (<10 cm) and have distinctive black markings, including a diagonal band from the middle of the dorsal fin to the base of the tail, and a black ring around the eye.

Common name: Three-banded butterflyfish

Scientific name: *Chaetodon humeralis*

Origin: Widespread

Description: Three-banded butterflyfish have three vertical black bands, a black border around the dorsal and anal fins, and three black bands on their tails.

Damselfish (Pomacentridae)

Although colorful and similar in size to many butterflyfish and surgeonfish, damselfish have a more perch-like (i.e., less oval) body, as well as spiny scales. Damselfish often aggressively defend a patch of algae in shallow waters.

Common name: Giant damselfish

Scientific name: *Microspathodon dorsalis*

Origin: Panamic

Description: The giant damselfish is a dusky-blue fish identifiable by its trailing dorsal and caudal fins, which are often highlighted with a fine pale border. The giant and bumphead damselfishes are the largest damselfish in the islands (reaching lengths greater than 30 cm). Juveniles are distinguished by iridescent blue markings on a deep-blue body.

Common name: Bumphead damselfish

Scientific name: *Microspathodon bairdii*

Origin: Panamic

Description: A drab, dark gray fish with an obvious bump on its forehead, the bumphead has bright blue eyes that are obvious in good lighting. Juveniles are similar but have an orange belly.

Common name: Galápagos ringtail damselfish (or Galápagos whitetail damselfish)

Scientific name: *Stegastes beebei*[2]

[2] The species epithet honors William Beebe, who named hundreds of species of fish. Beebe's *Galápagos: World's End* (1924) inspired many people to come to Galápagos (Chapter 2).

Origin: Panamic

Description: Another dark gray fish, the ringtail has a distinctive yellow tip on its pectoral fins and a faint orange crescent above the eye. Some ringtails have a light band around the caudal peduncle.

Common name: Panamic sergeant major

Scientific name: *Abudefduf troschelii*

Origin: Widespread

Description: The sergeant major is an abundant, small silver fish with a yellow back and five to six vertical black bands (Color Image 29). Breeding males turn dark blue while aggressively guarding their eggs.

Common name: Yellowtail damselfish

Scientific name: *Stegastes arcifrons*

Origin: Panamic

Description: A small fish with a black body and bright yellow tail, the yellowtail adults also have bright yellow lips (Color Image 30). Dark juveniles have dramatic blue scales and fin margins, and are abundant in shallow waters. They are territorial and aggressive, and you will probably see yellowtails chasing other fish from their territory. A yellowtail may even come after you if you linger too close to its spot.

Ecological drama at damselfish farms

Look closely at the yellowtail damselfish, and you will likely see how aggressively they farm and defend small patches of filamentous algae. This "farming" consists of removing undesirable algae, and "defending" involves constant vigilance, chasing away and attacking intruders. Working against these efforts are sea urchins, which are common, pointy invertebrates that you will encounter on most of your underwater forays. Urchins such as the pencil urchin (*Eucidaris galapagensis*) and the green urchin (*Lytechinus semituberculatus*) graze algal beds intensively, leaving behind bare rock or coral that is then colonized by encrusting algae. Uncontrolled grazing by urchins can lead to a habitat "switch," whereby the underwater landscape is completely altered and algal productivity is dramatically reduced. A similar habitat-switch occurred in formerly biodiverse kelp forests of the Pacific; with the decline of urchin-eating sea otters, urchin populations exploded and destroyed the kelp forests. Do the yellowtail damselfish in Galápagos likewise protect underwater productivity?

In 2009, biologists tested the hypothesis that damselfish protect algal turf beds from overgrazing by sea urchins. Investigators recorded the abundances of study organisms (yellowtail damselfish and urchins) and habitats (algal turfs—farmed or not—and encrusting algae) in Academy Bay at Santa Cruz Island. Habitats were manipulated to test the response of damselfish to sea urchins, and to evaluate how algal turfs recovered from sea urchin grazing, in the presence or absence of damselfish "farmers." When sea urchins were added to farmed algal turfs, the resident damselfish immediately attacked the sea urchins, until the intruder was removed from the farm (the intruder was often left upside-down!). When farmed algae were removed in the presence of damselfish, the algae recovered completely, returning to pre-removal conditions. In the absence of damselfish, the disturbed algae did not recover. Finally, when urchins were removed from encrusting algae, filamentous algal growth resumed. Thus, sea urchins alter the productivity of habitats in Galápagos nearshore environments. Aggressive yellowtail damselfish, farming their algal patches and evicting grazing urchins, preserve this productivity. These feisty little fish may help create habitats that enhance biodiversity in this fragile ecosystem.

Common name: Scissortail (or whitespot, or brown) chromis

Scientific name: *Chromis atrilobata*

Origin: Widespread

Description: You are most likely to encounter these small (<10 cm) fish in medium to large schools, feeding on plankton above reefs. The white or yellow spot just anterior to the caudal peduncle, and the deeply forked tail, allow easy identification of these chromis.

Sea Basses (Serranidae)

Sea basses are sturdy, large-mouth fish that are adapted for predation. Most are *protogynous hermaphrodites,* meaning that they begin life as females, after which a few become males as they mature.

Common name: Pacific creolefish (or five-spotted anthias)

Scientific name: *Paranthias colonus*

Origin: Widespread

Description: The creolefish may be the most common fish in the islands. Look for a midsized (10 to 20 cm) grayish-white body, with a pinkish caudal fin and three to five white or black spots on the back, leading to the caudal peduncle.

Parrotfish (Scaridae)

At lengths of up to 50 centimeters, the colorful parrotfish are larger than most other reef fish in Galápagos. Parrotfish are also distinguished by their strong jaws and beaklike mouths—used to scrape algae and soft corals from rock—and their reliance on their pectoral fins for swimming. Like many of the sea basses, parrotfish are *sequential hermaphrodites,* meaning that they can change their sex.

Common name: Bumphead parrotfish

Scientific name: *Scarus perrico*

Origin: Widespread

Description: Like the bumphead damselfish, the bumphead parrotfish is identifiable by its large, bulbous forehead. Adults and juveniles are green, with blue fins and blue irregular markings radiating from their eyes. Found alone or in small groups, bumphead parrotfish—the largest of all parrotfish—are associated with reef-building corals.

Why Do So Many Fish Undergo Sex-Changes?

Sexual reproduction requires male and female *gametes,* or sex cells. If an organism produces sperm, it is a male; if it produces eggs, it is a female. In *dioecious* organisms, the sexes are separate, and an individual can only make one type of sex cell (egg *or* sperm); this is the most common reproductive strategy for vertebrate animals, including Galápagos fish. A *hermaphrodite* individual can make both eggs and sperm, sometimes at the same time (*simultaneous hermaphroditism*), sometimes during different life stages, or under different environmental constraints (*sequential hermaphroditism*). Several fish you will see in Galápagos (e.g., the ubiquitous Cortez rainbow wrasse) are sequential hermaphrodites. Most sequential hermaphrodites are females first (*protogynous*), although some are males first (*protandrous*). How could this reproductive strategy have evolved?

One proposed explanation for changing sex with maturity hinges on size differences between males and females. All fish increase in size as they mature, but females are most successful when small, and males are most successful when large. Thus, protogynous wrasses should be able to maximize their reproductive output by reproducing as females when they are small, and reproducing as males when they are large. In some fish, a few large males monopolize most of the females in a population; thus, small males seldom reproduce at all. Given the alternative, it is adaptive for small individuals to be female, because females are rarely unable to reproduce. If size is a factor in sex-changing fish, a consequence of the female-first reproductive strategy should be a female-biased sex ratio. In fact, this is what occurs in natural populations of many female-first wrasses, parrotfishes, and sea basses.

Reducing the ratio of males to females in a population could have additional benefits—namely, *sperm competition* may be minimized with fewer large males. Sperm competition, which occurs when males compete to fertilize the available eggs in a population, can involve gamete-level competition. In other words, the most successful males have the best, or the most, sperm. In female-first, female-biased populations, fewer males should result in less competition for access to females. A recent comparison of gonad weight (a proxy for sperm competition) revealed that males in female-first species invest less in sperm production than do males in species that do not change sex. Thus, female-first sequential hermaphroditism seems adaptive on many levels. Perhaps instead of wondering why fish change sex, we should wonder why all fish are not hermaphrodites.

Common name: Bicolor parrotfish

Scientific name: *Scarus rubroviolaceus*

Origin: Indo-Pacific

Description: The mature male bicolor parrotfish is a distinctive bluish green, with a bright yellow eye and a blunt head (the blue-chin parrotfish, below, has a pointed or conical head). The female's body is divided by two distinct colors—brownish in the front, white or tan in back.

Common name: Blue-chin parrotfish

Scientific name: *Scarus ghobban*

Origin: Indo-Pacific

Description: These parrotfish exhibit pronounced sexual dimorphism, whereby the male and female can be mistaken for different species. Males are greenish-blue, with pink dorsal and anal fins bordered in blue, and blue markings under the mouth. Females have an orange or yellow body marked by five light blue, irregular vertical bars. You will probably see this striking fish almost everywhere, scraping algae off hard surfaces (rock and coral).

Wrasses (Labridae)

Although wrasses are similar to (and closely related to) parrotfishes, wrasses are typically smaller and longer. Like parrotfishes, many wrasses are sequential hermaphrodites—they begin life as one sex and can change sex as they mature. Many small wrasses follow larger fish, eating invertebrates disturbed by their passing.

Common name: Mexican hogfish

Scientific name: *Bodianus diplotaenia*

Origin: Widespread

Description: The Mexican hogfish is a large, gray-to-brown wrasse with distinct morphs depending on its life history stage. Adult males have a bulbous forehead and long, trailing filaments from their dorsal, anal, and caudal fins. Females and juveniles are pink-to-gray, more streamlined, and have two irregular black bars running horizontally along the sides of the body (Color Image 31).

Common name: Galápagos (or goldspot) sheephead

Scientific name: *Semicossyphus darwini*

Origin: Peruvian

Description: In 1842, Reverend Leonard Jenyns named this fish in honor of Charles Darwin, who had collected specimens during his visit to the islands in 1835 (Appendix 3). This wrasse has a slight bump on its forehead, and its body is gray-blue, with a large, irregular yellow patch above the pectoral fin. You are most likely to see the Galápagos sheephead in the cooler waters of the western archipelago.

Common name: Harlequin wrasse

Scientific name: *Bodianus eclancheri*

Origin: Peruvian

Description: A large wrasse that is a mixture of bright orange with black and white, the harlequin wrasse—with its many color variations—is difficult to describe. These fish exhibit *sequential protogynous hermaphroditism:* the juveniles are female, the adults are male.

Common name: Cortez rainbow wrasse

Scientific name: *Thalassoma lucasanum*

Origin: Panamic

Description: These wrasses are small, oblong, unmistakable fish that have different color-patterns, depending on their life history stage. Adult males (or *reverse-sexed* females) have a blue head, dark purple body and a blue tail, and are yellow above the pectoral fins. Females have dark backs and pink bellies with a broad yellow bar running horizontally on each side. In the terminal phase of their lives, adult males mate one-on-one with younger females. Prior to this stage of adulthood, in the intermediate phase, males and females engage in group spawning, surging upward en masse to form a large, gamete-releasing ball.

Common name: Chameleon wrasse

Scientific name: *Halichoeres dispilus*

Origin: Widespread

Description: The common name of this fish derives from its ability to change colors rapidly. Chameleon wrasses have a white, brown, or a yellow-red-green pattern of colors that is particularly striking. A pale stripe may run the length of its body. However, these fish are best identified as oblong, small (<15 cm long) fish with an irregular spot (black or otherwise) above each pectoral fin.

Squirrelfish (Holocentridae)

Squirrelfish have the large eyes that characterize nocturnal species. They can be seen tucked into crevices during the day, emerging at night to seek their invertebrate prey. Of the three species of squirrelfish in Galápagos, the Panamic soldierfish is the most common.

Common name: Panamic (or Crimson) soldierfish

Scientific name: *Myripristis leiognathus*

Origin: Panamic

Description: Panamic soldierfish are red fish with big eyes and obvious scales, but no distinctive markings or patterns.

Cardinalfish (Apogonidae)

Cardinalfish are small, reddish, nocturnal fish, most likely observed in large resting groups, tucked under rocky slopes or inside caves. Cardinalfish are active at night, emerging to eat tiny fish and small invertebrates. The two species described next often occur together.

Common name: Blacktip cardinalfish

Scientific name: *Apogon atradorsatus*

Origin: Panamic

Description: These fish have two separate dorsal fins and a distinguishing black tip on the rear dorsal fin; otherwise, they lack body markings.

Common name: Pink cardinalfish

Scientific name: *Apogon pacificus*

Origin: Widespread

Description: The pink cardinalfish is similar to the blacktip, but with a faint, black, vertical bar at the second dorsal fin.

Blennies (Labrisomidae, Chaenopsidae, Tripterygiidae, Blenniidae)

Collectively, the blennies are small, long-bodied fish with fleshy appendages (*cirri*) above the eyes. Blennies spend most of their time perched on a sandy bottom or tucked into small crevices with only their small heads protruding. Blennies have one- or three-part dorsal fins; the similar gobies have two-part dorsal fins.

Common name: Galápagos barnacle blenny

Scientific name: *Acanthemblemaria castroi* (Chaenopsidae, the tube blennies)

Origin: Endemic

Description: These blennies are reddish brown, with large white spots down their backs and along the middle of their bodies. Galápagos barnacle blennies can be distinguished by the many small blue spots on their heads. Look for them poking out of empty barnacle shells in shallow-water areas. They are shy, so approach them with care to get a good look.

Common name: Sabretooth blenny

Scientific name: *Plagiotremus azaleus* (Blenniidae, the true, or combtooth, blennies)

Origin: Widespread

Description: With its thin, elongate body, and its distinctive coloration—white underneath, with a black-yellow-black banding pattern running the length of its upper body—the sabretooth is not easily confused with other fish. Look for sabretooths associating with Cortez rainbow wrasses. As the rainbow wrasses clean client fish (such as blue-chin parrotfish), the sabretooth blennies may dart in and take bites out of the client's fins, skin, or mucus.

Common name: Panamic fanged blenny (or large-banded blenny)

Scientific name: *Ophioblennius steindachneri* (Blenniidae, the true, or comb-tooth, blennies)

Origin: Widespread

Description: These large (up to 20 cm long), olive-brown, robust blennies have several broad yellow bars running vertically from the dorsal fin. Panamic fanged blennies also have a large, irregular spot, often with a pale border, behind each eye. Juveniles lack the bars, but have the eyespots. These fish are abundant in rocky shallow waters throughout the archipelago, but they are also well camouflaged and easily missed (Color Image 32).

Common name: Galápagos four-eyed blenny

Scientific name: *Dialommus fuscus* (Labrisomidae)

Origin: Endemic

Description: This dark green fish can be identified by several pale blotches along the middle of its back. The Galápagos four-eyed blenny is unusual in that it can spend several minutes out of water, hunting for small arthropods (e.g., crabs and insects) among the lava rocks along the shore.

Adaptations for terrestrial forays in the four-eyed blenny

Organisms that can occupy both land and sea are puzzling, and adaptations for this dual life-history trait are intriguing. Sea lions, penguins, and marine iguanas move from land to sea with ease. Fish, on the other hand, seem confined to an aquatic existence, with adaptations to accommodate breathing (gills), moving (fins), and sensing (the lateral line system) underwater. So how does the Galápagos four-eyed blenny, a fish that hunts for crustaceans and insects in and out of tide pools, survive out of water?

An investigation of the four-eyed blenny, in a rocky intertidal zone near Tortuga Bay (Santa Cruz), suggested how this fish is adapted for terrestrial sojourns. First, the Galápagos four-eyed blenny has flattened corneas that converge in a central, vertical ridge bisecting each eye. This odd arrangement gives the blennies their four-eyed appearance and may enable clear vision both in and out of water. Juvenile blennies, which lack this split-eye configuration, do not leave the water to forage. Also, four-eyed blennies have long, flexible bodies that permit them to move out of water by bending and pushing themselves forward with their tails. In addition, these blennies move with the tides, ensuring that they are never too far from open water as they travel in and out of tide pools. While most terrestrial forays of Galápagos four-eyed blennies are short (less than 10 seconds), roughly 10% of these excursions exceed 1 minute, and a few are over 5 minutes.

Common name: Bravo clinid (or cheekspot labrisomid)

Scientific name: *Labrisomus dendriticus* (Labrisomidae)

Origin: Panamic

Description: These blennies come in many colors and can change colors rapidly, but they can be distinguished by three eye-area features: large, "goggle" eyes, or popeyes; two pairs of branched (bristle-like) cirri above the eyes; and a large spot on the gill cover, just anterior to the pectoral fin.

Common name: Galápagos triplefin blenny

Scientific name: *Lepidonectes corallicola* (Tripterygiidae)

Origin: Endemic

Description: This small (<8 cm long) reddish-brown blenny has three dorsal fins and an identifying black splotch at the base of its tail. Breeding males are marked by a black first dorsal fin, with two bright yellow spots.

Gobies (Gobiidae)

Gobies are similar in size (2–5 cm) and shape to the blennies, but they lack the fleshy cirri that blennies possess. Gobies have two-part dorsal fins and a more rigid perching posture than that of the blennies.

Common name: Marbled goby

Scientific name: *Eleotrica cableae*

Origin: Endemic

Description: Marbled gobies are light brown, with dark brown and white splotches and a dark irregular spot on the rear gill-cover. Look for these gobies darting about in tide pools and shallow, sandy areas.

Common name: Blue-banded goby

Scientific name: *Lythrypnus gilberti*

Origin: Endemic

Description: These gobies are small (<2 cm), red-to-maroon, with transparent fins and 9 to 10 vertical blue bars, outlined in black. The first spine of this goby's dorsal fin is much longer than the rest, giving the fin a sail-like appearance.

Batfish (Ogcocephalidae)

Batfish have an unusual, flattened shape, with a fleshy spike from the snout, and pectoral fins that look and function like limbs.

Common name: Red-lipped batfish

Scientific name: *Ogcocephalus darwini*

Origin: Peruvian

Description: The common name says it all! This small (<12 cm), drab-colored fish uses its pectoral fins to "fly" short distances along the bottom. Bright red lips and a dark, fleshy spike protruding from between the eyes make this fish unmistakable.

Lefteye Flounders (Bothidae)

These flat, bottom-dwelling fish lie on their right side and have pigment and both eyes on the left (i.e., upper) side. Dorsal and anal fins surround the body like fringing. Lefteye flounders can change colors to blend into the substrate.

Evolution and Adaptations of Flatfish

Flounders, with their flat bodies, odd fringing, and distinctive eyes—bulging, close-set, on one side of the body—are so freakish that it is difficult to imagine how natural selection could have produced such oddities. How, from a standard template of a vertical fish body, and eyes on each side, did flounders evolve? Flounders certainly are equipped with the standard bony fish architecture: juveniles have eyes on each side of their heads, and as they mature one eye migrates to join the other. Complicating matters has been a largely recalcitrant fossil record, from which no intermediary forms have arisen . . . until recently.

In 2008, University of Chicago graduate student Matt Friedman began scrutinizing a 50-million-year-old fossilized fish with fascinating anatomy: these fish appeared to have eyes positioned somewhere *in between* typical two-sided eyes and those of modern-day flatfish. Friedman has discovered many fish with this "transitional" eye-arrangement and has argued that even these small steps toward unilateral eye-position could have been adaptive. If these proto-flatfish propped themselves up on their pectoral fins, as many contemporary fish do, an eye moved even slightly forward would have given the fish better vision and possibly better hunting skills. Over time, mutations that moved the lower eye up and over would have been advantageous, leading to the flatfish bodies we see (if we look carefully) in Galápagos.

These discoveries are doubly interesting, for the evolution of flatfish has posed a problem for biologists for a long time. Not long after publication of *On the Origin of Species* (1859), Charles Darwin's critic George Mivart attacked Darwin with *On the Genesis of Species* (1871). As ammunition for his arguments, Mivart presented the problem of the flatfish: "how such transit of one eye a minute fraction of the journey towards the other side of the head could benefit the individual is indeed far from clear. It seems, even, that such an incipient transformation must rather have been injurious." This argument went largely unchallenged until Friedman's recent work, in which Friedman responded, "[t]he fossil record rejects the intuitive notion that incomplete orbital migration might be maladaptive."

Common name: Tropical flounder (or blue-eyed flounder, or flowery flounder)

Scientific name: *Bothus mancus*

Origin: Indo-Pacific

Description: Flat, unusual-looking fish with a pattern of blue spots and rings on their left, or upper, side. This flounder can be distinguished from the Pacific leopard flounder by its inter-eye distance: in the tropical flounder, the eyes are farther apart than the diameter of each eye; in the leopard flounder, the eyes are closer together (with only a single eye's space between them). Also, while the tropical flounder can grow to over 30 centimeters, the leopard flounder does not get longer than 20 centimeters.

Common name: Pacific leopard flounder (or leopard flounder)

Scientific name: *Bothus leopardinus*

Origin: Indo-Pacific

Description: Similar in shape and habitat to the tropical flounder, with the differences noted above. These fish are so well camouflaged that you may unknowingly swim right over them.

Scorpionfish (Scorpaenidae)

Many fish in this group possess venomous fin spines that can cause painful injuries. Scorpionfishes are stout, with large lips, tassels, and fleshy barbels that help with camouflage.

Common name: Rainbow scorpionfish

Scientific name: *Scorpaenodes xyris*

Origin: Widespread

Description: The small (<10 cm) rainbow scorpionfish is red with white mottling, has large black-and-red eyes, and lacks barbels under its mouth. Small rainbow scorpionfish often hover near long-spined sea urchins for protection.

Common name: Stone scorpionfish (or Pacific spotted scorpionfish)

Scientific name: *Scorpaena plumieri mystes*

Origin: Panamic

Description: With its stocky, midsize (20–40 cm) body and conspicuous, fringe-like barbels dangling below its mouth, the stone scorpionfish is easy to identify. Unfortunately, with its ability to change colors and tendency to remain motionless on the substrate, you are likely to swim right by this fish. Look for something that looks like lava encrusted with pink and orange lichens; if it has large, bulging eyes and fleshy appendages, it is probably a stone scorpionfish. Its ability to remain undetected, combined with its potent venom, makes this a dangerous fish. Keep your hands off the reef—it is certainly better for the ecosystem, and it may be better for you.

Common name: Player scorpionfish (or bandfin scorpionfish, or darkblotch scorpionfish)

Scientific name: *Scorpaena histrio*

Origin: Widespread

Description: Although similar in coloration to the stone scorpionfish (above), the player is smaller, lacks fleshy barbels below its mouth, and is distinguished by a dark red-to-brown circular splotch just posterior to the gill cover (at mid-body). The player scorpionfish is poisonous and can change colors to blend in with its surroundings.

Hawkfish (Cirrhitidae)

Hawkfish are stout fish characterized by fleshy cirri at the tips of their dorsal-fin rays. Hawkfish usually seek the highest areas of reefs, where they survey the area (much like hawks do on land). Hawkfish are also protogynous hermaphrodites.

Common name: Giant hawkfish (or hieroglyphic hawkfish, or chino mero)

Scientific name: *Cirrhitus rivulatus*

Origin: Panamic

Description: The giant hawkfish is unmistakable, but difficult to describe: its body is shades of olive-green and blue, with an irregular pattern (resembling topographic lines on a map) of dark bands outlined in white; two pale splotches occur just anterior to the caudal peduncle. You are likely to see these common fish perched on rocky slopes, darting into crevices when you get too close (Color Image 33).

Common name: Coral hawkfish

Scientific name: *Cirrhitichthys oxycephalus*

Origin: Indo-Pacific

Description: The coral hawkfish are white, but covered in irregular, large, red splotches. These small (4–7 cm) reef-dwelling fish can be distinguished by the yellow tassels at the tips of their dorsal fin rays (Color Image 34).

Puffers (Tetraodontidae)

Puffers are odd-shaped, boxy fish that can inflate their bodies by taking in extra air or water. Puffers "puff" out in self-defense when sufficiently agitated. If you have caused a puffer to inflate, you are bothering it! Although puffers can be covered in spines, the most common puffers in Galápagos—the guineafowl and the bullseye—are smooth. Puffers have powerful jaws and fused teeth that can crush hard-shelled invertebrates.

Common name: Guineafowl puffer

Scientific name: *Arothron meleagris*

Origin: Widespread

Description: The round guineafowl puffer appears in two distinct phases, both of them striking: in the spotted phase, this puffer is black with white spots; in the yellow phase, it is bright yellow.

Common name: Bullseye (or concentric) puffer

Scientific name: *Sphoeroides annulatus*

Origin: Indo-Pacific

Description: This common puffer gets its name from the dark concentric circles that form a "bullseye" on its back. These puffers spend a lot of time near the kitchen area of tour boats, awaiting any edible refuse, and they appear to eat just about anything. Bullseye puffers are also common in shallow, sandy bays.

Triggerfish (Balistidae)

Triggerfish are oval, laterally compressed fish, with dorsal and anal fins used for swimming. The pelvic fin is reduced, and the first spine of the first dorsal fin can be locked into place with the help of a triggerlike second spine. Triggerfish use this locking function as an antipredator device; a rigid dorsal fin helps to secure the triggerfish in protective holes or crevices.

Common name: Orangeside (or yellow-bellied) triggerfish

Scientific name: *Sufflamen verres*

Origin: Widespread

Description: Easily identified by its dark back, yellow sides, and dark abdomen, the orangeside is the most common triggerfish in Galápagos.

Common name: Blunthead triggerfish

Scientific name: *Pseudobalistes naufragium*

Origin: Indo-Pacific

Description: The largest of the triggerfish in the islands (up to 90 cm long), the blunthead is dark blue-gray, with darker, wide bands crossing the body vertically. Its prominent forehead superficially resembles the bumps characteristic of other, more common reef fish (e.g., the Mexican hogfish; see above). Like many triggerfish, the blunthead can sometimes be seen blowing on sand to reveal tasty prey items.

Goatfish (Mullidae)

Goatfish have two barbels under the chin, giving them a goatlike appearance.

Common name: Mexican (or yellow-tailed) goatfish

Scientific name: *Mulloidichthys dentatus*

Origin: Widespread

Description: Mexican goatfish are silvery-white, with yellow tails and a distinct yellow midbody stripe. White barbels under the chin are used to dig in the sand for food.

Jacks (Carangidae) and Tunas (Scombridae)

Jacks are silvery, open-sea fish characterized by their strong swimming ability and their forked caudal fins. Although it is easy to confuse jacks with tunas and their kin, jacks are alone in having a two-part dorsal fin—noticeably high in the front and low in the back. Jacks are prized in the Galápagos fishing industry.

Common name: Steel pompano

Scientific name: *Trachinotus stilbe*

Origin: Widespread

Description: The steel pompano is a small (<25 cm), silvery fish, distinguished by a white band running vertically between the head and the pectoral fin. Large schools of these fish are common near the surface of shallow water.

Common name: Bigeye trevalle (or bigeye jack)

Scientific name: *Caranx sexfasciatus*

Origin: Widespread

Description: Bigeye trevalles are large (20–60 cm), fast swimmers that can be seen occasionally in open water over reefs. These jacks can be identified by their white-tipped dorsal fins, large eyes, and a small black spot on the upper edge of the rear gill cover.

Common name: Indo-Pacific bonito (or Oriental bonito, striped bonito, Mexican bonito)

Scientific name: *Sarda orientalis*

Origin: Widespread

Description: These elongate, streamlined fish are excellent swimmers. The Indo-Pacific bonito is distinguished by a series of horizontal black lines running the length of its back. These fish live in small groups or large schools.

Common name: Yellowfin tuna

Scientific name: *Thunnus albacares*

Origin: Widespread

Description: Yellowfin tuna can be identified by their size (up to 2 m long), their yellow dorsal and anal fins, and the small yellow "finlets" between the dorsal and caudal fins. These fish are abundant in the archipelago, but after decades of being sought by the commercial fishing industry, selection has favored those that avoid nearshore environments.

Halfbeaks (Hemiramphidae)

The cylindrical halfbeaks are characterized by a long lower jaw that is unmistakable. You may see halfbeaks jumping out of the water to avoid hungry predators (e.g., sharks).

Common name: Longfin halfbeak

Scientific name: *Hemiramphus saltator*

Origin: Widespread

Description: The longfin halfbeak is a slender, silvery fish that can often be seen in small schools near the surface and close to shore. They have a long, needle-like lower jaw, which can be silvery to bluish purple; this jaw is punctuated by a bright orange tip.

Cornetfishes (Fistulariidae)

Common name: Reef cornetfish

Scientific name: *Fistularia commersonii*

Origin: Indo-Pacific

Description: Cornetfishes are long (up to 90 cm), silvery-white fish with a tubular snout and a long tail filament. Look for them hovering just below the surface, or drifting among submerged lava rock. They can be confused with trumpetfish, but trumpetfish are stockier and lack the long tail-filament.

Trumpetfish (Aulostomidae)

Common name: Trumpetfish

Scientific name: *Aulostomus chinensis*

Origin: Indo-Pacific

Description: Superficially similar to the cornetfish, the trumpetfish is a little smaller (25–70 cm) and has a mouth like a trumpet. Instead of a long, white, filamentous tail, the trumpetfish has a colorful, paddle-like caudal fin.

Pipefish and Seahorses (Syngnathidae)

The syngnathids are odd-shaped fish with tubelike snouts and paternal care, a behavior that is notable for its rarity.

Common name: Fantail pipefish

Scientific name: *Doryrhamphus excisus*

Origin: Widespread

Description: The fantail pipefish is small (<8 cm), timid, and well camouflaged. A key to properly identifying this fish is its dark paddle-like tail, marked with several orange splotches.

Common name: Pacific seahorse

Scientific name: *Hippocampus ingens*

Origin: Widespread

Description: Its tubular snout and flexible neck give the Pacific seahorse its horse-like appearance. This fish uses its flexible tail to curl around coral stalks; with its variable coloration in browns and reds, it is well camouflaged and easy to miss (Color Image 35). Seahorses are notable not only for their characteristic

Paternal Care in the Syngnathids

Most fish provide little if any parental care; instead, they release gametes into the water and move on. Some fish, however, care for their eggs (e.g., the Panamic sergeant major), keeping their newly hatched young safe inside a parent's mouth (e.g., the rare Galápagos jawfish) or, in the case of the seahorses and pipefishes, incubating the eggs inside an abdominal pouch.

In most animals, males' investment in offspring stops with the provisioning of sperm. In animals in which only one parent provides care, the mother is typically the caregiver. Females make eggs, which require a greater investment of resources, and therefore females generally have more to lose by abandoning their young. In the standard scenario, surplus males must compete for mating opportunities, and they make excessive amounts of sperm to compete for fertilizations. Interestingly, in most egg-laying fish with post-laying investment, it is the male that provides care.

Syngnathids (seahorses and pipefishes) take parental care to a level akin to gestation in mammals, although it is the male that carries the offspring. In seahorses, the female deposits her eggs in the male's specialized brood pouch, where they are fertilized. Then, in addition to protection, the male aerates his embryos and provides nutrients. When the young emerge from their eggs, they are released from the father's pouch, and there is no post-release parental care. Given that paternity is essentially assured, male investment itself is not surprising, but the *extent* of the care is extreme.

A logical prediction of full paternity assurance is a reduction in sperm competition. Without the threat of caring for another male's young, why invest resources in extra sperm? As might be expected, male seahorses have small testes (analogous to that seen in sex-changing wrasses) and a small sperm-to-egg ratio. While paternal care in seahorses is extreme relative to that of other animals, it does give the male a guarantee of fatherhood, and assurance of the ultimate reward in evolutionary currency (i.e., fitness).

shape, but also for exhibiting a rare form of parental care, in which the male provides most of the care.

Eels (Muraenidae, Congridae, Ophichthyidae)

Eels are fish with snakelike bodies. Many eels have continuous dorsal, caudal, and anal fins. Eels are typically reclusive, so consider yourself lucky if you see one. The most commonly encountered eels are the Galápagos garden eel, the tiger snake eel, the fine-spotted moray, and the jewel moray.

Common name: Galápagos garden eel

Scientific name: *Taenioconger klausewitzi* (Congridae)

Origin: Endemic

Description: This mottled gray-and-white fish can be seen extending its upper body from a sandy burrow, using its mouth to grasp planktonic food particles as they drift by. Dozens of individuals often occur together in a colony, and they will retreat in unison as you approach. If you are patient, you can see them pop back up from their burrows when they sense danger has passed.

Common name: Tiger snake eel

Scientific name: *Myrichthys tigrinus* (Ophichthyidae)

Origin: Widespread

Description: This eel is easily identified by its cream-colored body, which is covered by two rows of large dark spots. Look for it hiding in recesses in rocks or coral.

Common name: Fine-spotted moray

Scientific name: *Gymnothorax dovii* (Muraenidae)

Origin: Panamic

Description: This dark gray, long (>125 cm) eel can be distinguished by the small white spots that cover its body. Like all morays, the fine-spotted moray has sharp teeth in powerful jaws, which the eel will use to protect itself if you bother it. Morays are active at night, but you may see a head poking out from holes and crevices in rocky areas. They are always opening and closing their mouths; with their exposed teeth, this behavior can be intimidating, but it is necessary to move oxygenated water for respiration.

Common name: Jewel (or lentil) moray

Scientific name: *Muraena lentiginosa* (Muraenidae)

Origin: Indo-Pacific

Description: The jewel moray is brown, with yellowish-cream spots covering the head and body (Color Image 36). This moray can be distinguished by its dorsal fin, which begins right behind the head and is continuous with the anal fin. Also, the jewel moray has two pairs of tubelike nostrils on its head—one pair just above the mouth, the other pair between the eyes.

Sharks and Rays (Carcharhinidae, Sphyrnidae, and Rhincodontidae)

Sharks and rays belong to a group of fish called Chondricthyes. Within the Chondrichthyes, the Elasmobranchii includes the sharks, skates, and rays; these fish have a skeleton composed primarily of flexible cartilage, rather than rigid bone. Sharks and rays have toothlike scales, giving their skin a rough texture. These fish lack swim bladders and lungs, and they breathe through exposed gill slits rather than the operculum-covered gills of bony fish. All of the elasmobranchs have separate sexes and internal fertilization, whereby sperm are delivered to the eggs inside the female. By not releasing gametes into the water, each sperm has a greater probability of fertilizing an egg.

An estimated 30 species of sharks, from nine different families, have been observed in Galápagos. The most commonly encountered species are the white-tipped reef shark, the Galápagos shark, and, for divers, the scalloped hammerhead. If you are lucky, you will also see the world's largest fish, the whale shark. At least a dozen species of skates and rays have been recorded in the islands. The most commonly encountered members of this group are the diamond (or whiptail) stingray, the golden cowray, the spotted eagle ray, and the manta ray.

Common name: Galápagos shark

Scientific name: *Carcharhinus galapagensis* (Carcharhinidae)

Origin: Indo-Pacific

Description: The Galápagos shark, like many other sharks in the islands, has a pointed nose and two dorsal fins; the first dorsal fin is noticeably larger than the second, is tall and pointy, and begins just above the pectoral fin. This shark can exceed 3 meters in length. It is often observed in pairs or small groups, circling tour boats for discarded food.

Common name: White-tipped reef shark

Scientific name: *Triaenodon obesus* (Carcharhinidae)

Origin: Indo-Pacific

Description: The white-tipped reef shark is a long (up to 2.5 m), slender, blue-gray fish, easily identified by the bright white tips of the first dorsal and caudal fins. These sharks are often encountered in groups of five to nine individuals, resting under rocky overhangs during the day.

Common name: Whale shark

Scientific name: *Rhincodon typus* (Rhincodontidae)

Origin: Widespread

Description: The world's largest fish, the whale shark can exceed 20 meters in length. With its size, deep ridges running the length of its body, and large white spots, the whale shark is unmistakable. Whale sharks filter-feed offshore, and are harmless to divers and swimmers.

Common name: Hammerhead shark

Scientific name: *Sphyrna lewini*

Origin: Widespread

Description: The well-named hammerhead shark has a flattened head, with eyes set on the edge of each side; rounded lobes on its front give the head a scalloped appearance. The smooth hammerhead, *Sphyrna zygaena,* lacks the scalloped head and is rarely seen in Galápagos. Few snorkelers see hammerhead sharks, because these sharks are most common in the scuba-diving areas around Darwin and Wolf islands. However, it is possible to see hammerheads in shallow-water snorkeling areas around the central islands. Most of the diet of hammerhead sharks is squid.

The large schools of hammerhead sharks in Galápagos are a primary attraction for divers from around the world. However, illegal fishing, fueled largely by a demand for shark fin soup, led the IUCN (in 2008) to list the hammerheads as Endangered.

Shark Attacks in Galápagos

Sharks are famously unpredictable, and sometimes they attack without warning. However, shark attacks are globally rare (fewer than 100 reported incidents per year) and not always fatal. In Galápagos, only 7 or 8 (depending on the source) shark attacks have been reported in the past 20 years. Thus, your likelihood of being attacked by a shark is slim. Simple, commonsense precautions can make this likelihood infinitesimal: do not swim (or snorkel, or dive) near a fish-cleaning station; do not swim with an open wound; do not swim near sea lion colonies that have many babies; do not swim at night, when most sharks feed; and do not threaten a shark by swimming too close. In sum, a healthy fear of sharks is probably smart, but you are still more likely to get struck by lightning than you are to get attacked by a shark.

Common name: Diamond (or whiptail) stingray

Scientific name: *Dasyatis brevis*

Origin: Panamic

Description: The brown-to-gray stingray has a slightly pointed snout and wing tips. This ray can be identified by its size (between 1 and 2 m long), its tail (shorter than the rest of its body), and the paddle-like tip of the tail. Like many rays, the diamond stingray appears to fly along the bottom, burrowing into sand to hide or feed. Look for small opportunistic fish (e.g., Pacific creolefish and various wrasses) trailing the diamond stingray, feeding on small sand-dwelling invertebrates as they are agitated by the ray. Given that the small fish get food, and the ray appears unaffected, this relationship is probably commensalism (Chapter 1).

Common name: Golden cowray (or golden cownose ray)

Scientific name: *Rhinoptera steindachneri*

Origin: Widespread

Description: The small (up to 1 m in diameter) golden cowray can be distinguished by its color (white on the bottom, brownish yellow or golden on top) and its square, fleshy head that is divided into two distinct lobes (Color Image 37). These fish prefer to swim in schools in mangrove-ringed lagoons, but they can also be seen along the cliffs in deepwater areas (e.g., South Plaza and Daphne Major).

Common name: Spotted eagle ray

Scientific name: *Aetobatus narinari*

Origin: Widespread

Description: With its size (1 to 2 m in diameter), pointed snout, and black back covered in white spots, the spotted eagle ray is easy to identify. These rays cruise over reefs and sandy bottoms, feeding on mollusks and crustaceans in the northern and central islands of the archipelago.

Common name: Manta ray

Scientific name: *Manta hamiltoni*

Origin: Widespread

Description: The graceful manta ray can be identified by its size (up to 6 m in width), coloration (black on the back, white on the front), and the fleshy, pointed pectoral fins that are anterior to its head. Primarily oceanic, these filter feeders are usually observed at a distance, with individuals or small groups flopping in the water.

CHAPTER 9

Invertebrates

The terrestrial invertebrates of Galápagos are the largest group of organisms [and have the] highest species diversity. They are present in all habitats and represent an estimated 51% of the total biodiversity.

—Charles Darwin Foundation, 2012

ALTHOUGH THE ICONIC ANIMALS OF GALÁPAGOS ARE PREDOMINANTLY those with backbones—boobies, iguanas, sea lions—fascinating invertebrates also live throughout the islands. These animals range from postcard-friendly Sally lightfoot crabs and sea stars to ecologically important carpenter bees and green sea urchins, to toxic scorpions and centipedes. As is true for other wildlife in Galápagos, the diversity of invertebrates in the islands is relatively low compared with that of the mainland. Nevertheless, the invertebrates of Galápagos account for most of the islands' biodiversity. You will see them everywhere.

Despite their relatively low diversity, the invertebrates of Galápagos are far too numerous to describe in detail. Indeed, there are more than 1,800 species of insects, 600 species of mollusks, 100 species of crabs, and 20 species of sea urchins in Galápagos.[1] Here, we discuss many of the echinoderms, cnidarians, crustaceans, and insects you will encounter in Galápagos.

There have been relatively few studies of the archipelago's invertebrates, and the conservation status has not been established for most of these organisms. However, several are disappearing. For example, only one living specimen of the snail *Bulimulus achaetellinus* has been collected since 1890, scarab beetles (*Neoryctes*) have become increasingly rare, the endemic sunstar (*Heliaster solaris*) has not been seen since the 1982–1983 El Niño,[2] and local populations of spiders, scorpions, and native ants have disappeared after their habitats have been

[1] Approximately half of the 3,000 species of land invertebrates reported in Galápagos are endemic.

[2] El Niños have also brought new species to Galápagos. For example, the Moorish idol (*Zanclus cornutus*) entered Galápagos with the 1982–1983 El Niño, and Indo-Pacific wrasses (e.g., *Thalassoma* spp.) entered with the 1997–1998 El Niño.

invaded by introduced fire ants (*Wasmannia auropunctata*). Several of the more than 500 introduced invertebrates—many of which are highly invasive—are discussed in Chapter 11.

CNIDARIA

Cnidarians are named for their *cnidae,* which are cells that can sting, penetrate, or adhere to prey and perceived threats (including snorkeling humans who get too close). Cnidarians include a diverse array of aquatic animals, from the Scyphozoans (jellies) to the Anthozoans (corals and anemones). One of the most commonly encountered, and easily identified, cnidarians is the orange cup coral.

Common name: Orange cup coral

Scientific name: *Tubastrea coccinea*

See it: Throughout the islands

Residence status: Native

What you will see and what it means: Orange cups live as soft, tentacled polyps inside calcareous cups that these animals secrete (Color Image 38). Although similar to other stony corals, the orange cups do not build reefs, nor do they depend on photosynthetic symbionts for food. Instead, orange cups use their long, fleshy tentacles to capture passing zooplankton, typically at night.

ECHINODERMATA

Echinoderms, named for their spiny skin, are invertebrates that have an internal skeleton of bony plates (ossicles) and an adult body-plan based on five radiating points (i.e., radial symmetry). Examples of echinoderms include the sea stars (Asteroidea), sea cucumbers (Holothuroidea), and sea urchins and sand dollars (Echinoidea); these groups are abundant in Galápagos tide pools and nearshore environments.

Common name: Green sea urchin

Scientific name: *Lytechinus semituberculatus*

See it: Tide pools and shallow nearshore waters

Residence status: Endemic

What you will see and what it means: You may see green sea urchins in tide pools, and you will definitely encounter these urchins while snorkeling. These animals have bright green spines, and they often adorn themselves with leafy bits of algae or fragments of clamshells. Sea urchins are ecological engineers that shape underwater habitats by controlling algal growth, which in turn allows many other organisms to proliferate. When urchins are removed, unrestrained algal growth can have cascading effects on entire underwater communities. This ability of sea urchins to alter nearshore productivity in Galápagos is discussed further in Chapter 8.

Why do sea urchins decorate themselves? Many sea urchins, such as the endemic green sea urchins of Galápagos, adorn their bodies with various materials from their environments. You may see green urchins covered with a motley assemblage of clamshell fragments, algal "leaves," spines of pencil sea urchins, and dead coral. This is similar to the covering behavior seen in decorator crabs, polychaete worms, and many insect larvae. Several hypotheses have been suggested for the adaptive significance of this behavior. For example, external layers may protect the animal, making it less palatable to predators; this seems to be the case for many insect larvae, such as caddisflies that build and live in tubes of gravel or sand. Decorations may also make an animal less visible to predators; this type of *crypsis* is gained by decorator crabs and hermit crabs that adorn their shells. There must be, or must have been, an advantage to this behavior, for there is clearly a cost: decorated sea urchins are not as nimble as their unencumbered counterparts. So why, then, do urchins decorate themselves? Is it for camouflage, external defense, or something else?

Sea urchins are sensitive to ultraviolet (UV) radiation from sunlight, and this sensitivity might make covering adaptive. There have been several studies supporting the benefits of covering to protect against UV rays; however, some sea urchins that live in complete darkness also cover themselves. Other suggested benefits of covering include: food storage, in the case of urchins that cover themselves with edible pieces of algae; protection from water loss, for those urchins that find themselves above water during low tide; or defense against water-borne particles. Recent experimental work with a different species of green sea urchin, *Strongylocentrotus droebachiensis,* revealed that juveniles cover themselves much more than adult urchins, and multiple environmental factors appear to regulate covering. That is, instead of a single cause for decorating themselves, covered sea urchins can gain protection against sunlight, wave activity, and predators. When you see decorated green sea urchins in Galápagos, consider what environmental forces may have produced this unusual behavior.

Common name: Pencil-spined sea urchin (or slate-pencil urchin)

Scientific name: *Eucidaris thouarsii*

See it: Tide pools and shallow nearshore waters

Residence status: Native

What you will see and what it means: The pencil-spined sea urchin is darker than the green sea urchin, with short, stubby spines that range in color from brown to purple. Its thick spines not only resemble pencils, they can also be used to draw on lava rock—try it yourself with the stray spines you will see on many beaches.

Common name: Crowned sea urchin (or hatpin sea urchin)

Scientific name: *Centrostephanus coronatus*

See it: Shallow nearshore waters

Residence status: Native

What you will see and what it means: With its long (10–15 cm), thin, black spines, the crowned sea urchin is unmistakable. It will be doubly unmistakable should you accidentally step on or grab one of these urchins; the spines dislodge easily and are toxic and aggravating. This urchin lives in holes and crevices in shallow, rocky water. It forages within a meter of its shelter, to which it returns at night.

Common name: Galápagos sand dollar

Scientific name: *Encope micropora galapagensis*

See it: Rarely encountered underwater, but often seen washed up on beaches just after high tide

Residence status: Endemic subspecies

What you will see and what it means: Living sand dollars are rarely encountered because they are burrowers that feed on organisms disturbed by moving the sand. Their burrowing explains their shape: although closely related to the sea urchins, sand dollars have evolved a flattened body that facilitates sliding just beneath the sand. Although sand dollar larvae are food for many marine creatures, adult sand dollars have few predators.

Common name: Galápagos sea cucumber ("pepino")

Scientific name: *Isostichopus fuscus*

See it: Sandy bottoms throughout the archipelago

Residence status: Native

What you will see and what it means: This large, dark, cucumber-shaped echinoderm is often covered with brown or orange bumps, and lives throughout the archipelago. However, sea cucumbers are declining in number, due to an intense harvest—legal and illegal—of this animal (Chapter 11).

Common name: Chocolate-chip sea star

Scientific name: *Nidorellia armata*

See it: Clinging to rocks or coral in shallow waters

Residence status: Native

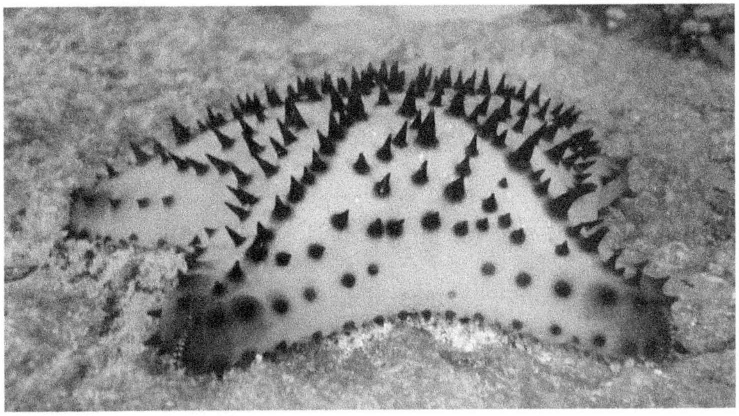

FIGURE 9.1
You will encounter the aptly named chocolate-chip sea star in shallow waters throughout Galápagos.

What you will see and what it means: This pale orange sea star, with its short black spines reminiscent of chocolate chips, is unmistakable. Sea stars such as the chocolate-chip sea star are related to other bilaterally symmetrical organisms (e.g., vertebrates, mollusks, and arthropods). Like the cnidarians (e.g., anemones and corals), sea stars exhibit radial symmetry as adults, yet they are bilaterally symmetrical as larvae and are not closely related to anemones and their kin.

Common name: Panamic cushion star

Scientific name: *Pentaceraster cumingi*

See it: Sandy bottoms throughout the archipelago

Residence status: Native

What you will see and what it means: You are likely to see these large, colorful (with elaborate patterns of red, orange, and white) sea stars in shallow-water areas with sandy bottoms. Panamic cushion stars graze on other animals, especially bivalves, pulling their shells apart with slow and steady hydrostatic pressure. Sea stars then evert their stomachs and lay them on the clam's soft internal tissues. After some external digestion, the sea stars pull their stomachs back into their bodies to complete digestion.

Common name: Variable sea star

Scientific name: *Linckia columbiae*

See it: Clinging to lava rock in intertidal zones

Residence status: Native

What you will see and what it means: The variable sea stars, as their name implies, vary greatly in color and number of rays. While many of these long-rayed sea stars are fairly dull, others are vibrant blue, orange or red (Color Image 39). Variable sea stars undergo *autotomy*, a process in which one (or more) of their rays detaches itself from the sea star's body. During the next six months, the separated ray grows replacement rays and a new body.

Are "Darwin's Finches" Really Snails?

Mollusks are soft-bodied invertebrates common to terrestrial, freshwater, and marine environments. Mollusks in Galápagos include cephalopods (squid and octopus), bivalves (clams), and gastropods (snails). Recent work suggests that, relative to adaptive radiation, land snails in the genus *Bulimulus* are to mollusks what Galápagos finches are to birds, or lava lizards are to reptiles. In fact, bulimulid snails represent the most diverse of these adaptive radiations, with over 70 endemic species distributed throughout most of the islands (and in five of the six vegetative zones). No other group of organisms in the archipelago comes close to these land snails in number of species. They caught Darwin's eye—he collected 15 species of these snails.

The distribution of bulimulid snails satisfies many predictions of *island biogeography,* a theory that explains the distribution of species on islands (or "island-like" habitats such as desert oases or lakes). According to island biogeography, species diversity is a function of inter-island colonization, within-island speciation, and extinction. Inter-island colonization should depend on distance to the nearest island or mainland, and within-island diversity should increase with increasing island size and habitat types (often measured as plant diversity). A recent evaluation of habitat characteristics associated with high bulimulid diversity confirmed that species richness of snails increases with that of plants, increased elevation (itself associated with plant diversity; Chapter 10), island area, and distance to the nearest mainland. Thus, large islands with elevational changes, such as Santa Cruz, house the greatest diversity of these snails. The land snails of Galápagos are an adaptive radiation that exceeds the others (finches, cactus, lizards) in species diversity. However, *why* this group of snails could radiate throughout the archipelago in this way is still a mystery.

ARTHROPODA

Arthropods are animals characterized by a molting exoskeleton, jointed appendages in myriad permutations, and complex life histories. Globally, arthropods rule the land, water, and sky, yet Galápagos—having never been connected to the mainland—is relatively arthropod-poor. However, the islands are home to a few myriapods (centipedes and millipedes) and crustaceans, and over 1,600 described species of insects (about 500 of which are beetles).

Crustacea

Common name: Sally lightfoot crab

Scientific name: *Grapsus grapsus*

See it: Intertidal zones throughout the islands

Residence status: Native

What you will see and what it means: Adult Sally lightfoot crabs, with their bright red and orange shells against black lava rock, are unmistakable (Color

Image 40). And although these crabs are not endemic to the archipelago, they are iconic and occur here in numbers that exceed what you will encounter elsewhere. Sallys scoot gingerly across the rock, in a light-on-their-feet manner that presumably gave them their common name. These crabs, like neighboring marine iguanas, eat algae fringing the rocks along the shore, but they will also scavenge on carcasses and feces. The algae they prefer are typically so abundant that there appears to be little, if any, competition between crabs and iguanas. Sallys groom marine iguanas, removing encrusted algae, parasites, and dead skin, in a mutualism unique to these islands.

Why are Sally lightfoot crabs so brightly colored? Adult Sally lightfoot crabs are so dramatic, in striking combinations of red, yellow, blue, and purple, it begs the question: what is the significance of this coloration? Bright pigmentation typically exacts energetic and ecological costs on the bearer: pigments are metabolically expensive to produce, and bright colors make prey species, such as the Sallys, more visible to predators with color vision (e.g., herons and other shorebirds, octopuses). So, what are the possible benefits of this feature in Sallys?

Bright coloration is often used to attract mates through sexual selection (e.g., the bright red throats of frigatebird males) or signal toxicity to predators (e.g., the striking orange and black colors of monarch butterflies). However, both male and female Sallys mature from dark, lava-colored juveniles into brightly colored adults, and there is little evidence that color variation affects mate choice in these crabs. In addition, Sallys are eaten by a range of animals, none of which seems harmed by any of the crabs' metabolic compounds. So, the adaptive significance of Sally lightfoot coloration, if any exists, remains a mystery.

Common name: Ghost crab

Scientific name: *Ocypode gaudichaudii*

See it: Sandy beaches throughout the islands

Residence status: Native

What you will see and what it means: Ghost crabs resemble Sally lightfoot crabs in size, and are often a pinkish color that, from a distance, can cause ghost crabs to resemble Sally lightfoot crabs. However, ghost crabs are not nearly as vibrant as Sallys, are more often cream-colored, and can be distinguished by their eyes, which are on long, pointed, retractable stalks. Ghost crabs are active mainly at night, although you may see a few emerging from their burrows near sunset. They are skittish and will vanish quickly if you get too close.

On many beaches in Galápagos, you may see an array of round balls of sand, radiating from a hole in the sand. This likely marks the home of a ghost crab that has created the sandy balls while excavating the burrow. While building the burrow, a ghost crab will feed on organic matter deposited on the grains of sand, thereby forming a ball of clean sand that is placed at the entrance to the burrow.

Common name: Fiddler crab

Scientific name: *Uca helleri*

See it: Mangrove swamps

Residence status: Native

What you will see and what it means: Fiddler crabs are small (<5 cm), well camouflaged inhabitants of mangrove swamps and intertidal zones. Perhaps their most distinctive feature is the male's single large claw, which can be used to fight rival males or, via a series of stereotypical waving motions, solicit females.

Common name: Galápagos hermit crab

Scientific name: *Calcinus explorator*

See it: Rocky intertidal zones

Residence status: Endemic

What you will see and what it means: If you see a gastropod shell moving clumsily across rocks along the shoreline, it is likely a hermit crab, dragging its home—an abandoned snail shell. Hermit crabs have an asymmetrical, coiled abdomen, which accommodates the spiral shells that snails produce.

Myriapoda

Common name: Galápagos giant centipede

Scientific name: *Scolopendra galapagensis*

See it: Arid zones, especially near nightfall

Residence status: Endemic

What you will see and what it means: One of the world's largest centipedes, reaching lengths greater than 30 centimeters, the Galápagos centipede has large fangs for injecting poison into lizards, invertebrates, and small birds and mammals. Fortunately for humans, the centipede is furtive and primarily nocturnal. A 1986 report documented an unusual instance of kleptoparasitism (*a la* frigatebirds; Chapter 5), whereby a Galápagos mockingbird stalked a giant centipede and stole its prey, a wingless cricket.

Chelicerata

Arachnida (scorpions, ticks, and spiders)

Common name: Galápagos scorpion

Scientific name: *Hadruroides maculatus galapagoensis*

See it: Arid zones of most islands

Residence status: Endemic

What you will see and what it means: This light brown, small (<10 cm) scorpion is a nocturnal predator of many other, smaller arthropods.

Common name: Galápagos black widow spider

Scientific name: *Latrodectus apicalis*

See it: Lowlands of all major islands

Residence status: Endemic

What you will see and what it means: This web-spinning spider is related to the venomous continental black widow, sharing a similar body shape and color—a large black abdomen with red markings on its underside. Although the bites of other black widow spiders can be fatal, no humans have been seriously injured by the Galápagos black widow spider.

Common name: Zig-zag spider

Scientific name: *Neoscona cooksoni*

See it: Arid zones of most islands

Residence status: Endemic

What you will see and what it means: The common zig-zag spider is often encountered by walking into its fine web, strung across a trail. If your guide sees the zig-zag's web, he or she may divert the group to avoid disrupting this harmless, cream-and-brown spider.

Insecta

Lepidoptera (butterflies and moths)

Common name: Galápagos blue butterfly

Scientific name: *Leptodes parrhasioides*

See it: Throughout the islands

Residence status: Endemic

What you will see and what it means: This is a small (<3 cm), blue butterfly, with gray along the margins of its wings. It is easy to identify the Galápagos blue butterfly because it is so small and is the only bluish butterfly on the islands. These butterflies are most abundant after a rain.

Common name: Galápagos sulfur butterfly

Scientific name: *Phoebis sennae marcellina*

See it: Throughout the islands

Residence status: Endemic subspecies

What you will see and what it means: This bright yellow butterfly is commonly encountered on the main islands, feeding on the nectar of muyuyo and *Parkinsonia* flowers. *Phoebis* is the only yellow butterfly in Galápagos.

Common name: Monarch butterfly

Scientific name: *Danaus plexippus*

See it: Throughout the islands, in areas with milkweed

Residence status: Native

What you will see and what it means: Monarchs, the largest (>10 cm across the wings) butterflies on the islands, are similar in coloration (orange, black, and white) to queen butterflies (see below). However, monarchs are distinguished by the bold, black venation on their orange wings. Bright white spots against this black background set off their bodies and surround their wings.

Monarch butterflies are extreme specialists, developing only on leaves of milkweed (*Asclepias curassavica*). The monarchs' degree of specialization is unusual for Galápagos. Given the islands' age and distance from the mainland, one expects insect colonists to be generalists, capable of feeding on a variety of plants and animals. But in addition to milkweed-dependent monarchs, Galápagos is home to another extreme specialist: small *Gerstaeckeria* weevils, which feed exclusively on *Opuntia* cactus.

How Did Insects Get to Galápagos?

Insects, as well as many other fauna, have a limited number of ways in which they can colonize distant islands: they may be carried by a nonhuman animal; they can be transported, intentionally or not, by humans; they can fly; or they can arrive on floating rafts (of debris, logs, etc.). For some of the archipelago's insects, it is easy to identify their mode of entry. For example, carpenter bees nest in wood, so the ancestor of today's Galápagos bees probably arrived aboard floating logs from the mainland. However, the arrival mode of most Galápagos insects is unclear.

The monarch butterfly found on the islands (*Danaus plexippus*) is the same migrating monarch that occurs in North America. Like its continental counterparts, the Galápagos monarch depends on milkweed, which was introduced to the islands. Milkweed was first documented by the California Academy of Sciences during their 1905–1906 expedition to the islands, but the Academy's thorough surveys did not record any monarchs. In 1924, William Beebe (Chapter 2) first documented the monarch's presence in the archipelago, noting, "*Danais* (sic) *plexippus* . . . is definitely added to the fauna of the Galápagos Archipelago. On several islands I had seen a milkweed in blossom, so this wide-ranging butterfly should have no difficulty in establishing itself." So, it appears the monarch arrived to the islands sometime between 1906 and 1924, but how?

A 1998 report in *Noticias de Galápagos* concluded that monarchs are too fragile to have reached the islands through rafting. In addition, humans have not intentionally brought monarchs to the islands. Thus, given this migrant's famed ability to travel great distances, it seems most likely to have dispersed to the archipelago naturally, by wind-assisted flight. Monarchs have probably had numerous failed colonizations of Galápagos, and were finally successful after the establishment of their host plant.

Common name: Queen butterfly

Scientific name: *Danaus gilippus*

See it: On Isabela and Baltra

Residence status: Native

What you will see and what it means: The queen butterfly is easily confused with the monarch, with similar size and coloration. Queens lack the bold black venation that characterizes monarch wings, and the queen is not exclusive to milkweed. Queens, like monarchs, produce noxious compounds that make them unpalatable to many insectivores. When two toxic, but unrelated, species share predators and resemble each other, they gain increased protection from their predators' learned aversion to their appearance. Selection, through coevolution, favors this resemblance, in a type of mimicry known as *Mullerian mimicry.*

Common name: Green hawk moth

Scientific name: *Eumorpha labruscae yupanquii*

See it: Widespread throughout the islands, especially at night near lights

Residence status: Endemic subspecies

What you will see and what it means: This large (8–12 cm), common, bright green moth is often seen trailing the lights of ships, far out to sea. The proboscis of green hawk moths is almost twice as long as the insect's body. When feeding at flowers, these hawk moths resemble hummingbirds.

Hymenoptera (bees, ants, and wasps)

Common name: Galápagos carpenter bee

Scientific name: *Xylocopa darwini*

See it: Throughout the central and southern islands

Residence status: Endemic

What you will see and what it means: Male carpenter bees are small, brown, and rare. The females are more common, black, and larger (>2 cm). Carpenter bees are the archipelago's only bee, and they pollinate many of the islands' endemic flowering plants (e.g., the columnar cactus *Jasminocereus,* arrow-leafed morning glory, cordia, and lantana). Carpenter bees nest in wood, and they may have gotten to the islands via a floating log. Male carpenter bees establish territories and fly in figure-eight patterns to attract females.

Common name: Galápagos carpenter ants

Scientific name: *Camponotus* spp.

See it: Three species of *Camponotus* occur throughout the islands

Residence status: Endemic

What you will see and what it means: These light brown ants are named for their wood-boring tendencies. You may detect their presence by the holes they construct, or by the piles of chewed, rotten wood ("frass") they evict from their nests.

Common name: Little red fire-ant

Scientific name: *Wasmannia auropunctata*

See it: All of the inhabited islands, plus several frequented by tourists (e.g., Santiago, Isabela) and, surprisingly, more isolated islands such as Pinta and Marchena

Residence status: Introduced

What you will see and what it means: The aptly named little red fire-ant is a small (<2 mm), aggressive, biting pest that has the dubious distinction of being named one of "the 100 worst invaders in the world" by the IUCN. These ants are generalists, feeding on many different foods; this, combined with their hyper-aggressive nature, makes the little red fire-ant a potent colonizer of novel habitats. This species of ant has colonized islands throughout the Pacific (e.g., Hawaii, Tahiti), and was brought to Galápagos in the last century. In Galápagos, the fire ant has harmed native reptiles, birds, and invertebrates, and is a serious concern for the archipelago's future biodiversity (Chapter 11).

Orthoptera (crickets, grasshoppers, and locusts)

Common name: Galápagos field cricket

Scientific name: *Gryllus abditus*

See it: Several species occur throughout the islands

Residence status: Endemic

What you will see and what it means: There are several species of *Gryllus* in Galápagos, all large (about 5 cm) and dark, and identified primarily by island.

Common name: Large painted locust

Scientific name: *Schistocerca melanocera*

See it: All islands except Española

Residence status: Endemic

What you will see and what it means: You can't miss this large (6–10 cm), elaborately painted locust, which displays striking patterns of yellow, black, and red.

Common name: Small painted locust

Scientific name: *Schistocerca literosa*

See it: All islands

Residence status: Endemic

What you will see and what it means: The small painted locust is harder to detect than its larger relatives. This locust is also less colorful and blends in with the rocks and sand on most trails.

Common name: Galápagos flightless grasshopper

Scientific name: *Halemus robustus*

See it: Throughout the islands

Residence status: Endemic

What you will see and what it means: This grasshopper is difficult to find, due to its small size and drab, light brown coloration. Similar to other island endemics, this grasshopper has lost the ability to fly and, consequently, no longer has wings. In contrast to their continental kin, island insects may not enjoy a benefit from flight, because oceanic winds may carry them out to sea.

Coleoptera (beetles)

Common name: Spotless ladybug

Scientific name: *Cycloneda galapagoensis*

See it: Transition and humid forest zones

Residence status: Endemic

What you will see and what it means: The spotless ladybug is similar to the introduced vedalia beetle, but lacks the black markings on its back.

Common name: Vedalia beetle, or cardinal ladybug

Scientific name: *Rodolia cardinalis*

See it: This small (about 4 mm) red-and-black beetle lives on all the major islands in the archipelago

Residence status: Introduced

What you will see and what it means: While many (23%) of the insects of Galápagos were unintentionally introduced, the vedalia beetle was intentionally brought to the islands in 2002 to control another invasive species, the cottony cushion scale insect. So far, the project appears to have been successful (Chapter 11).

Diptera (flies)

Common name: Galápagos green-eyed horsefly

Scientific name: *Tabanus vittiger*

See it: Throughout the archipelago

Residence status: Endemic

What you will see and what it means: The largest fly in the islands, these horseflies are notorious pests to many of the islands' fauna, including humans. With its size, loud buzzing noise, and painful bite, you are unlikely to miss this horsefly.

CHAPTER 10

Plants

Most of the organic productions [of Galápagos] are aboriginal creations, found nowhere else; there is even a difference between the inhabitants of the different islands; yet all show a marked relationship with those of America, though separated from that continent by an open space of ocean . . .

—Charles Darwin, *Journal of Researches,* 1845

ALTHOUGH CHARLES DARWIN'S VISIT TO GALÁPAGOS HELPED make the archipelago famous, he was not initially overwhelmed by the islands' plants—for example, he described the flora of San Cristóbal as "wretched-looking little weeds." However, his impressions of the archipelago's plants must have improved during his next five weeks in the islands, for he eventually collected 173 different plants, many of which were new to science.[1] Darwin sent these plants to his Cambridge University mentor John Stephens Henslow, and he hoped that Henslow—who taught "botanical geography" at Cambridge—would examine them quickly.[2] In 1837, Henslow published descriptions of two *Opuntia* that Darwin had collected (which Henslow named after Darwin). Although Henslow had told Darwin that "most of the plants are very desirable to me," Henslow began devoting more time to his clerical duties and made little more progress with Darwin's plants. Meanwhile, Darwin was wondering if the

[1] Darwin collected more plants from Galápagos than any other group of organisms; his was the largest and most diverse collection of plants that had ever been gathered in the archipelago. Unlike when he collected birds in Galápagos, Darwin recorded on which island he collected each of the plants. Darwin visited Galápagos during the cool, dry season (Chapter 3), when many of the plants are leafless. Some of the prominent endemic species of plants he collected include Darwin's cotton (*Gossypium darwinii*), Galápagos tomato (*Lycopersicon cheesmanii*), cutleaf daisy (*Lecocarpus pinnatifidus*), radiate-headed scalesia (*Scalesia affinis*), macrea (*Macraea laricifolia*), thin-leafed Darwin's shrub (*Darwiniothamnus tenuifolius*), Galápagos carpetweed (*Sesuvium edmonstonei*), and castela (*Castela galapageia*). Darwin also collected many native species, including matazarno (*Piscidia carthagenensis*) and yellow cordia (*Cordia lutea*). These plants are described later in this chapter.

[2] Darwin, who had collected plants for Henslow in Wales in 1831 while an undergraduate, sent Henslow more than 2,600 plants (including duplicates) collected during his years aboard the *Beagle*. Henslow prepared more than 900 herbarium sheets containing Darwin's plants. In all, Darwin contributed 211 specimens from Galápagos, of which 74% had been documented for the first time in the archipelago.

islands' plants, like their animals, resembled those of the mainland. As he noted in 1835, "I certainly recognize South America in ornithology; would a botanist?"

Late in 1838, a frustrated Darwin pleaded for Henslow not "to take any trouble in giving me names" and that "all I want to know is whether . . . there are of near species, of the same genus; —one species coming from an island, & the other from a second island." Five years later, Darwin transferred his collection of plants to Joseph Dalton Hooker, son of the then-director of Kew Gardens. Hooker quickly recognized the importance of these plants: Most of them lived nowhere else on Earth except in Galápagos (i.e., were endemic) and were closely related to plants of western South America, a finding that Hooker summarized as "a fact that quite overturns all our preconceived notions of species radiating from a centre & migrating to any extent from one focus of greater development." Hooker was especially impressed by both the large number of endemic genera (35 species, 29 of which are endemic) and their treelike forms unique to Galápagos.

Hooker's analysis thrilled Darwin, and Darwin responded that he was "astonished at the amount of creative force, if such an expression may be used, displayed on these small, barren, and rocky islands; and still more so, at its diverse yet analogous action on points so near each other. . . . I cannot tell you how delighted & astonished I am at the results of your examination; how wonderfully they support my assertion on the differences in the animals of the different islands, about which I have always been fearful." Although Darwin had already published the first edition of his *Journal of Researches* without this information about the plants of Galápagos, he used the information from Hooker and others to expand his descriptions of Galápagos in the second edition (Chapter 2).

In 1847, Hooker's *An Enumeration of the Plants of the Galápagos Archipelago; with Descriptions of Those Which Are New* described many of Darwin's plants; this was the first complete analysis of the flora of Galápagos.[3] Hooker—who became a lifelong friend of Darwin's—eventually described 78 new species of Galápagos plants, most of which were endemic, and some of which were restricted to individual islands. Hooker was also the first person to whom Darwin confided his ideas about evolution, noting in his famous letter in 1844, "at last gleams of light have come, & I am almost convinced (quite contrary to the opinion I started with) that species are not (it is like confessing a murder) immutable."

Hooker realized that the flora of Galápagos is *disharmonic;* that is, it is not merely a smaller replica of that of nearby South America, but instead, because of the archipelago's isolation, includes only a few genera and species from the mainland. Virtually all of these species—which may have resulted from fewer than 400 chance arrivals—are adapted for long-distance transport; for example, they have lightweight seeds, buoyant and salt-tolerant fruits for dispersal in the

[3] Hooker also included plants collected by people who preceded Darwin to Galápagos (e.g., David Douglas and John Scouler; see Chapter 2).

ocean, fruits that are eaten by birds, or fruits adorned with hooks or adhesive for attaching to birds. But the odds of getting to and colonizing Galápagos have been slim; although there are more than 20,000 native species of vascular plants in mainland Ecuador, there are fewer than 600 such species in Galápagos.

WHO ARE THE PLANTS?

Plants are multicellular organisms having cellulose-rich cell walls and plastids, which are used for photosynthesis and the storage of starch. Approximately 30% of the more than 560 species of vascular plants of Galápagos are endemic. Endemism is highest among the flowering plants. Other groups—for example, ferns, mosses, and liverworts—have small spores that are efficiently dispersed, thereby reducing the likelihood of the isolation necessary for the evolution of new species.[4] Most of the native and endemic plants of Galápagos came from western South America, with smaller percentages from Mexico and Central America. More than 40% of the native plants are endemic to Galápagos, and many of the endemic plants of Galápagos are endemic to a single island. The presumed ancestral species (i.e., the closest relative from South America) of only about 15% of the endemic vascular plants of Galápagos now lives in the archipelago. Thus, for the remaining 85% of these plants, the colonizing population has evolved into taxa different from the colonizers (or the colonizers evolved and the original species are now extinct).

VEGETATIVE ZONES OF HIGH ISLANDS

Visitors to Galápagos quickly see that particular species of plants often live in particular parts of islands, and that this distribution is strongly influenced by the availability of water. Because Galápagos is in the Pacific Dry Belt, most of the low islands receive 0–300 millimeters of rain per year, and are therefore inhabited by desert-like vegetation.[5] Indeed, arid plants such as cacti and palo santo dominate the small, flat islands such as Baltra. However, some islands (e.g., Santa Cruz, Floreana, and San Cristóbal) also have damp highlands that receive 300–2,000 millimeters of rain, and these areas are covered by lush vegetation. Between the arid and moist regions are several intergrading vegetative zones, some of which are named for their dominant plants (e.g., *Scalesia*, *Miconia*), and others for their location (e.g., coastal). These zones, which help

[4] The same generalization applies to the animals of Galápagos: Reduced mobility is associated with increased rates of endemism. For example, reptiles and terrestrial mammals have higher rates of endemism than terrestrial birds, which, in turn, have higher rates of endemism than marine birds and marine mammals.

[5] Lowland plants of Galápagos are closely related to those of the semidesert along the Gulf of Guayaquil and in northern Perú, whose climates resemble that of Galápagos.

define the distributions of animals in the archipelago, are affected by elevation, the type of soil, and moisture, which changes with elevation and exposure to prevailing winds. The southern slope of Santa Cruz demonstrates the vegetative zones of the high islands of Galápagos (Figure 10.1):[6]

Name of Zone	Altitude[7] (m)	Vegetation
Coastal (littoral)	0–10	Mangroves and other salt-tolerant plants
Arid	10–120	Cacti, *Cordia, Fursera,* and *Gossypium*
Transition	100–200	Semideciduous forest of *Pisonia* and *Psidium*
Humid		
Scalesia	200–400	Dense evergreen forest of *Scalesia*
Zanthoxylum	400–500	Mossy open evergreen forest
Miconia	500–600	Mossy evergreen shrub
Pampa	>600	Sedges, ferns, grasses

Although animals such as tortoises can be found in all of these vegetative zones, others live only in particular zones (e.g., vermilion flycatchers live only in the higher zones). Some introduced plants have invaded several vegetative zones; for example, poison apple (*Hippomane mancinella*) grows from sea level to the *Scalesia* zone, especially near ponds and clearings. Regardless of where they live, plants help determine the distribution of animals.

Coastal (Littoral) Zone: 0–10 meters

The coastal zone is a narrow belt extending from just below the low-tide line to a few meters above the high-tide line on all islands of the archipelago.[8] Substrates range from sandy beaches to lava rocks, both of which may be submerged during parts of the day, but exposed at others. Plants growing in this zone are mostly evergreen plants that have a high tolerance to salt (i.e., they can survive in brackish or seawater); mangroves are common along sheltered coves, whereas vines, grasses, and succulent shrubs (e.g., *Scutia spicata*) inhabit beaches. Plants living in the coastal zone are readily dispersed by the ocean. As a result, these species are rarely isolated enough for the evolution of new species, thus explaining why this zone has few endemic plants.

---

[6] These zones are similar on other large, high islands such as San Cristóbal, Floreana, Santiago, Isabela, and Fernandina. The smaller, low islands have only a coastal and arid zone. On some islands, the same species is found in different ecological zones. For example, the native flame tree (*Erythrina velutina*) grows at mid-elevations of high islands such as Santa Cruz and Isabela, but in the arid zone of low islands such as Wolf and Genovesa. The flame-red flowers of these trees are easily spotted in the otherwise drab arid zone.

[7] These heights and characterizations were originally proposed by Alban Stewart after the California Academy of Sciences expedition to Galápagos in 1905–1906 (Chapter 2). Although this classification system is informative, it often belies the complexity of the islands. For example, there are usually significant differences at any one height on different islands. Moreover, because the northern (i.e., lee) sides of islands get less rain than the southern sides, the vegetative zones on the northern sides of islands often extend higher than the indicated heights.

[8] There are two high tides and two low tides per day in Galápagos, and the tidal range is 2.5 meters.

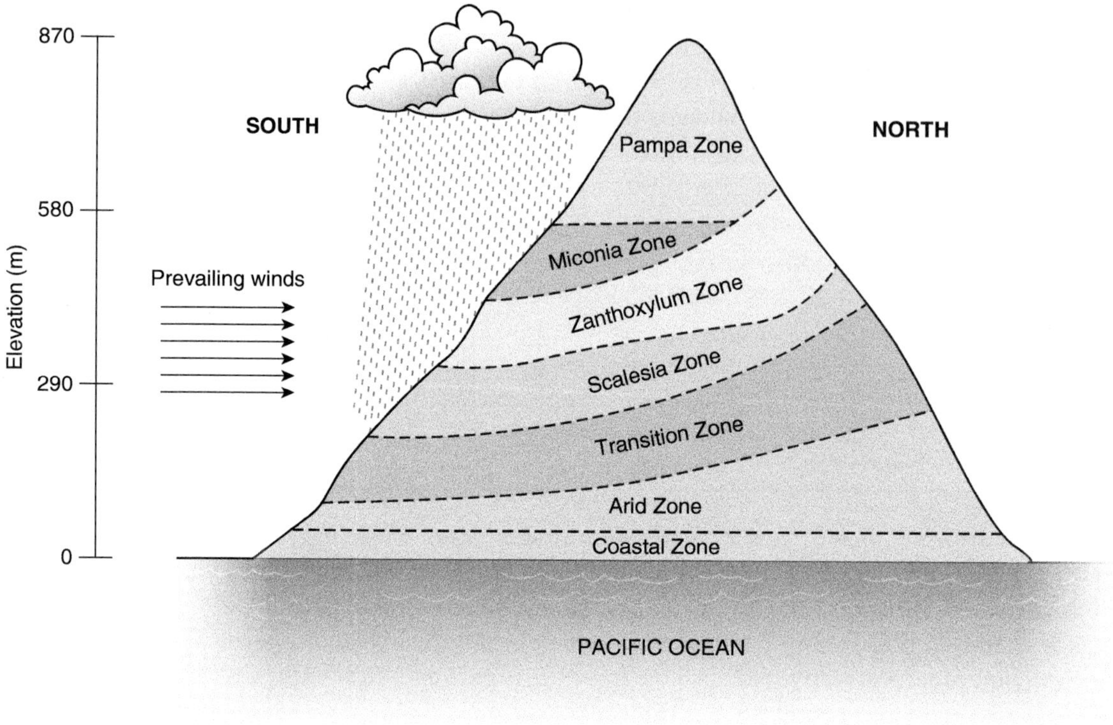

FIGURE 10.1 High islands in Galápagos such as Santa Cruz have several ecological zones. The southern sides of islands and higher altitudes are wetter because moisture-carrying winds are usually from the southeast. The northern sides of high islands are rain shadows.

Plants: Saltbush (*Cryptocarpus pyriformis*), *Sesuvium portulacastrum, Scutia spicata, Cryptocarpus pyriformis,* mangroves (red, *Rhizophora mangle;* black, *Avicennia germinans;* white, *Laguncularia racemosa;* button, *Conocarpus erecta*), beach morning glory (*Ipomoea pes-caprae*), *Nolana galapagensis, Heliotropium curassavicum, Atriplex peruviana, Lycium minimum, Sesuvium edmostonei,* buckthorn (*Scutia pauciflora*), *Sporobolus virginicus, Maytenus octogona, Scaevola plumieri, Mollugo, Tiquilia, Amaranthus sclerantoides, Polygala, Salicornia fruticosa, Batis maritima*

Arid Zone: 10–120 meters

The arid zone, which is just inland from the coast, is the largest ecological zone in Galápagos. This semidesert is dominated by cacti (Color Image 41), shrubs, other evergreen drought-resistant species (e.g., *Cordia, Prosopis, Gossypium, Opuntia, Croton*), and deciduous trees such as palo santo (*Bursera graveolens*), which drop their leaves during the cool, dry season to minimize water loss. As a result, plants in the arid zone often appear dead. Because water is usually the limiting resource for growth, plants in this zone are often uniformly spaced, such as the endemic shrub *Tiquilia* on the volcanic cinders of Bartolomé (Figure 10.4), and palo santo on the hillsides of several islands (see below). Many plants in the arid zone have small leaves and deep roots; annual plants growing in the arid zone survive the dry season as seeds. Drought-resistant lichens that can absorb garúa are often abundant in the arid zone,[9] and shrubs growing here often form impenetrable thickets.

The arid zone houses the largest number of endemic species of plants; more than 60% of the archipelago's endemic plants live here. This zone is also home to lava lizards, land iguanas, endemic rats, many species of land birds, and some seabirds. On islands such as Bartolomé, the only sign of plant life in the arid zone is often *Tiquilia* and endemic lava cacti.

Plants: Palo verde (*Parkinsonia aculeata*), often covered with lichens; leatherleaf (*Maytenus octogona*), prickly pear (*Opuntia*), palo santo (*Bursera graveolens*), candelabra cactus (*Jasminocereus*), chala (*Croton scouleri*), lava cactus (*Brachycereus*), cutleaf daisy (*Lecocarpus*), *Tiquilia,* poison apple (*Hippomane*), acacia (*Acacia*), *Lignum vitae,* passion flower, muyuyo (*Cordia lutea*), lava morning glory, algarrobo (*Prosopis juliflora*), buckthorn (*Scutia pauciflora*), *Parkinsonia, Mollugo, Chamaesyce, Erythrina velutina, Cryptocarpus pyriformis,* Galápagos cotton (*Gossypium darwinii*), *Macraea laricifolia*

Transition Zone: 100–200 meters

Just above the arid zone is the more diverse and dense transition zone, which receives more moisture than the arid zone. Trees such as *Bursera, Cordia, Pisonia, Psidium,* and *Piscidia* are abundant and taller here than in the arid zone. Perennial herbs and tangled shrubs are common in the transition zone, but the *Opuntia* common in the arid zone is much less abundant in the transition zone. Because the transition zone usually has few dominant indicator plants, this zone is often hard to pinpoint.

Plants: Pega-pega (*Pisonia floribunda*), guayabillo (*Pisidium galapageium*), matazarno (*Piscidia cathagenensis*), palo santo (*Bursera graveolens*), Galápagos tomato, *Croton scouleri,* lechoso (*Scalesia*), epiphytes (especially lichens such as the pendant yellow-green *Ramalina* and *Usnea*), *Chiococca alba, Cordia lutea, Mormordica charantia, Tournefortia*

Humid Zone: 200–600 meters

Scalesia zone: 200–400 meters

The transition zone merges into the evergreen *Scalesia* zone, which is a lush, humid, and diverse cloud forest dominated by the endemic lechoso (*Scalesia pedunculata,*[10] a member of the Asteraceae) with an understory of *Psychotria*

[9] Although lichens contain few nutrients, they condense garúa and, when saturated, the excess water drips to the ground. Puddles of water that form beneath the lichens are a significant part of the water cycle on some islands.

[10] *S. pedunculata* forms the *Scalesia* zone on Floreana, San Cristóbal, Santiago, and Santa Cruz. On Fernandina, the *Scalesia* zone is dominated by *S. microcephala.*

and *Alternanthera.* The *Scalesia* zone, which is the lowest of the humid zones, is present only on higher-elevation islands; for example, on Santa Cruz the transition zone is bathed in garúa, even while the lowlands are dry. Trunks and branches of trees in the transition zone are covered with liverworts, mosses, ferns, and other epiphytes. Introduced plants such as elephant grass (*Pennisetum purpureum*) have invaded the *Scalesia* zone on Santa Cruz.

Habitat clearance has been most damaging in the humid zone, and especially in the *Scalesia* zone. Indeed, more than 70% of the *Scalesia* zone on Santa Cruz is gone, and on San Cristóbal virtually *all* of this habitat is gone. Because of its humidity, rain,[11] and exceptionally fertile soil (up to a meter deep), most of the native vegetation in this zone has been cut down, and the area is now used for agriculture.

Plants: Lechoso (*Scalesia pedunculata*), *Pisonia, Psidium, Psychotria, Zanthoxylum fagara, Miconia robinsoniana, Croton scouleri, Peperomia,* bracken ferns, bromeliads (e.g., *Tillandsia*), liverworts (especially the brown *Frullania*), mosses, and other epiphytes such as *Epidendrum spicatum* (buttonhole orchid)

Zanthoxylum (Brown) zone: 400–500 meters

Between the dense forest of the *Scalesia* zone and the shrub vegetation of the *Miconia* zone is the *Zanthoxylum* zone. This zone is dominated by the small evergreen tree *Zanthoxylum fagara,* as well as by *Tournefortia pubescens* and *Acnistus ellipticus.* Branches of trees in this zone are often covered with epiphytes such as liverworts, mosses, and ferns. During the cool, dry season, vegetation here turns brown, thus accounting for the name "brown zone." Farming has largely eliminated the *Zanthoxylum* zone.[12]

Plants: Cat's claw (*Zanthoxylum fagara*), *Tournefortia pubescens, Acnistus ellipticus,* mosses, ferns, and liverworts (especially *Frullania*)

Miconia zone: 500–600 meters

The *Miconia* zone, which is named for the dense cacaotillo shrub (*Miconia robinsoniana*) that once dominated this zone, occurs on the southern slopes of San Cristóbal and Santa Cruz, where rain averages 1.7 meters per year. There are few native trees in this zone, but there are many liverworts and ferns. This zone is easily observed if you visit Laguna El Junco on San Cristóbal. On Santa Cruz, the *Miconia* zone—a breeding ground for dark-rumped petrels—is threatened by invasive elephant grass (*Pennisetum purpureum*), quinine (*Cinchona succirubra*), and grazing by introduced animals. Other plants common in the *Miconia* zone include *Acnistus* and *Croton.*

Plants: Cacaotillo (*Miconia robinsoniana*), *Dicranopteris flexuosus, Hypericum, Verbena, Darwiniothamnus,* epiphytes

---

[11] The town of Bellavista on Santa Cruz, which is in the *Scalesia* zone, gets about 1 meter of rain per year.

[12] *Zanthoxylum* lives in several vegetative zones, but only reaches tree sizes in the upper *Scalesia* and *Zanthoxylum* zones.

Pampa Zone: >600 meters

The pampa zone is the highest and wettest vegetative zone,[13] especially in garúa season (when it gets more than 2 m of rain). This zone, which is dominated by ferns, grasses, club mosses, and sedges, has no trees except the endemic tree fern (*Cyathea weatherbyana*), which grows to 10 meters tall. The pampa zone covers large areas of Volcán Cerro Azul, Volcán Sierra Negra, Santa Cruz, and San Cristóbal. In some areas, introduced plants such as cinchona and guava have invaded the pampa zone and now threaten the ecosystem. During the cool season, the pampa zone is continually wet, and the ground is littered with puddles of water.

Plants: *Cyathea weatherbyana, Habenaria monorrhiza, Cyperus, Jaegeria gracilis, Lycopodium,* ferns, sedges, grasses

PLANTS OF GALÁPAGOS

Studies of the plants in Galápagos have uncovered many endemic species, interesting stories, and amazing evolutionary tales, some of which have been described in earlier chapters of this book. Here, we describe the endemic and native plants you are most likely to encounter in Galápagos. Many of these species are displayed at several places in towns (e.g., at both ends of Malecón Charles Darwin and north of Playa de Oro in Puerto Baquerizo Moreno). We have not listed all of the endemic plants, for that is beyond the scope of this book. Readers wanting a more comprehensive list can consult Charles Darwin Research Station (CDRS), Galápagos National Park Service (GNPS), and *Flowering Plants of the Galápagos* (McMullen, 1999).

What Breeding Systems Typify Galápagos Plants?

Most native and endemic plants of Galápagos are *monoecious* (i.e., hermaphroditic), meaning that eggs and sperm are produced on the same plant. Most of these plants are self-compatible (i.e., sperm from pollen fertilize eggs on the same plant), which is adaptive for plants that are colonizing islands such as Galápagos.

About 3% of the native and endemic plants of Galápagos are *dioecious*, meaning that individual plants are male or female (i.e., no plants produce both eggs and sperm on the same plant).[14] As a result, dioecious plants depend on wind, insects, and other pollinators to transfer pollen from one plant to another. Once established, one might assume that natural selection would favor dioecy, which would increase the probability of outcrossing leading to

[13] Atop the tallest volcanoes—that is, Isabela's Volcán Cerro Azul and Volcán Wolf—there is an upper arid zone that penetrates the inversion layer. These areas are usually dominated by *Opuntia* and *Scalesia*.

[14] A similar percentage of dioecy occurs among flora on continental Ecuador. Interestingly, the normally dioecious *Croton scouleri* (whose relatives on mainland Ecuador are dioecious) is sometimes monoecious in Galápagos.

increased genetic variation. However, none of the endemic dioecious species in Galápagos evolved this breeding system after its arrival in the archipelago (all of the related genera on the mainland are also dioecious). Examples of dioecious plants in Galápagos include *Batis maritima, Zanthoxylum fagara, Bursera graveolens, Bursera malacophyla,* and *Pisonia floribunda.*

Coastal Plants

Common name: Black mangrove

Scientific name: *Avicennia germinans*

Resident status: Native

Black mangroves are the largest and most salt-tolerant of the four species of mangroves in Galápagos and are named for their dark, grayish-black bark. Individuals can be up to 20 meters tall. Black mangroves live along sandy beaches and brackish lagoons throughout Galápagos and produce small, dense clusters of white flowers. They also often grow in waterlogged soil, where they produce spongy, upward-growing roots called *pneumatophores* that help aerate the underground root system. The oval-shaped leaves of black mangroves are smaller than those of red mangroves, and their large, hairy, buoyant fruits are dispersed by the ocean.

Where you will see it: Santa Cruz (Puerto Ayora, Tortuga Bay, Las Bachas), Santiago (Espumilla Beach), Floreana (Punta Cormorán), Fernandina, Isabela (Elizabeth Bay), and Rábida

Common name: Red mangrove

Scientific name: *Rhizophora mangle*

Resident status: Native

Red mangroves, which are the most common mangrove in Galápagos, are named for their hard, reddish wood. These plants grow as shrubs and small trees throughout Galápagos along sheltered beaches, coves, and lagoons. Their many prop roots, which can form impenetrable thickets, anchor the trees to rocks, dissipate waves, and accumulate organic debris that supports a variety of organisms (e.g., oysters) in the intertidal zone. Individuals are usually about 7 meters tall, but they can (especially on Isabela) reach heights exceeding 15 meters. Red mangroves have white flowers and large, thick, dark-green, waxy leaves.

Red mangroves are *viviparous,* meaning that their offspring develop while still attached to the parent plant (rather than after becoming dormant in seeds, as in most other flowering plants). In red mangroves, these torpedo-shaped propagules are 15 to 25 centimeters long and often "plant" themselves when they fall into the wet, sandy soil in which the parent plants live. They are often dispersed by waves to other places, where they anchor themselves between rocks. In some instances, the propagules of red mangroves must be exposed to water for at least

6 weeks, during which time the embryo continues to develop (for black mangroves, this period is 2 weeks, whereas in white mangroves it is only 1 week).

Where you will see it: Santa Cruz (Puerto Ayora, Tortuga Bay, Black Turtle Cove, Caleta Negra), Isabela (Elizabeth Bay, Punta Garcia, Puerto Villamil), Fernandina, Genovesa, and Bartolomé

Common name: White mangrove

Scientific name: *Laguncularia racemosa*

Resident status: Native

This shrub grows up to 8 meters tall and produces clusters of small, greenish-white flowers. It is best distinguished from other mangroves in Galápagos by its pale, oval leaves that have spots on their undersides. White mangroves, which often grow with red mangroves in brackish lagoons, have flask-shaped, pale green fruits.

Where you will see it: Santa Cruz (Puerto Ayora, Tortuga Bay, Las Bachas), Santiago (Espumilla Beach), Floreana (Punta Cormorán), Fernandina, Isabela (Elizabeth Bay), Rábida

Common name: Button mangrove

Scientific name: *Conocarpus erectus*

Resident status: Native

Although button mangroves are not actually mangroves, they resemble mangroves enough to share the name. Button mangroves, which are the least common mangroves in Galápagos, have dark gray bark, green flowers, alternate leaves, clusters of small flowers, and elliptical brownish-purple fruit. Button mangroves grow up to 10 meters tall on sandy beaches and near lagoons. You will see button mangroves on your walk to CDRS in Puerto Ayora. Button mangroves are so common near Puerto Villamil that a street is named for the genus.

Where you will see it: Santa Cruz (Puerto Ayora), Santiago, Floreana, Fernandina, Isabela (Puerto Villamil)

Common name: Beach morning glory

Scientific name: *Ipomoea pes-caprae*

Resident status: Native

Beach morning glory is a perennial that grows as a creeping vine (up to 10 m long) on dunes. Its distinctive trumpet-shaped flowers, which are among the largest in Galápagos, are purple.

Where you will see it: Española (Gardner Bay), Genovesa, Santa Cruz (Puerto Ayora, Tortuga Bay), Bartolomé, Isabela (Puerto Villamil, Punta Albemarle, Elizabeth Bay)

Common name: Lava morning glory

Scientific name: *Ipomoea habeliana*

Resident status: Endemic

These perennial vines, which can be up to 10 meters long, have alternate leaves, white flowers, and brown capsule fruit. The large, white, and fragrant trumpet-like flowers, which are often pollinated by hawk moths, open late in the day and last only until about noon the following day. As its name suggests, lava morning glory is common along lava cliffs. Lava morning glory, which colonizes lava, is also called *soguilla.*

Where you will see it: Española, Floreana, Genovesa, Rábida, Santa Cruz

Common name: Leatherleaf

Scientific name: *Maytenus octogona*

Resident status: Native

Leatherleaf (Figure 10.2) is a low-growing evergreen shrub or small tree that is typically about 5 meters tall (but can reach heights of 8 m). It has cracked, dark gray bark, small flowers, and thick, waxy, alternate leaves that are usually oriented vertically. Its fruit are red berries. Leatherleaf is common inland and on dunes and beaches; its seeds are a favorite food of finches. Leatherleaf is also called *arrayancillo.*

Where you will see it: Santa Cruz (Puerto Ayora), Baltra, Seymour, Rábida, Santiago (Espumilla Beach, Puerto Egas), Santa Fé, Floreana, South Plaza, Bartolomé

Common name: Saltbush

Scientific name: *Cryptocarpus pyriformis*

Resident status: Native

Saltbush (Color Image 42) is a low, creeping evergreen shrub that often forms dense thickets near beaches. It has thick, waxy, opposite leaves and small, inconspicuous green flowers at the tips of its branches. As its name suggests, saltbush is extremely salt-tolerant. Saltbush, which is one of the most ubiquitous plants in Galápagos, often houses nests of pelicans and frigatebirds. Residents of Galápagos often refer to saltbush as *monte salado.*

FIGURE 10.2 Leatherleaf (*Maytenus octogona*) is common on several islands of Galápagos. The plant shown here grows on Santa Cruz.

Where you will see it: Santa Cruz (Tortuga Bay, Puerto Ayora, along the road to CDRS), North Seymour, Española (Punta Suárez, Gardner Bay), Rábida, Genovesa, Floreana (Punta Cormorán, Post Office Bay), Bartolomé, Santiago (Sulivan Bay, Espumilla Beach), Isabela (Tagus Cove), Santa Fé

How did the ancestors of endemic plants get to Galápagos?

Thanks to the work by Joseph Hooker (see above), Charles Darwin knew that most of the endemic plants of Galápagos came from western South America, especially Peru and mainland Ecuador. Indeed, 90% of the native and endemic plants of Galápagos came from this region; the rest came from North and Central America, the Caribbean, and the rest of South America. But how did they get to Galápagos?

We have no precise record of how colonization occurred, but studies suggest that birds brought 60–80% of the endemic plants, ocean currents brought another 8–12%, and winds brought 8–30%. The more than 500 native vascular plants endemic to Galápagos were produced by an average colonization rate of one species every 10,000 years; for comparison, the more than 800 introduced plants have arrived at an average rate of 1.3 species per year. The often-quoted claim that the introduction rate of introduced species is increasing exponentially is confounded by the fact that the detection rate may be influenced more by research, awareness, and sampling techniques than their actual rate of introduction.

Common name: Saltwort

Scientific name: *Batis maritima*

Resident status: Native

Saltwort is a pale green shrub that grows in dense clusters near beaches. Saltwort is usually about a meter high and has opposite leaves. Whereas leaves of saltbush are flat and waxy, those of saltwort are fleshy and cylindrical.

Where you will see it: Santa Cruz (Tortuga Bay), Floreana (Punta Cormorán)

Common name: Scorpion weed

Scientific name: *Heliotropium curassavicum*

Resident status: Native

Scorpion weed is a dense, branched herb that grows in mats usually less than a meter wide. It grows along coastal lagoons and has long, gray leaves and spikes (resembling the curved tail of a scorpion) of white flowers.

Where you will find it: North Seymour, South Plaza, Rábida, Santiago (Sulivan Bay), Española (Punta Suárez), Genovesa

Common name: Salt sage

Scientific name: *Atriplex peruviana*

Resident status: Native

Salt sage is a small shrub (usually less than a meter tall) with pale, leathery, wrinkled leaves and clusters of small, brownish-yellow flowers at the tips of its branches. Salt sage often grows in mats.

Where you will see it: North Seymour, Española (Punta Suárez)

Common name: Galápagos clubleaf

Scientific name: *Nolana galapagensis*

Resident status: Endemic

Galápagos clubleaf is a low-growing shrub that you will see near the shore at several Visitor Sites. It has yellow-green, fleshy leaves and small white flowers.

Where you will see it: Santa Cruz (Tortuga Bay), San Cristóbal (Punta Pitt)

Common name: Galápagos carpetweed

Scientific name: *Sesuvium edmonstonei*

Resident status: Endemic

Galápagos carpetweed is a low-growing perennial herb having cylindrical, opposite, fleshy leaves and small, white, star-shaped flowers. Plants typically grow in sandy and rocky areas. During the rainy season, its leaves are green, and during the dry season they become a bright reddish-orange. Carpets of *S. edmonstonei* plants help produce the colorful landscapes of South Plaza (Color Image 6).

Where you will see it: South Plaza, Genovesa, Santiago (Sulivan Bay), Sombrero Chino, Española, North Seymour

Common name: Common carpetweed

Scientific name: *Sesuvium portulacastrum*

Resident status: Native

Common carpetweed is similar to Galápagos carpetweed, but its leaves are flatter and its flowers are pink. These flowers are often visited by the Galápagos carpenter bee and the Galápagos sulfur butterfly. *S. portulacastrum* is also referred to as *sea purslane*.

Where you will see it: Santa Cruz (Puerto Ayora, Tortuga Bay), Isabela (Puerto Moreno), North Seymour, Santa Fé, Fernandina

Common name: Galápagos shore petunia

Scientific name: *Exedeconus miersii*

Resident status: Endemic

Galápagos shore petunia is an annual herb that grows as dense mats in the rainy season. It has distinctive white, trumpet-shaped flowers.

Where you will see it: Genovesa (Darwin Bay), Fernandina, Española (Punta Suárez), Floreana (Punta Cormorán, Post Office Bay), Isabela (Elizabeth Bay, Tagus Cove)

Arid and Transition Zone

Cacti

Cacti, the most conspicuous plants in Galápagos, dominate the areas that are most easily visited by tourists (e.g., coastal zone, arid zone, young lava).

Common name: Lava cactus

Scientific name: *Brachycereus nesioticus*

Resident status: Endemic

Lava cactus (Color Image 43) is one of the first plants to colonize lava, but is the smallest cactus that grows in Galápagos. Lava cactus is a perennial plant that grows only on barren lava fields and produces succulent, cylindrical, yellow-tipped stems in clumps about 0.7 meters high and up to 2 meters wide. As the stems age, they darken to brown and gray. The stems are covered with soft spines (i.e., modified leaves), and the white flowers—which last only a few hours—are often visited by Galápagos carpenter bees. Fruits are red berries. *Brachycereus,* one of seven endemic genera of flowering plants in Galápagos (see below), disappears when plant succession diversifies islands. *Brachycereus,* which includes only one species, is derived from the Greek *brachys,* meaning *short,* and the Latin *cereus,* meaning *candle,* both of which refer to the distinctive shape of the plant's stems (Color Image 43).

Where you will see it: Fernandina (Punta Espinosa), Genovesa (Prince Philip's Steps), Isabela (Punta Moreno), Pinta, Santiago (Sulivan Bay), Bartolomé, Sombrero Chino

Common name: Candelabra cactus

Scientific name: *Jasminocereus thouarsii*

Resident status: Endemic

Candelabra cacti, which can be up to 7 meters tall, have fluted stems and reddish-orange flowers (2–6 cm wide) that open just before dawn and shrivel a few hours later. The fleshy, reddish-green fruits are edible. Candelabra cacti are prominent on cliffs above Academy Bay in Puerto Ayora (east of Angermeyer Point) and along the road to CDRS. *Jasminocereus thouarsii* is also referred to as *cacto esbelto* and *cirio.*

Where you will see it: The three varieties of candelabra cacti are *J. thouarsii* var. *thouarsii* (Floreana, San Cristóbal, and Devil's Crown), *J. thouarsii* var. *delicatus* (South Plaza, Santiago, and the cliffs along the southwest side of Academy Bay at Santa Cruz), and *J. thouarsii* var. *sclerocarpus* (Fernandina, Isabela).

Common name: Prickly pear cactus

Scientific name: *Opuntia* spp.

Resident status: Endemic

Opuntia (Figure 4.9, Color Image 44), the most abundant and widely spread cactus in Galápagos, is—like the more famous "Darwin's finches" (Chapters 2, 5)—an excellent example of adaptive radiation in the archipelago. The 6 species and 14 varieties of *Opuntia* vary on different islands; some are prostrate, but most are erect trees having thick trunks covered by flaky, reddish bark. On islands such as North Seymour, *Opuntia* is usually less than 0.5 meters high, whereas on others (e.g., Santa Fé and Santa Cruz) it can be more than 10 meters tall. Clusters of spines (i.e., modified leaves) often form beautiful patterns on the fleshy pads (i.e., stems) of *Opuntia*. *O. galapageia*, the only genus of cactus collected by Darwin while he was in Galápagos, was the first endemic plant species from the archipelago to be given a scientific name.

The tallest *Opuntia* usually live on islands such as Santa Fé, where there are (or were) tortoises. On islands such as Marchena and Rábida, where there have been no tortoises, *Opuntia* are usually short and have small, soft spines (Chapter 6). Stems, flowers, seeds, and the reddish-purple fruits of *Opuntia* are important foods for tortoises, land iguanas, doves, mockingbirds, and finches (especially cactus finches).[15] The plants' yellow flowers, which form throughout the year, open for just a few days, but are visited by a variety of insects. *Opuntia* are slow-growing and typically live 30 to 50 years before flowering, but then often live another 150 years. *Opuntia* has pads, but candelabra cactus (*Jasminocereus thouarsii;* see above) does not.

Where you will see it: In the arid and transition zones throughout the archipelago

O. echios var. *barringtonensis* (Figure 4.9) lives only on small, uninhabited Santa Fé, where you will also find land iguanas but no tortoises.[16] *Opuntia* here are up to 13 meters tall and have the thickest trunks in the islands. The towering *Opuntia* along the cliffs of Santa Fé form one of the most striking vistas in Galápagos (see Figure 4.9).

O. echios var. *echios* lives on Baltra, Daphne, South Plaza, and northern Santa Cruz. On most of these islands, this cactus is a low-growing shrub, but on South Plaza it grows as a tree up to 4 meters tall. Except for Santa Cruz, none of these islands has tortoises, but land iguanas live on South Plaza and Santa Cruz.

O. echios var. *gigantea* lives only near Puerto Ayora (Santa Cruz) and is one of the tallest species of *Opuntia*. Individuals, which often grow

[15] A statue of an iconic tortoise reaching for a pad of *Opuntia* is in Puerto Baquerizo Moreno where Avenida Quito intersects the south end of Malecón Charles Darwin.

[16] Tortoise bones (but no shells) have been found here; these bones were probably left by visitors camping here because of the island's good anchorage.

near candelabra cacti, can be up to 12 meters tall. Flowers and fruits of this variety of *Opuntia* are important food sources for finches and mockingbirds.

O. *echios* var. *inermis* lives only along Isabela's Volcán Sierra Negra and in lava fields near Puerto Villamil. It can be up to 5 meters tall. Isabela is home to tortoises and land iguanas.

O. *echios* var. *zacana* is a shrub that grows only on North Seymour. There are land iguanas here, but no tortoises. This variety of *Opuntia* often grows near dwarf palo santo.

O. *galapageia* var. *galapageia* lives on Pinta and Santiago, both of which have (or had) tortoises and land iguanas. Individuals on Pinta have a tapering trunk, atop which is a dense group of pads.

O. *galapageia* var. *macrocarpa* grows only on Pinzón, where there are tortoises but no land iguanas.

O. *galapageia* var. *profusa* is a soft-spined shrub that grows only on Rábida, which has never hosted tortoises or land iguanas. The epithet *profusa* refers to this variety's abundant production of flowers and fruit.

O. *helleri* is a low-growing, soft-spined shrub living on the northern islands of Galápagos (i.e., Wolf, Darwin, Marchena, and Genovesa), none of which has tortoises or land iguanas. O. *helleri* is the only *Opuntia* on Genovesa, where its flowers, fruits, and seeds are eaten by ground finches and doves. The lack of stiff spines on O. *helleri* in the northern islands may be an adaptation to encourage pollination by birds; Galápagos carpenter bees, a common pollinator of *Opuntia* on other islands, are absent on the northern islands of Galápagos. On Wolf Island, frigatebirds nest in clusters of O. *helleri.*

O. *insularis* is a treelike (up to 6 m tall) variety that grows on Fernandina and Isabela, both of which have (or had) tortoises and land iguanas. Trunks of this variety are covered by reddish-orange "bark." At Volcán Sierra Negra, O. *insularis* lives with O. *echios inermis,* and at Volcán Cerro Azul it lives with O. *saxicola.* If you are on Fernandina and you see *Opuntia,* you are seeing O. *insularis.*

O. *saxicola* grows only on Isabela's Volcán Cerro Azul, where there are tortoises and land iguanas.

O. *megasperma* var. *megasperma* grows only on Floreana and nearby Devil's Crown. Floreana was home to tortoises and land iguanas. As its *megasperma* epithet suggests, this species has the largest seeds of all the species of *Opuntia.*

O. *megasperma* var. *orientalis* is a thick-trunked variety of *Opuntia* that lives only on the eastern islands of Galápagos (i.e., Española and San Cristóbal; *orientalis* comes from the Latin word for *eastern*). Both of these islands have tortoises, but Española has no land iguanas. On Española, *O. megasperma* was endangered until the island's goats were eradicated in 1978.

O. megasperma var. *mesophytica* is a treelike variety of *Opuntia* that grows only at higher elevations of San Cristóbal and its nearby islands. San Cristóbal has tortoises and land iguanas.

How are endemic flowering plants of Galápagos pollinated?

After seeing a few of the endemic flowering plants of Galápagos, many visitors to Galápagos assume that these plants are wind pollinated because their flowers typify those of other wind-pollinated plants. Indeed, with few exceptions (e.g., *Opuntia,* Galápagos cotton), most endemic flowering plants of Galápagos have small, plain-looking white or yellow flowers that produce small amounts of pollen. Because there are relatively few pollinators in Galápagos, there has been little selection for the large, showy, pollen-laden flowers that often typify insect-pollinated plants. As a result, most endemic plants of Galápagos have few or no adaptations to attract pollinators. Native plants of Galápagos have larger, more attractive flowers that evolved on the mainland. By the time these plants arrived in Galápagos, there were enough insects present for pollination.

Wind pollination is relatively unimportant in Galápagos. For example, air currents in Galápagos carry relatively little pollen of endemic plants, and the flowers of most endemic plants of Galápagos (1) lack adaptations for wind pollination, such as elaborate stigmas and protruding stamens, (2) produce small amounts of pollen, and (3) produce flowers at times least suited for wind pollination (e.g., during garúa, when pollen would be washed out of the air). That is, the same environmental factors (rain and garúa) that stimulate flowering of many species of endemic plants in Galápagos also prevent pollination by wind. Moreover, except for Galápagos carpenter bees (the most common pollinator in Galápagos), there are few insects to help with pollination.[17] If neither insects nor wind is responsible for pollination, how does pollination occur in Galápagos' endemic plants?

Flowering plants arrived at Galápagos before there were insects with which to establish mutualisms to ensure pollination. These colonizers were *self-pollinating,* meaning that pollen rubs against the flower's stigma as the flower develops, or falls against the stigma when the flower opens. Sperm from the pollen then fertilizes the plant's eggs (i.e., the plant is self-compatible). These traits are selected for when plants colonize oceanic islands. This strategy would have allowed a single plant to establish a sexually reproducing population.

[17] Galápagos carpenter bees are abundant and visit several different plants each day. However, few other insects visit flowers in Galápagos. Other pollinators in Galápagos, especially on islands lacking Galápagos carpenter bees, include Galápagos sulfur butterfly (*Phoebis sennae),* Galápagos blue butterfly (*Leptotes parrhasioides*), and hawk moths (e.g., *Eumorpha labruscae yupanquii;* Chapter 9).

Trees

Common name: Palo santo

Scientific name: *Bursera graveolens*

Resident status: Native

Palo santo, the most abundant and easily identifiable tree in the arid and transitional zones, lives in the lowlands of most islands in Galápagos. Trees produce leaves in the wet season near Christmas, and their wood is used as incense,[18] thereby accounting for its name (*palo santo* means *holy stick*). Unless you are in Galápagos near Christmas, these trees will be leafless. Individuals are seldom more than 12 meters high, and their brownish-red bark is usually covered by gray, crustose lichens (especially *Ramalina usnea*). Leaves are alternate, and the plants' white flowers produce small berries that are eaten by land iguanas, endemic rats, and finches.

Where you will see it: Floreana (Punta Cormorán, Post Office Bay), Santa Cruz, Isabela (Tagus Cove, Puerto Villamil, Punta Moreno, Volcán Alcedo, Punta Garcia, Urbina Bay, Elizabeth Bay), Genovesa (Darwin Bay, Prince Philip's Steps), Rábida, Baltra, Santa Fé, Santiago (Puerto Egas, Sulivan Bay, Espumilla Beach), San Cristóbal (Frigatebird Hill). You will find forests of palo santo around Whale Bay Hill (Santiago) and Mirador de la Baronesa (Floreana; Color Image 45).

Common name: Dwarf palo santo

Scientific name: *Bursera malacophylla*

Resident status: Endemic

Dwarf palo santo, also known as Galápagos palo santo, is more rare and shorter than, but otherwise similar to, *B. graveolens*. This species evolved from palo santo on North Seymour and Baltra, and then spread back to Santa Cruz.

Where you will see it: North Seymour, Baltra, Daphne, and Santa Cruz

Common name: Matazarno

Scientific name: *Piscidia carthagenensis*

Resident status: Native

Matazarno is distinguished by its large, alternate leaves, pinkish-purple flowers, and winged seeds. These slow-growing trees are the tallest plants in the arid and transition zones, and their hard, resistant wood is used to make buildings, docks, piers, and the black-and-white stakes that mark GNP trails (see Figure 3.4). Unlike palo santo, matazarno has leaves during the dry season, making the plants easily identifiable. On all islands where it grows, most of the

[18] Palo santo, which is also referred to as "Galápagos Incense Tree," is related to the plants that produce myrrh (*Commiphora myrrha*) and frankincense (*Boswellia* spp.). *Graveolens,* the specific epithet, means *strong smelling,* and refers to the trees' aromatic resin.

largest trees have been harvested, and residents have turned to the endemic guayabillo (*Psidium galapageium*) for wood.

Where you will see it: San Cristóbal, Santa Cruz (on the trail to Tortuga Bay, and between Bellavista and Puerto Ayora)

Common name: Pega-pega

Scientific name: *Pisonia floribunda*

Resident status: Endemic

Pega-pega, which means *sticky-sticky,* is a tree that produces sticky fruits that are often dispersed by birds. Individual plants, which are usually cloaked with mosses and lichens, have opposite, oblong leaves, inconspicuous green flowers, and are 10 to 15 meters tall. Pega-pega is also referred to as *Galápagos pisonia.*

Where you will see it: Santa Cruz (between Bellavista and Puerto Ayora), Fernandina, Floreana, Isabela, Santiago

Common name: Guayabillo

Scientific name: *Psidium galapageium*

Resident status: Endemic

These shrubs and small trees are up to 7 meters tall and are distinguished by their opposite, leathery, elliptical leaves; large white flowers (up to 12 cm in diameter); and smooth, pinkish-gray bark, which is often decorated by lichens and mosses. Guayabillo, which is also known as *Galápagos guava,* produces small guava-like fruit, and its wood is used for fences and houses. A relative of guayabillo, guayaba (*Psidium guajava*), is an invasive species that covers large areas of the archipelago (Chapter 11).

Where you will see it: Isabela (Volcán Sierra Negra), Fernandina, Santiago, Santa Cruz (between Puerto Ayora and Itabaca Channel)

Common name: Poison apple

Scientific name: *Hippomane mancinella*

Resident status: Native

These large evergreen trees, which are also referred to as *manzanillo,* are distinguished by their small, green, poisonous, apple-like fruits (about 4 cm in diameter) and milky sap.[19] Trees are up to 10 meters tall and have small, elliptical, shiny, alternate leaves and spikes of pale green flowers at the tips of branches. As is true for most members of the spurge family, the sap of poison apple can be extremely irritating. Do not touch this plant.

Where you will see it: Santiago (Espumilla Beach), Santa Cruz (Puerto Ayora, Academy Bay), Floreana

[19] If you are exposed to the alkaline sap, rub the area with lemon juice or orange juice.

Common name: Muyuyo

Scientific name: *Cordia lutea*

Resident status: Native

This shrub (Color Image 46), which is also known as *yellow cordia,* grows up to 7 meters tall. It produces light gray bark; dark green, hairy, round, alternate leaves; trumpet-shaped, lilac-scented yellow flowers; and whitish berries containing sticky seeds that are favorites of mockingbirds. Flowers of muyuyo are frequently visited by the Galápagos sulfur butterfly and Galápagos carpenter bee. If you see a flower of muyuyo visited by a carpenter bee, you may witness "nectar robbing," in which the bee opens the base of the flower and gets the nectar without ever entering the flower. This is referred to as "robbing" because the bee is rewarded, but the plant's pollen is not dispersed to another flower. Wood of muyuyo is often used to make the carvings sold at souvenir shops in Galápagos.

Where you will see it: Floreana (Puerto Velasco Ibarra, Post Office Bay), Santa Cruz (Puerto Ayora), Española (Punta Suárez), Rábida, Baltra, Santa Fé, Isabela (Tagus Cove), North Seymour

Common name: Parkinsonia

Scientific name: *Parkinsonia aculeata*

Resident status: Native

This tree (Color Image 47) gets up to 10 meters tall, has thin green bark on its young stems (and darker bark on its older stems), and produces branches having hooked spines, yellow flowers, pods of seeds up to 20 centimeters long, drooping branches, and long, spikelike alternate leaves each containing 20 to 25 pairs of leaflets. The sap of *Parkinsonia* is consumed by scale insects, which are nurtured by ants who eat their honeydew. *Parkinsonia,* which is often pollinated by the Galápagos carpenter bee, is also referred to as *Jerusalem thorn* and *palo verde.* Finches and land iguanas eat the flowers and leaflets of *Parkinsonia.*

Where you will see it: Santa Cruz (Puerto Ayora), Floreana (Punta Cormorán, Post Office Bay), Baltra, North Seymour, Isabela (Puerto Villamil), Española (Gardner Bay), San Cristóbal (Puerto Baquerizo Moreno)

Common name: Galápagos acacia

Scientific name: *Acacia rorudiana*

Resident status: Endemic

These spreading trees (up to 8 m tall) have compound alternate leaves, each with 5 to 25 pairs of delicate leaflets (Figure 10.3). The ball-like flowers are yellowish-orange and produce long seed pods. *A. rorudiana* is also referred to as *algarrobo;* related species include *A. insulae-iacobi* and *A. nilotica* (Nile acacia, a cultivated plant common on Santa Cruz). Many visitors to Galápagos confuse *A. rorudiana* with the invasive lead tree (*Leucaena leucocephala*), both of which

FIGURE 10.3 Galápagos acacia (*Acacia* sp.) growing on San Cristóbal. Branches of *Acacia* have long (up to 5 cm) thorns, thus explaining why residents often refer to it as "thorn tree."

have similar leaves and fruit. However, unlike *Acacia,* lead tree has white flowers and lacks thorns.

Where you will see it: Española, Isabela, Santiago (Espumilla Beach), Floreana (Post Office Bay), Santa Cruz (Puerto Ayora)

Common name: Mesquite

Scientific name: *Prosopis juliflora*

Resident status: Native

These large trees resemble *Acacia,* but have shorter thorns at the bases of their leaves. They also have spikes of yellow flowers (rather than ball-like flowers in *Acacia*).

Where you will see it: Española (Gardner Bay, Punta Suárez), Isabela (Elizabeth Bay), Floreana (Post Office Bay, Punta Cormorán), Santa Cruz (Tortuga Bay)

Shrubs

Common name: Galápagos croton

Scientific name: *Croton scouleri*

Resident status: Endemic

This abundant shrub (up to 5 m tall), which is also called *chala,* has greenish-gray leaves, gray bark, small yellowish-white flowers at the tips of its branches, and spherical fruit. The sap of Galápagos croton will leave a brown stain on your clothes. Several species of finches eat the seeds of *Croton scouleri.*

Where to see it: Santa Fé, Santa Cruz, Genovesa, Fernandina, Floreana, Española, Isabela (Tagus Cove), Santiago (Puerto Egas)

Common name: Darwin's cotton

Scientific name: *Gossypium darwinii*

Resident status: Endemic

This shrub (Color Image 48), which is 1–3 meters tall, has purple stems and alternate leaves covered by hairs and black spots. Petals of the large, yellow flowers—the largest and showiest of all endemic and native plants in Galápagos—have a purple center, and the lint (i.e., cotton) is used by some finches to build nests. Darwin's cotton, which is not grown as a commercial crop in Galápagos, grows in relatively open rocky areas; its seeds are eaten by various birds. Galápagos cotton is also known as *algodoncillo.*

Where you will see it: Española, Fernandina, Floreana (on the trail to Black Beach), Isabela, Rábida, Santa Cruz (e.g., at Cerro Dragón and on the way to the lava tunnel at Post Office Bay), San Cristóbal, Santa Fé, North Seymour

Common name: Lance-leafed cordia

Scientific name: *Cordia leucophylctis*

Resident status: Endemic

This shrub has lancelike alternate leaves, clusters of small white flowers, and red oblong fruit. Like palo santo, lance-leafed cordia is usually leafless in the dry season. Individual plants are up to 3 meters tall. Other species include *C. scouleri* and *C. anderssonii.*

Where you will see it: Santa Cruz (Puerto Ayora), Fernandina, Santa Fé, Isabela (Puerto Villamil)

Common name: Galápagos lantana

Scientific name: *Lantana peduncularis*

Resident status: Endemic

Galápagos lantana, which often forms dense thickets on lava and cinder fields, has thin, pointed leaves and clusters of small, yellow-centered white flowers. Individual shrubs are up to 2 meters tall. Leaves of Galápagos lantana are eaten by a variety of caterpillars, which are in turn eaten by finches. The invasive *L. camara,* a relative of *L. peduncularis* that was brought to Floreana in 1938 as an ornamental plant, is outcompeting local flora on Floreana and elsewhere.

Where you will see it: Santa Cruz (Puerto Ayora), Santa Fé, Española (Punta Suárez), Isabela (Tagus Cove), Floreana (Post Office Bay, Punta Cormorán)

Common name: Tournefortia

Scientific name: *Tournefortia* spp.

Resident status: Endemic

Tournefortia pubescens, which is less than 4 meters tall, has elliptical alternate leaves, white flowers, and white, berry-like fruit. Leaves and young branches are covered with white hairs. A related endemic species, *T. rufo-sericea* (red-haired tournefortia), has brownish-red bark, alternate leaves, white flowers, and white, berry-like fruit. Its leaves and young branches are covered with red hairs. Flowers are usually pollinated at night, and its fruit is a favorite food of Galápagos mockingbirds.

Where you will see it: Santa Cruz (road from Puerto Ayora to Itabaca Channel)

Common name: Spiny bush

Scientific name: *Scutia spicata*

Resident status: Endemic

This dark green bush grows up to 3 meters high and produces alternate leaves, small green flowers, and red berries that are eaten by land iguanas, finches, and hungry children. Its brown spines are up to 5 centimeters long. Spiny bush is also known as *thorn shrub* and *espino.*

Where you will see it: At most Visitor Sites throughout Galápagos

Common name: Bitterbush

Scientific name: *Castela galapageia*

Resident status: Endemic

This erect, spiny shrub produces small, dark green, waxy, oval leaves; small yellow or red flowers; and red fruits. Individuals get up to 2 meters tall. Young leaves are covered with white hairs. Bitterbush, which is also referred to as *amargo,* branches profusely and usually grows on rocky lava.

Where you will see it: Baltra, Floreana (Punta Cormorán), Santa Cruz (Puerto Ayora), Santa Fé

Common name: Desert thorn

Scientific name: *Lycium minimum*

Resident status: Native

This dense, spiny shrub grows up to 3 meters high and produces spines at the tips of each of its branches (but nowhere else). It has club-shaped leaves, small white flowers, and small red fruits.

Where you will see it: Española (Punta Suárez), Baltra, Santa Fé

Common name: Pearl berry

Scientific name: *Vallesia glabra*

Resident status: Endemic

This shrub (up to 6 m tall) has white flowers, white berry-like fruit, and alternate, pointed, long leaves that hang from branches. Pearl berry is also referred to as *peralillo* and *sauquillo.*

Where you will see it: Floreana (Post Office Bay), Isabela (Puerto Villamil, Urbina Bay, and all six volcanoes), Santiago (Espumilla Beach), Santa Cruz (by the road from Puerto Ayora to Itabaca Channel), San Cristóbal (Puerto Baquerizo Moreno)

Common name: Thread-leafed chaff flower

Scientific name: *Alternanthera* spp.

Resident status: Endemic

These shrubs, which are usually about 1.5 meters tall, have many upright branches, opposite leaves, and white or greenish flowers at the tips of their branches. Six of the 14 species of *Alternanthera* in Galápagos are endemic. A native species, *A. echinocephala,* is easily identified by its spiny, large flower-heads.

Where you will see it: *A. galapagensis* is found on satellite islets of Floreana, but not on Floreana itself

Common name: Thin-leafed Darwin's shrub

Scientific name: *Darwiniothamnus tenuifolius*

Resident status: Endemic

This dense, low-growing shrub has pointed, alternate leaves and fragrant, yellowish-white, daisylike flowers. When crushed, its leaves release a sweet odor. Related species include lance-leafed Darwin's shrub (*D. lancifolius*) and the endangered alternate-leafed Darwin's shrub (*D. alternifolius*), whose leaves are not clumped at the tips of branches.

Where you will see it: Isabela (Volcán Sierra Negra), Fernandina

Endemic Families of Plants in Galápagos

Although 19 genera of vascular plants in Galápagos have undergone adaptive radiation, three families have been isolated long enough to have evolved endemic genera:

Genus	Family	Number of Taxa
Darwiniothamnus	Asteraceae	4
Lecocarpus	Asteraceae	3
Macraea	Asteraceae	1
Scalesia	Asteraceae	20
Brachycereus	Cactaceae	1
Jasminocereus	Cactaceae	3
Sicyocaulis	Cucurbitaceae	1

In contrast, Hawaii, another archipelago of volcanic islands, hosts 32 endemic genera of plants, presumably because Hawaii is older and more isolated than Galápagos. While in Galápagos, Darwin collected representatives of four of these genera (*Macraea, Darwiniothamnus, Scalesia,* and *Lecocarpus*) and described a fifth (*Brachycereus*).

Common name: Cutleaf daisy

Scientific name: *Lecocarpus pinnatifidus*

Resident status: Endemic

Cutleaf daisies are small (usually about 1 m tall), evergreen bushes that grow as pioneers on cinder and bare lava. They have a single stem supporting a head of lobed leaves and large, yellow daisylike flowers. Cutleaf daisy, which is also called *Floreana daisy* and *wing-fruited lecocarpus,* is threatened with extinction. *L. darwinii* (Darwin's daisy) is an endemic plant similar to cutleaf daisy, but it lives only on San Cristóbal. *L. lecocarpoides* (FitzRoy's daisy) also resembles cutleaf daisy, but it lives only on Española.

Where you will see it: Floreana (Punta Cormorán)

Common name: Needle-leafed daisy

Scientific name: *Macraea laricifolia*

Resident status: Endemic

This shrub, a member of the sunflower family, grows up to 2.5 meters high and has short, narrow leaves and yellow composite flowers.

Where you will see it: Floreana, Isabela, Rábida, San Cristóbal, Santa Cruz, Santiago

Herbs and vines

Common name: Galápagos passion flower

Scientific name: *Passiflora foetida*

Resident status: Endemic

This vine (Color Image 49) produces hairy stems and large, ivy-shaped leaves that are up to 5 centimeters long. Its flowers are white with a purple center, and the vine—thanks to tendrils that wrap around objects that the plant touches—grows over rocks and other shrubs. The greenish-orange fruits are 1.5 centimeters long and covered in a protective sheath. *P. foetida* resembles *P. colinvauxii* (Indefatigable passion flower; see below), but leaves of *P. colinvauxii* are crescent-shaped and its flowers are less colorful.

Where you will see it: Isabela, Floreana, Santa Cruz (Puerto Ayora)

Common name: Grey matplant

Scientific name: *Tiquilia* spp.

Resident status: Endemic

These small herbs form regularly spaced mats having small gray-green leaves and small white flowers, which are eaten by lava lizards. Although *T. nesiotica* is the most common species of grey matplant (e.g., at Bartolomé; Figure 10.4), other species include *T. darwinii* and *T. gusca* (both of which grow as a mat) and the rare *T. galapagoa* (which grows as a shrub up to 40 cm high) on Santiago and Bartolomé. On the north side of Bartolomé, you will see many regularly spaced *Tiquilia.*

Where you will see it: Santiago (Sulivan Bay), Bartolomé, Isabela, San Cristóbal, Floreana, Rábida, Santa Fé, Plazas, Baltra

FIGURE 10.4 Grey matplant (*Tiquilia nesiotica*) growing on Bartolomé.

Common name: Spurge

Scientific name: *Chamaesyce* spp.

Resident status: Endemic

This genus, like *Scalesia* and *Opuntia,* includes several species that have radiated into a variety of niches in Galápagos. All of the 10 species produce small white flowers and white sap, but they range from annual herbs to small shrubs that are up to 2 meters tall. Flowers and fruits of *Chamaesyce* are eaten by finches.

Where you will see it: *C. amplexicaulis* is a small shrub that grows on Bartolomé and Genovesa; *C. punctulata* grows on Santiago and Isabela; *C. viminea* grows on Santa Cruz, Isabela, Floreana, and Genovesa.

Common name: Galápagos milkwort

Scientific name: *Polygala* spp.

Resident status: Endemic

The three species (and several subspecies) of Galápagos milkwort are small, erect herbs having orange stems, elliptical or rounded leaves, and clusters of yellow-centered flowers at the tips of their yellowish stems. *P. anderssonii* and *P. galapageia* are perennial bushes that get about 1 meter tall, whereas *P. sancti-georgii* (St. George's milkwort) is an annual herb that is about 0.5 meters tall. *Polygala* is derived from the Greek *poly,* meaning *much,* and *gala,* meaning *milk,*

FIGURE 10.5 Mollugo (*Mollugo crockeri*) growing on lava on Santiago Island.

denoting the belief that cows who eat this plant will produce more milk. It has also been used to increase the flow of a human nursing mother's milk.

Where you will see it: Santiago, Santa Cruz, Rábida, San Cristóbal, Floreana, Isabela

Common name: Mollugo

Scientific name: *Mollugo* spp.

Resident status: Endemic

These annual, low-growing plants are some of the first plants to colonize lava (Figure 10.5). One of the most common of the five species (and seven subspecies) of mollugo is *M. flavescens,* which has red stems, elongate leaves, and white, green-veined flowers each having five petals.

Where you will see it: Genovesa, Fernandina, Isabela (Punta Moreno), San Cristóbal (Frigatebird Hill), Santa Cruz, Bartolomé, Santiago

Common name: Galápagos sarcostemma

Scientific name: *Sarcostemma angustissimum*

Resident status: Endemic

These slender, perennial vines have opposite leaves, purple and yellow flowers, and flask-shaped fruits covered with white hairs. Galápagos sarcostemma, which climbs over rocks and other plants, is common in lava fields near Puerto Villamil.

Where you will see it: Baltra, Fernandina, Española, Floreana, Rábida, San Cristóbal, Santa Cruz, Isabela

Common name: Galápagos tomato

Scientific name: *Solanum cheesmaniae* and *S. galapagense*

Resident status: Endemic

Of the 13 species of tomatoes worldwide, only 4 are edible, and 2 of these edible species are endemic to Galápagos: *Solanum cheesmaniae* and *S. galapagense*.[20] These highly branched herbs have alternate pear-shaped leaves, hairy stems, yellow flowers, and small, edible fruits. Rates of seed germination increase significantly after seeds pass through the digestive tracts of animals such as land iguanas, tortoises, and mockingbirds. Galápagos tomato is self-pollinated, thus explaining why clusters of plants in the wild are remarkably homogenous. Galápagos tomato has tasty, edible fruit, but it is not grown commercially in Galápagos. Both endemic species of Galápagos tomato resist many pests and are salt-tolerant, and they are being cross-bred to confer desirable traits to commercial varieties of tomato. Galápagos tomato is being outcompeted by invasive species of tomato (*S. lycopersicum and S. pimpinellifolium*), which grow in dense clusters at El Chato, Sombrero Chino, and elsewhere in the archipelago. Unlike the endemic species, which have yellow or bright orange fruit, both of the invasive species have red fruit.

Where you will see it: Isabela, Santa Cruz, San Cristóbal. Although tomatoes grow on 19 islands, only on Isabela and Fernandina do the two endemic species co-occur.

Common name: Goat's head

Scientific name: *Tribulus cistoides*

Resident status: Native

This low-growing herb produces silver-gray, compound leaves; yellow flowers; and spiked fruit. These piercing spikes account for another of the species' common names: *puncture weed*. (If you walk around Galápagos without shoes, you will get a painful introduction to this plant.) Another species (*T. terrestris*) is distinguished by its larger seeds. Like many native and endemic flowering plants of Galápagos, *Tribulus* is self-pollinated; even plants growing side-by-side seldom cross-pollinate. Goat's head is also referred to as *caltrop*.

Where you will see it: Santa Cruz, Daphne, Baltra, Bartolomé, South Plaza, Santiago (Puerto Egas) Sombrero Chino, Española (Gardner Bay, Punta Suárez)

Common name: Stickleaf

Scientific name: *Mentzelia aspera*

Resident status: Native

This annual herb produces alternate leaves, yellow flowers, and barbed seed-capsules that stick to virtually anything they touch. Stickleaf is widespread; after

[20] The 13 species of tomatoes include the cultivated tomato (*Solanum lycopersicum*) and 12 wild relatives (Darwin, 2009; Peralta et al., 2008, and references therein). *Solanum cheesmaniae* was formerly called *Lycopersicon cheesmaniae; S. galapagense* was also formerly called *L. cheesmaniae,* but it is now categorized as a separate species. *S. lypersicum* was formerly called *L. esculentum,* and *S. pimpinellifolium* was formerly called *L. pimpinellifolium.*

a day of hiking, you may have fruits of this plant attached to your socks, shoes, or pants.

Where you will see it: Española, Baltra, Floreana, Isabela, Genovesa, Rábida, San Cristóbal, Santa Fé, Santa Cruz, Santiago

Common name: Galápagos purslane

Scientific name: *Portulaca howellii*

Resident status: Endemic

This succulent-leaved perennial—a relative of Galápagos native *Portulaca oleraceae*—grows only stems during the dry season, but it quickly produces leaves and large, nocturnal, yellow flowers (up to 5 cm across) after rains. Grazing by goats has restricted Galápagos purslane to offshore islets. A related species (*P. oleraceae*, or common purslane) is native to Galápagos; it has smaller flowers, thinner stems, and persistent leaves. Leaves of *P. howellii* are short-lived, thus making Galápagos purslane leafless most of the year.

Where you will see it: Daphne, Santa Fé, South Plaza, Santa Cruz (Puerto Ayora), Genovesa, North Seymour, Sombrero Chino

Common name: Arrow-leafed morning glory

Scientific name: *Ipomoea linearifolia*

Resident status: Endemic

This perennial vine has alternate, arrow-shaped leaves, large white or pink flowers, and brown seeds. Leaves are thin, long, and straight (i.e., linear). Vines typically cover lava and other plants, and blooms can persist for several weeks after rains.

Where you will see it: Genovesa, Fernandina, Isabela, Santa Cruz, Rábida, Seymour

Common name: Oily pectis

Scientific name: *Pectis tenuifolia*

Resident status: Endemic

This small, perennial herb produces yellow flowers and opposite, purplish, needle-like leaves that have distinctive oil-glands on their undersides. These plants often grow on bare rocks, with their roots extending into crevices. Less-common species of pectis in Galápagos include *P. linifolia* and *P. subsquarrosa* (both native). All members of this genus release a distinctive odor when their leaves are crushed.

Where you will see it: Fernandina, Floreana (Post Office Bay), Genovesa, Isabela (Elizabeth Bay), San Cristóbal, Santa Cruz, Santa Fé, Bartolomé

Common name: Galápagos bean

Scientific name: *Phaseolus mollis*

Resident status: Endemic

This herbaceous, spreading vine, which is usually about 4 meters long, has hairy stems, alternate leaves, pinkish-purple flowers, and short pods that hold two to four seeds. A cultivated species, *P. vulgaris,* grows on Santa Cruz and San Cristóbal and is referred to as green bean, kidney bean, and snap bean. *P. adenanthus,* which has white flowers, is a native plant found only on Española.

Where you will see it: Isabela, Fernandina, Santa Cruz

Common name: Scalesia

Scientific name: *Scalesia* spp.

Resident status: Endemic

The adaptive radiation of *Scalesia* in Galápagos has earned this genus the name, "The Darwin's Finches of the Plant World." Indeed, the 15 species (and several subspecies) of *Scalesia*—which range from shrubs less than a meter tall to trees more than 13 meters tall—have adapted to several different ecological zones on several different islands, ranging from dry lowlands to humid upland forests. The largest species, *S. pedunculata* (Figure 10.6), defines and dominates the *Scalesia* zone. On Santa Cruz, much of the popular Lava Java coffee is harvested from plants that grow in the forest understories of *S. pedunculata.*

Scalesia is a member of the sunflower family (Asteraceae), thus accounting for its common name of *sunflower tree. Scalesia* forests host the bright red male (and pale yellow female) vermilion flycatchers. In well-established forests of *Scalesia,* most of the trees have similar ages, and environmental changes such as an El Niño (Chapter 3) often produce synchronous deaths, after which new generations develop only after the older generation dies. On Santa Cruz (e.g., at El Chato), these new generations of *Scalesia* are being outcompeted by invasive guayaba (*Psidium guajava*) and Spanish cedar (*Cedrela odorata*),[21] resulting in *Cedrela* forests with understories of guayaba trees.

Leaves of *Scalesia* cluster near the tips of branches, often with the dead leaves of the previous season just behind the tips (Figure 10.6). Its wood is soft and pithy, and its small white flowers are aggregated into daisylike flower heads (i.e., inflorescences) at the tips of branches. *Scalesia,* which is also called *lechoso,* has a short life-cycle; it grows fast, produces flowers within a few years, and dies after a few decades. Most shrublike species of *Scalesia* grow in the arid and transition zones, while treelike species (e.g., *S. pedunculata* and *S. cordata*) are the dominant species of dense forests in the highlands.

A single ancestor that reached Galápagos long ago gave rise to the different species of *Scalesia.* On some islands, *Scalesia*—which first colonized the lowlands, and then radiated to the wetter regions—is threatened by grazing by feral goats. Agriculture has destroyed the entire *Scalesia* forest on San Cristóbal.

[21] Despite its common name, Spanish cedar is not a cedar; rather it is an introduced flowering plant. There are no gymnosperms in Galápagos.

FIGURE 10.6 The endangered *Scalesia pedunculata* growing at Los Gemelos on Santa Cruz.

Some species of *Scalesia* are endangered; for example, *S. retroflexa* is rare and known only from southeastern Santa Cruz, and *S. atractyloides* was presumed extinct until it was rediscovered in 1990. Some species of *Scalesia* are widely distributed, whereas others have limited distributions; for example, *S. baurii* lives on Pinta, Wolf, and Pinzón, islands which have no Terrestrial Visitor Sites. Here are the species of *Scalesia* you are most likely to see in Galápagos:

S. affinis (radiate-headed *Scalesia*) is a treelike shrub (usually less than 3 m tall) that often colonizes bare lava. It is characterized by alternate, serrated leaves, and white, daisylike flowers at the tips of its branches. Subspecies of *S. affinis,* which grow from sea level to about 500 meters, live on Santa Cruz (*brachyloba* subspecies), Floreana (*affinis* subspecies), and Isabela and Fernandina (*gummifera* subspecies). *S. affinis,* along with *S. helleri* and *S. villosa* (see below), are the most common species of *Scalesia* in Galápagos.

Where you will see it: Isabela (Puerto Villamil, Punta Moreno, Tagus Cove, Sierra Negra, Punta Albemarle), Santa Cruz

S. microcephala (small-headed *Scalesia*) grows as a small shrub or tree that is up to 4 meters tall. It has alternate leaves and white flowers.

Where you will see it: Wolf, Isabela, Fernandina

S. divisia grows as a shrub or small tree having a maximum height of about 4 meters. It is characterized by opposite leaves and flower heads having from 30 to 110 white flowers.

Where you will see it: *S. divisia* grows only on San Cristóbal; you are most likely to see it at Sappho Cove.

S. helleri (Heller's *Scalesia*) is a shrub (usually less than 3 m tall) that usually grows near shorelines and along cliffs, possibly because of grazing pressures by feral goats. This species has oval, divided, opposite or alternate leaves, and flower heads consisting of up to 100 flowers. Leaves release a pleasant odor when crushed. The *helleri* subspecies lives on Santa Fé and Santa Cruz, and the *santcruziana* subspecies lives only on Santa Cruz. Hairs on stems and leaves of *S. helleri* gather water from garúa.

Where you will see it: Santa Cruz (Tortuga Bay), Santa Fé

S. pedunculata (tree *Scalesia*), the largest of the sunflower trees, is often more than 12 meters tall and is the dominant species of the *Scalesia* zone on Floreana, Santiago, Santa Cruz, and San Cristóbal. These unique trees have a moss-covered, unbranched trunk topped by an umbrella-like cluster of branches; alternate, hairy, and elliptical leaves; and white flowers. The largest remaining stand of *S. pedunculata* is at Los Gemelos on Santa Cruz (altitude = 400 to 500 m), but this is only about 20% of the original population.[22]

Where you will see it: Santa Cruz (Los Gemelos, on the road between Puerto Ayora and Itabaca Channel), Floreana (highlands), San Cristóbal, Santiago

---

[22] Several populations of *Scalesia* have fared poorly. For example, on Santiago, goats destroyed more than 99% of the original population; the remaining individuals, which were protected by several fenced-in areas, cover less than 0.02 square kilometers. On Volcán Sierra Negra, more than 99% of the population of *S. cordata* is gone; fires there between 1985 and 1995 resulted in *S. cordata* being replaced by invasive *Psidium guajava,* whose seeds germinate quickly after fires.

S. villosa (longhaired *Scalesia*) is a shrub that grows up to 3 meters tall and has hairy, narrow, alternate leaves at the tips of its stems, and white flower heads comprised of up to 300 flowers. This species is easily identified by its abundant hairs, which gather water from garúa.

Where you will see it: Only on Floreana (Punta Cormorán) and nearby islets

S. crockeri (Crocker's *Scalesia*) is a small (less than 1 m tall) shrub with opposite, elliptical leaves at the tips of its branches, and white flower heads made of 20 to 50 flowers. *S. crockeri* is the only *Scalesia* on Baltra.

Where you will see it: Baltra, Santa Cruz

S. stewartii (Stewart's *Scalesia*) is a 3-meters-tall shrub with alternate, lancelike leaves whose undersides are coated with white hairs. Flower heads consist of 30 to 80 flowers.

Where you will see it: Santiago (Sulivan Bay), Bartolomé

S. cordata (heart-leafed *Scalesia*) is a tall (up to 10 m high) tree with alternate, heart-shaped leaves. Flower heads include 10 to 30 flowers.

Where you will see it: Isabela (Sierra Negra)

S. incisa (cut-leafed *Scalesia*) is a small shrub (less than a meter high) with opposite and divided leaves, and flower heads made of 20 to 50 flowers. This is one of the rarest *Scalesia* species; it lives only in northern San Cristóbal.

Where you will see it: San Cristóbal (Punta Pitt)

Humid Zone

Trees and shrubs

Common name: Soapberry

Scientific name: *Sapindus saponaria*

Resident status: Native

Soapberry is a tall tree (sometimes exceeding 20 m) having alternate leaves, small (less than 5 mm wide) white flowers at the tips of its branches, and fruits up to 4 centimeters in diameter. The scientific and common names of this plant result from the use of the plants' pulp as soap. Soapberry is also called *jaboncillo*.

Where you will see it: Santa Cruz, Isabela, San Cristóbal

Common name: Miconia

Scientific name: *Miconia robinsoniana*

Resident status: Endemic

This shrub (Color Image 50), which can get up to 6 meters tall, forms dense thickets above the *Scalesia* zone on the southern slopes of Santa Cruz and San

Cristóbal. These thickets often house nests of petrels during the garúa season. Its long leaves are opposite, pointed, and red, and its pink or red flowers form at the tips of branches. Fruits are dark blue berries. *Miconia,* which is also called *cacaotillo* (its leaves resemble those of cacao), is threatened by invasive blackberries, guava, and quinine.

Where you will see it: Santa Cruz, San Cristóbal (Laguna El Junco)

Common name: Cat's claw

Scientific name: *Zanthoxylum fagara*

Resident status: Native

This shrub or small tree gets its common name from the sharp, hooked spines that adorn its branches. The genus is named for the Greek words *xanthos,* meaning *yellow,* and *xylon,* meaning *wood,* referring to the plants' yellowish wood. Cat's claw (*uña de gato*) produces compound alternate leaves, inconspicuous pale green flowers, and small dark berries. It dominates the brown zone (where it forms impenetrable thickets) and is often covered by epiphytes such as lichens, liverworts, and mosses.

Where you will see it: Santa Cruz (by the road between Puerta Ayora and Itabaca Channel), Floreana, Isabela

Common name: Galápagos mistletoe

Scientific name: *Phoradendron henslowii*

Resident status: Endemic

This shrub parasitizes a variety of woody plants—especially cat's claw (*Zanthoxylum fagara*), Galápagos croton (*Croton scouleri*), and needle-leaf daisy (*Macraea laricifolia*)—throughout Galápagos. Galápagos mistletoe has elliptical, green, leathery leaves; small green flowers; and translucent, berrylike fruits. Unlike other members of this genus, fruits of *P. henslowii* are not sticky.

Where you will see it: Fernandina, Floreana, Isabela, Pinzón, San Cristóbal, Santiago, Santa Cruz

Common name: Galápagos pilea

Scientific name: *Pilea baurii*

Resident status: Endemic

This herb has red stems, opposite oval-shaped leaves with serrated edges, green flowers, and small seeds (0.3 mm). Most individuals are less than 70 centimeters tall and grow in moist, shaded areas. *P. baurii* is also referred to as *ortiga*. Tortoises at Santa Cruz's El Chato eat these plants. *P. microphylla,* an introduced species known only from Santa Cruz, shoots its pollen from its tiny flowers.

Where you will see it: Isabela, Floreana, Fernandina, San Cristóbal, Santa Cruz, Santiago

Herbs and vines

Common name: Galápagos peperomia

Scientific name: *Peperomia galapagensis*

Resident status: Endemic

This perennial herb has bunched stems; whorled, fleshy leaves with red edges; small, green flowers; and stems covered with dense, small hairs.

Where you will see it: Santa Cruz (Los Gemelos), Floreana, Isabela, Pinta, San Cristóbal, Santiago

Common name: Galápagos orchid

Scientific name: *Epidendrum spicatum*

Resident status: Endemic

Galápagos orchid is an epiphytic perennial herb typically found hanging from the stems of other plants on which it grows. It has alternate, fleshy leaves up to 12 centimeters long, whose bases wrap around the stem. Small, yellowish-green flowers form at the tips of stems, and its tiny seeds are powdery. This plant is also called *buttonhole orchid*.

Where you will see it: Santa Cruz (Los Gemelos), Floreana, Isabela, San Cristóbal, Santiago

Why aren't there more orchids in Galápagos?

Ecuador's mainland is an orchid lover's paradise; throughout the country you will find more than 4,000 species of orchids. However, although orchids' tiny seeds are lightweight and easily dispersed by wind, there are only 14 known species of orchids in Galápagos. Why aren't there more?

Orchids are an evolutionarily successful family of plants that have coevolved with two other groups of organisms: fungi and insects. For example, the germination and development of many orchids depend on specific fungal partners. In these mutualisms, the orchid provides the fungus with sugars, and the fungus provides the orchid with nutrients from the soil. If these fungal partners were not present when the orchid seeds landed on Galápagos, the seeds would either not have germinated, or the germinated seedlings would not have developed.

Many species of orchids are also pollinated by specific insects, and they have evolved traits that attract those particular helpers. In these mutualisms, the visiting insect gets nutrient-rich nectar in exchange for dispersing the orchid's pollen to other flowers. However, when seeds of these orchids reached Galápagos, their insect pollinators were almost surely not present, so the plants could not reproduce.

Like virtually all endemic plants of Galápagos, orchids are well adapted for long-distance transport, and their seeds have probably landed in the archipelago for millennia. However, only 14 species have become established because their evolutionary partners—that is, particular insects and fungi—have not always preceded them.

Common name: Galápagos tillandsia

Scientific name: *Tillandsia insularis*

Resident status: Endemic

This relative of Spanish moss and pineapple grows as an epiphyte throughout the humid zone. The bases of its long (up to 50 cm), pointed, greenish-red leaves wrap around the stem, thereby forming flowerpot-like structures that collect water and debris in which invertebrates live.

Where you will see it: Santa Cruz, Fernandina, Floreana, Isabela, San Cristóbal, Santiago

Common name: Galápagos justicia

Scientific name: *Justicia galapagana*

Resident status: Endemic

This perennial herb has hairy stems, pear-shaped and opposite leaves, and purple flowers with a white base. Most plants are less than a meter tall, and their flowers are visited by syrphid flies (*Toxomerus crockeri*).

Where you will see it: Santa Cruz, Santiago, Fernandina, Isabela

Common name: Galápagos jaegeria

Scientific name: *Jaegeria gracilis*

Resident status: Endemic

This delicate, annual, low-growing herb has brownish-purple stems, oval and opposite leaves, and yellow, daisylike flowers at the ends of its stems.

Where you will see it: Fernandina, Isabela, Floreana, San Cristóbal, Santa Cruz, Santiago

Common name: Galápagos three-awn grass

Scientific name: *Aristida subspicata*

Resident status: Endemic

This perennial grass, which is less than 60 centimeters tall, has alternate leaves and yellowish-brown spikes of flowers.

Where you will see it: Baltra, Fernandina, Floreana, Genovesa, Isabela, Marchena, San Cristóbal, Santa Fé, North Seymour

Common name: Indefatigable passion flower

Scientific name: *Passiflora colinvauxii*

Resident status: Endemic

This herbaceous climbing vine, which is one of seven species of *Passiflora* in Galápagos, resembles *P. foetida* (see above) but has crescent-shaped leaves and larger flowers. These plants often climb over rocks and other plants, regardless of size. *Passiflora* refers to how various parts of these plants have been used to symbolize the suffering and death of Jesus (i.e., Christ's Passion).[23] A similar species (*P. suberosa*) also lives in Galápagos, but is rare.

Where you will see it: Santa Cruz (Los Gemelos)

Sedges

Common name: Galápagos sedge

Scientific name: *Cyperus anderssonii*

Resident status: Endemic

Galápagos sedge, which is also referred to as *Andersson's sedge,* is a perennial, upright herb found in most vegetative zones in Galápagos. Galápagos sedge has alternate leaves and clumped stems less than a meter long. Only 3 of the 18 species of *Cyperus* in Galápagos are endemic (4 are introduced, and 11 are native). The solid, triangular stems of *Cyperus* distinguish Galápagos sedge and other sedges from grasses, which have hollow, rounded stems.

Where you will see it: Fernandina, Baltra, Isabela, Genovesa, Rábida, Santa Fé, Santa Cruz, San Cristóbal

Ferns

Common name: Galápagos tree fern

Scientific name: *Cyathea weatherbyana*

Resident status: Endemic

There are more than 85 species of ferns in Galápagos, one of the most interesting of which is *Cyathea weatherbyana*. These ferns, which grow up to 4 meters tall, have thick trunks atop which are clusters of fronds. *Cyathea,* with *Tournefortia rufo-sericea* and *Zanthoxylum fagara,* are important sources of shade for tortoises. *C. weatherbyana* is common in the pampa zone. Like all ferns, *Cyathea* is dispersed by spores.

Where you will see it: Santa Cruz (Los Gemelos, Mount Crocker)

Mosses and liverworts

More than 90 species of mosses, and 110 species of liverworts, live in Galápagos. One of these liverworts, *Frullania aculeata,* covers trees in the *Scalesia* and brown zones and is responsible for much of the color of the brown zone. Like ferns, mosses and liverworts are dispersed by spores.

Common name: Club mosses

Scientific name: *Lycopodium* spp.

Resident status: Native

[23] For example, the plant's tendrils are said to represent whips used by Roman soldiers, the 5 stamens represent Christ's 5 wounds, the 3 styles represent the 3 spikes in Christ's body, the corona represents the crown of thorns, and the 10 petals and sepals represent the 10 disciples who attended the crucifixion.

Despite their name, club mosses are not mosses; instead, they are vascular plants that lack flowers and reproduce by spores formed in specialized structures called strobili. There are at least six species of club mosses in Galápagos, each of which is a small, creeping vascular plant having small (less than 1 cm long) yellow-green leaves: *L. reflexum,* which grows on rocks and has branch tips that curve upward; *L. cernuum,* which has an upright, conical shape and grows primarily in the *Miconia* zone; *L. dichotomum* and *L. passerinoides,* which hang from branches throughout the humid zone; *L. clavatum,* whose branch tips grow upward and produce clublike strobili; and *L. thyoides,* a terrestrial species that prefers shade.

Where you will see it: Santa Cruz (highlands), San Cristóbal (highlands)

Lichens

Lichens are some of the most abundant organisms in Galápagos. Indeed, during the 1905–1906 California Academy of Science expedition to Galápagos (Chapter 2), Alban Stewart noted, "when one lands for the first time on almost any of the islands, one is immediately struck by the great abundance of lichens. . . ."[24] Although lichens are less abundant today in Galápagos than during Stewart's visit, the almost 600 species of lichens in Galápagos today will catch your attention.

Lichens are not plants, but are instead a symbiosis between a fungus and an alga or cyanobacterium. The fungus protects and nourishes the alga, and the alga provides the fungus with sugars. Lichens are especially common in the arid and transition zones, where they often cover rocks, trees, and even tortoises. The most colorful lichens that you will see in Galápagos include *Arthothelium galapagosensis* (white), *Lecanora pseudopinguis* and *Anthracothecium ochraceoflavum* (yellow), *Caloplaca isidiosa* (orange), and *Pyrenula cerina* (red). *Roccella* spp. (dyer's moss), an epiphytic, fruticose lichen common in the transition zone, is a source of litmus and was the basis of an early business venture in Galápagos (Chapter 2).[25] When *Roccella* is fermented in ammonia or urine from cattle, it turns deep blue to purple, depending on the pH.

INTRODUCED PLANTS

Humans have introduced hundreds of species of plants to Galápagos. More than 90% of these species were brought intentionally for agriculture, medicine, and aesthetic reasons, and include crop plants such as bananas, coffee, and onion that pose no environmental threats to native organisms. However, many other introduced plants—for example, quinine and blackberry—are now outcompeting native vegetation and changing the Galápagos ecosystem. Some of these plants are discussed in Chapter 11.

[24] Stewart's collection of more than 3,000 plants from Galápagos was the basis for his *A Botanical Survey of the Galápagos Islands* (1911).

[25] Dyes from *Roccella,* which are not permanent, are often used to enhance the tints of other dyes.

CHAPTER 11

The Future of Galápagos

While the magnitude of the problems facing Galápagos may seem overwhelming, we have an unparalleled opportunity here. The world—not just Ecuador, the Galápagos National Park Service, the Charles Darwin Foundation, but the world—has the opportunity to preserve biological diversity in a pristine state.

—Howard Snell, Director of Science Programs,
Charles Darwin Research Station, 2001

GALÁPAGOS—THE CRADLE OF OUR UNDERSTANDING OF EVOLUTION—is Earth's only tropical archipelago that has not been dramatically changed by humans. Almost 95% of the islands' native and endemic species still live in Galápagos, and it is these creatures and their unique evolutionary stories that draw visitors to the archipelago. As ornithologist Robert Bowman noted in 1983, "No area on earth of comparable size has inspired more fundamental changes in Man's perspective of himself and his environment than the Galápagos Islands."

However, humans' interactions with island ecosystems throughout the world have been consistently catastrophic, as documented by the extinctions of dodos on Mauritius, elephant birds on Madagascar, moas on New Zealand, and countless other species elsewhere by humans. In many ways, Galápagos has been no different. Although some species were extinct before humans arrived, humans have caused—and continue to cause—a wealth of problems for the islands' plants and animals.

Galápagos has been attacked since its earliest visitors—whalers, pirates, scientists,[1] and others—arrived, slaughtering native and endemic species and introducing pathogens, pests, and predators that continue to alter the islands'

[1] Scientists justified their collections of hundreds of tortoises by claiming that the tortoises were the "last survivors" of a "dying race" destined for extinction, and that their killing of the remaining tortoises for museums and other scientific collections was a service to science. When later expeditions found more tortoises on the islands, scientists invoked the same justifications to *again* kill all of the tortoises they could find. For example, on Pinzón, scientists took all of the tortoises they could find in 1897, 1898, 1900, and 1901; five years later, scientists from the California Academy of Sciences took another 86 tortoises from Pinzón that the earlier collectors had missed. Scientists collected the last tortoise from Fernandina in 1905.

ecosystems. Some visitors killed the archipelago's animals for profits, whereas others killed animals merely for entertainment. As biologist Irenäus Eibl-Eibesfeldt noted in 1960, American sailors "beat several hundred iguanas to death. . . . Sea-lions with bashed in skulls lay rotting on the shores. Birds with broken wings and smashed beaks lay about. . . . [We were] openly offered for sale living or dead specimens of 'protected' animals—penguins, young tortoises. . . . It was obvious that the laws [protecting wildlife] were inoperative . . ." Today these slaughters, combined with other factors such as introduced species and habitat destruction, have placed large percentages of the archipelago's plants and animals at risk:

IUCN Status	Vertebrates	Plants	Marine Organisms
Critically Endangered	7%	12%	6%
Endangered	9	26	13
Vulnerable	23	32	31
Near Threatened	13	8	11
Least Concern	46	32	38

Virtually all of the threats to Galápagos can be traced to a single source: humans. Many of these threats have been direct, such as our killing hundreds of thousands of tortoises, our killing tens of thousands of fur sea lions, and the destruction and fragmentation of habitat by citizens, tourists, developers, and governmental agencies (e.g., as during the U.S. occupation of Baltra and other islands during World War II; see Chapter 2). Other threats have been indirect, such as climate change and the invasive species (e.g., cats, pigs, black rats) that damaged native and endemic wildlife long after the earlier visitors left the islands (Figure 11.1). Together, these threats have changed Galápagos. Indeed, in 2007, Ecuador's President Rafael Correa declared Galápagos "at risk" and refocused attention on conservation in the archipelago. That same year, UNESCO—noting the environmental threats posed to Galápagos by tourism, immigration, and introduced species—designated Galápagos a World Heritage Site in Danger. This designation was removed three years later when Ecuador created a new $15 million Invasive Species Fund and stricter measures for immigration and quarantine. Nevertheless, threats remain. When this book went to press in late 2012, there were renewed concerns about the declining populations of blue-footed boobies, vermilion flycatchers, and other iconic species of the archipelago. Of the taxa fully evaluated, almost 10% are Critically Endangered (i.e., are on the brink of extinction), and another 15% are Endangered.

The biggest threat to marine organisms in Galápagos is overfishing. Indeed, overfishing has depleted populations of sea cucumbers, spiny lobsters, and several other organisms in the archipelago. For example, fishing for sea cucumbers was banned in 1994, but in 1997, GNP rangers discovered 40,000 dried sea cucumbers aboard *Magdalena,* and tens of thousands more on other boats and at makeshift

FIGURE 11.1 Invasive species have damaged populations of native and endemic species on several islands in Galápagos. This feral dog on Isabela has killed a marine iguana.

campsites along the coast of Fernandina (where fishermen had cut down mangroves and burned the wood to dry the sea cucumbers).[2] As a result of this overfishing, harvests of sea cucumbers began to drop; in 1999, divers collected an average of 35 kilograms per day, but by 2003 the average harvest had dropped to 25 kilograms per day. This reflected the ongoing shrinkage of the sea cucumber population; whereas in 2001 the density of sea cucumbers along Fernandina was 1.61 individuals per square meter, the following year it dropped to 1.00 per square meter, and to only 0.45 per square meter in 2003. GNPS closed the fishery in 2006, but reopened it in 2007 to appease angry fishermen. Fishermen harvested more than 8 million sea cucumbers in 2002, but only 1.2 million cucumbers in 2007. In 2011, the Participatory Management Board[3] of the Galápagos Marine Reserve authorized a 60-day fishing season for one million sea cucumbers by artisanal fishermen, but only 40% (i.e., 489 of 1,209) of the registered fishermen participated in the harvest. Since 2008, harvests have been allowed only when densities of sea cucumbers reach 0.11 individuals per square meter.[4]

Populations of spiny lobsters have also been damaged by overfishing. In the early 1960s, when commercial lobster fishing began, spiny lobsters were so abundant in some parts of Galápagos that fishermen could gather them without getting wet. Within a few years, however, overfishing reduced harvests from 11 kilograms per day per diver in 1978 to 7 kilograms per day in 2002, and to 4.7 kilograms per day in 2004. Industrial harvests of spiny lobsters were banned in 1984, but illegal

[2] One of the instigators of the illegal harvests of sea cucumbers was Luis Copiano, a businessman who managed the Coca-Cola distributorship in Galápagos. When caught, Copiano blamed the government for being insensitive "to the people living [in Galápagos], born here," despite the fact that he was a recent arrival from Guayaquil (where he was linked to illegal fishing connected with Asian markets). Soon after the illegal fishing camps were exposed, GNPS guides found 81 butchered tortoises at five sites on Isabela.

[3] The Participatory Management Board consists of representatives from five major stakeholder groups in Galápagos: tourism, fishing, park guides, CDRS, and GNPS.

[4] Previous limits were not based on scientific data, but instead resulted from political negotiations. The limit of 0.11 individuals per square meter was the first limit based on scientific data.

harvests continue. More than one-third of marine species in Galápagos are now threatened, and as many as 20% of these species may already be extinct.

The situation is no better on land, where humans and introduced species have significantly altered the biological diversity of several islands. Although Charles Darwin collected and wrote of "great beds" of the vine *Sicyos villosus* (a relative of cucumbers and pumpkins) when he visited Floreana in 1835, no one has seen the plant since. Two other endemic species of plants—Galápagos amaranth (*Blutaparon rigidum*) on Santiago, and the daisy *Delilia inelegans* on Floreana—are also extinct, and others such as rock purslane (e.g., *Calandrinia galapagosa* on San Cristóbal) are threatened. Other extinct species include tortoises on Pinta, Fernandina, and Floreana, and rice rats on Santa Cruz and San Cristóbal. Mangrove finches have been extinct on Floreana since the 1970s; on Isabela, there were five original populations, but only two remain.

Nearly 60% of the 168 endemic plant species of Galápagos are threatened, and more than 40 other endemic species of plants and animals are on the brink of extinction, including the Floreana mockingbird (i.e., the species which helped inspire Darwin's theory of evolution by natural selection). As a result of climate change and overfishing, three species of corals unique to Galápagos—Floreana coral (*Tubastraea floreana*), Wellington's solitary coral (*Rhizopsammia wellingtonia*), and Galápagos coral (*Polycyathus isabela*)—are near extinction. More than 40 species of endemic land snails (*Bulimulus*; Chapter 9) and 13 species of moths and butterflies (Lepidoptera) are threatened with extinction, as are more than 70 species of algae (10 of which are Critically Endangered). The list of vanished and vanishing plants and animals grows every year.

Other problems abound. Even with a renewed emphasis on recycling, the production of trash in the islands—more than 12 tons per day in 2009—is increasing at a rate of 7% per year. The more than 2,000 registered vehicles in Galápagos jam streets in Puerto Ayora, pollute the air, and damage wildlife. Cars on Santa Cruz kill hundreds of birds each month,[5] and there are no viable long-term plans to restore the highlands of colonized islands. In recent years, we have seen feral cats at Tagus Cove, feral goats on Santa Cruz, invasive wasps and plants at countless sites, and feral cats in tortoise pens at CDRS. Conservation challenges are everywhere.

INTRODUCED SPECIES

When Charles Darwin visited Galápagos in 1835, he saw evidence of human impact on each of the four islands he visited. He also noted pigs, goats, and other introduced species and the damage they could do to the native plants and animals of the islands: "What havoc the introduction of any new beast of prey

[5] In 1990 there was one taxi per 350 residents of Puerto Ayora, but in 2012 there was one taxi per 40 residents. On average, each taxi consumes about 10 gallons of gasoline per day. Collectively, taxis consume more than 60% of the fuel used for terrestrial transportation in Galápagos.

must cause in a country, before the instincts of the indigenous inhabitants have become adapted to the stranger's craft or power." Today, this havoc continues.

Introduced species are the biggest threat to Galápagos. Introduced species occur throughout most of the archipelago (especially on inhabited islands), are abundant (there are now more than 1,400 known introduced species in the archipelago),[6] and are remarkably diverse. For example:

> There are 36 introduced species of vertebrates, 30 of which are now established: 1 freshwater fish, 2 amphibians (both frogs), 4 reptiles (all geckos), 10 birds (e.g., ducks, chickens, cattle egrets, smooth-billed anis), and 13 mammals (e.g., goats, cats, rats, pigs, dogs; Table 11.1). Goats, which compete with native and endemic species for vegetation, have been disastrous for many native species (Figure 11.2). At Volcán Alcedo on Isabela, goats transformed *Cyathea* forests into barren landscapes. By 1987, grazing goats had destroyed the habitat and reduced the population of Galápagos rails on Santiago to only 25 birds. Similar environmental changes on Santa Cruz reduced the population of *Scalesia affinis* to fewer than 10 individuals.
>
> There are more than 540 introduced species of insects, most of which arrived in fruits, vegetables, and wood. The most invasive of these insects are fire ants (*Wasmannia auropunctata, Solenopsis geminata*), wasps (*Brachygastra lecheguana, Polistes versicolor;* Figure 11.3), cottony cushion scale insects (*Icerya purchasi*), and botfly (*Philornis downsi*), an ectoparasite of birds (Table 11.1; also see discussion in Chapter 5). Another 55 or so

TABLE 11.1 Invasive Animals with the Greatest Current or Potential Impact and Their Distribution on the Larger Islands of Galápagos

	Santiago	Floreana	San Cristóbal	Santa Cruz	Isabela	Pinzón
Pigs (*Sus scrofa*)	●	●	●	●	●	
Dogs (*Canis lupus familiaris*)	●	●	●	●	●	
Cattle (*Bos primigenius taurus*)		●	●	●	●	
Donkeys (*Equus africanus asinus*)	●	●	●	●	●	
Horses (*Equus ferus caballus*)		●	●	●	●	
Cats (*Felix catus*)		●	●	●	●	
Rats (*Rattus* spp.)	●	●	●	●	●	●
Ants (*Wasmannia, Solenopsis*)	●	●	●	●	●	
Cottony cushion scale (*Icerya* spp.)	●	●	●	●	●	
Wasps (*Polistes*)	●	●	●	●	●	
Ectoparasitic fly (*Philornis*)		●	●	●	●	

[6] Since the discovery of Galápagos in 1535 (Chapter 2), the introduction rate for vascular plants has averaged more than one species per year; this is approximately 10,000 times greater than the rate of natural arrivals in the archipelago. Although several sources claim that the introduction rate of vascular plants is increasing exponentially, it may be an artifact of increased awareness, more research, more sampling, and better detection.

FIGURE 11.2 Grazing by giant herds of feral goats (shown here on Santiago Island) has transformed lush vegetation to barren areas on several islands in Galápagos.

introduced species of insects in Galápagos may be invasive. These invaders change ecosystems; for example, *Philornis downsi* (whose larvae are obligate small-bird parasites) arrived in Galápagos in the 1960s and now kills birds (including endemic species of finches) on several islands. The mortality of botfly-infected chicks averages 76%, and sometimes reaches 100% (surviving nestlings often grow slowly, have deformed beaks, and suffer from anemia). Similarly, wasps such as *Polistes,* which prey on native invertebrates, can consume more insect biomass than *all* of the finches on the same island. Wasps are now spreading throughout the archipelago, including at several Visitor Sites and throughout several towns. Invasive giant African land snails (*Achatina fulica* = *Lissachatina fulica*), which mainland Ecuador exports as "Chinese escargot," are now an agricultural

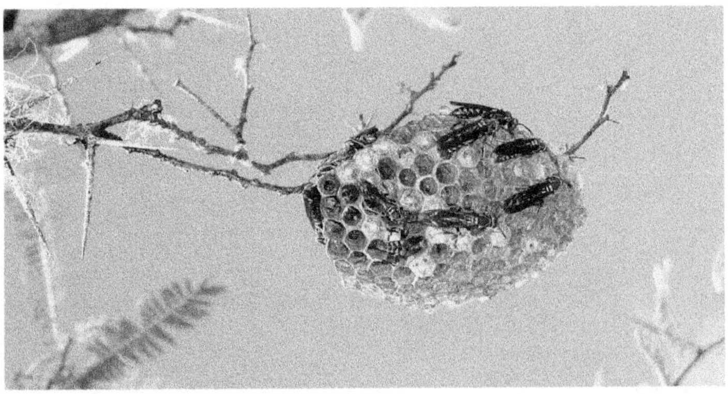

FIGURE 11.3 Wasps such as *Polistes versicolor* have invaded several islands and are killing native invertebrates. This nest was photographed on San Cristóbal.

pest in Galápagos.[7] Although the effects of introduced mammals such as cats and goats are obvious, introduced insects and plants are changing Galápagos more slowly and just as profoundly.

There are more than 870 known species of introduced plants, more than 90% of which were introduced intentionally for aesthetic, agricultural, or medicinal uses. You will see invasive plants such as supirosa,[8] elephant grass, and cedrela on your way from Puerto Ayora to Itabaca Channel, as well as at several Visitor Sites; blackberries, passion fruit, and elephant grass populate Los Gemelos; cedrela, blackberry, and guava have invaded El Chato; and blackberry is abundant at Laguna El Junco and elsewhere on San Cristóbal (Table 11.2). Introduced species of plants far outnumber native and endemic species of plants in Galápagos; at least 25% of the introduced species are now naturalized (i.e., are established and reproduce without the help of humans), and more than 130 of the introduced species of plants have invaded natural areas of the islands. For example, on San Cristóbal, only 23% of the 29 square kilometers of natural vegetation present in 1987 remained in 2009. On Santa Cruz, the

TABLE 11.2 The Most Aggressively Invasive Plants in Galápagos and the Islands on which They Are Most Problematic

	Santiago	Floreana	San Cristóbal	Santa Cruz	Isabela
Blackberry (*Rubus* spp.)	●	●	●	●	●
Guava (*Psidium guajava*)		●	●	●	●
Cascarilla (*Cinchona pubescens*)					●
Cedrela (*Cedrela odorata*)				●	●
Supirosa (*Lantana camara*)		●		●	●
Sauco (*Cestrum auriculatum*)				●	
Maracuyá (*Passiflora* sp.)					●
Pomarosa (*Syzgium jambos*)				●	
Hola del aire (*Bryophyllum pinnatum*)			●		●
Pasto elefante (*Pennisetum purpureum*)					●
Cabuya (*Furcraea hexapetala*)					●
Sauco macho (*Citharexylum* sp.)				●	
Tulipán africano (*Sphatodea campanulata*)				●	
Zaragoza (*Aristolochia odoratissima*)					●
Higuerilla (*Ricinus communis*)				●	●
Poleo (*Hyptis pectinata*)	●			●	
Saboya (*Panicum maximum*)				●	
Floripondio (*Datura* spp., *Brugmansia* spp.)				●	

[7] IUCN classifies *A. fulica* (which often harbors parasites) as one of the world's top 100 invasive species.
[8] Supirosa (*Lantana camara*), a shrub introduced on Floreana in 1832, now covers more than 3,000 hectares of the island and outcompetes local vegetation.

situation is even worse: only 14% (about 5 km^2) of the natural vegetation remains. Humans have replaced birds as the most important means for disseminating plants in Galápagos.

Fewer than 5% of introduced species become invasive. For example, plants such as pineapples and carrots were brought to Galápagos for food and do not pose a serious threat to native and endemic plants of Galápagos. However, other invasive species *are* damaging the islands. For example, invasive species (along with agricultural practices such as the clearing of land to grow crops) have degraded large areas of the humid zones of the inhabited islands (Table 11.3).

GNPS currently lacks the resources to stop invasive plants. For example, in 2007 the GNPS budget for weed control ($630,000) was spent to control invasive plants on less than 5 square kilometers. For comparison, guava alone covers more than 850 square kilometers in the archipelago.

Who Are the Invaders, and What Do They Do?

Invasive species damage the native and endemic plants and animals of Galápagos in a variety of ways:

Black rats and Norwegian rats have destroyed most of the archipelago's (generally smaller) endemic rodents. Black rats live on all islands that have lost their native rats. Native rats survive on Fernandina and Santa Fé, which have not yet been infested by black rats; only on Santiago do native rats co-occur with invasive rats. Black rats eat juvenile marine iguanas, birds, and tortoises, especially on dry islands such as Pinzón, where they have killed *all* tortoise hatchlings for several decades (Chapter 2). Along with disease and feral cats, black rats drove endemic mice (*Nesoryzomys indefessus*) on Santa Cruz to extinction (the last endemic mouse there was captured in 1934). Rats (with dogs, cats, and pigs) now threaten the dark-rumped petrel and other animals.

Cats and dogs, which came to populated islands as pets, dig up eggs and eat juvenile iguanas and tortoises (Figure 11.1). Cats were eradicated

TABLE 11.3 Percentages of Land in Various Vegetative Zones that have been Degraded by Invasive Species and Agriculture

	Vegetative Zone		
Island	**Arid**	**Transition**	**Humid**
San Cristóbal	2	23	>95
Santa Cruz	0.4	25	>90
Floreana	0.5	2	>80
South Isabela	0.2	4	>70

from Baltra in 2003, but they remain common on several other islands. Cats eat invasive black rats, but they also eat native and endemic birds, iguanas, and lizards, none of which has evolved behaviors to counter a feline predator. Feral dogs and other introduced animals eat juvenile birds and iguanas, who often do not flee or defend themselves. As Victor von Hagen noted in 1934, feral dogs "attack tortoises, rob eggs from flamingo and cormorant nests, slaughter penguins and cormorants, and even swim out to sea after the sea iguanas."

Introduced birds and other animals have brought diseases to native and endemic birds of Galápagos. For example, avian pox now affects several species of birds in Galápagos, and a *Plasmodium* blood parasite has infected Galápagos penguins. Smooth-billed anis (*Crotophaga ani*), which were introduced on Galápagos in 1962, are suspected of killing chicks of native vermilion flycatchers.

Pigs were brought to Galápagos as a source of food, but they soon escaped captivity and began digging up eggs and eating juvenile tortoises and marine iguanas. Pigs can be remarkably destructive; for example, in one month, one pair of pigs located and destroyed 23 tortoise nests.

Donkeys trample nests and crush the eggs of tortoises and marine iguanas while transforming forested areas into grassy—and often dusty—landscapes, thereby leaving little food, shelter, or nesting sites for native and endemic species. Donkeys first appeared in Galápagos in 1834 on Floreana, where they were used for chores such as moving kegs of tortoise oil to settlements and visiting boats. By the 1890s, large populations of donkeys roamed Floreana, Santa Cruz, San Cristóbal, Isabela, and Santiago.

Goats compete with tortoises and other animals for food. Although goats were brought to Galápagos for milk and meat (after tortoises became rare because of overhunting), their populations on some islands were in the tens of thousands (Figure 11.2).

Invasive ants, which now live on all cultivated islands in Galápagos, kill native invertebrates and reduce the breeding success of birds, iguanas, tortoises, and other wildlife. Local populations of invertebrates such as spiders, scorpions, and native ants have disappeared in areas infested by introduced ants (especially *Wasmannia auropunctata*). The first of these invasive ants were accidentally introduced on Santa Cruz in 1930, and today have spread to many other islands. Ants are among the most aggressive invertebrates in Galápagos; on Champion Island, which is home to one of the last two populations of Floreana mockingbirds (*Mimus trifasciatus*), GNPS's poison-based traps have failed to control fire ants (*Solenopsis geminata*), presumably because new populations of ants continually fly over from nearby Floreana (700 m away; fire ants have also island-hopped from Isabela to the nearby Marielas islets). In 2012, the population of Floreana mockingbirds on Champion Island had dropped to between 20 and 50 birds, depending on environmental conditions (the population of the birds on nearby Gardner Island had increased to 400 to 500 birds). GNPS officials are considering biological controls involving phorid flies (e.g., *Pseudacteon*), which prey on South American fire ants.[9]

Introduced plants—many of which outcompete native and endemic vegetation—outnumber endemic plants on San Cristóbal, Santa Cruz, Floreana, and Isabela. Many of these introduced plants are aggressively invasive. For example, hill blackberry (*Rubus niveus*) was introduced on San Cristóbal in 1980 for its edible fruits and its ability to form dense, fence-like thickets. Today, these plants cover more than 7,500 hectares of San Cristóbal and elsewhere, and often form thickets up to 4 meters high that block farming. It is difficult to stop the spread of blackberries; plants produce fruits within six months after germination, and their seeds are dispersed by birds. In some areas, soil contains more than 20,000 blackberry seeds square meters, many of which can remain viable for more than 20 years.[10] Other invaders include Spanish cedar (*Cedrela odorata*), whose aromatic wood is used to build ships, closets, and furniture. This species, which was introduced on Santa Cruz in 1950 as a source of timber, covers more than 1,000 hectares of GNP, where it inhibits the growth of several native and endemic species such as *Tournefortia pubescens* and *Zanthoxylum fagara*. Quinine (*Cinchona pubescens*), which was introduced in Galápagos in 1946 from South America as a cure for malaria (a nonexistent disease in Galápagos), now covers more than 11,000 hectares in the highlands of some islands. Quinine spreads quickly; when they are only three years old, plants begin releasing large numbers of winged seeds that are wind-dispersed throughout the region. (The control and subsequent eradication of quinine would take 10 to 15 years and almost $10 million.) Other invasive plants in Galápagos include elephant grass (*Pennisetum purpureum*) and guayaba (*Psidium guajava;* Table 11.2).[11]

The impacts of several other introduced species—for example, frogs on Isabela—have not been evaluated. Officials are especially concerned about invasive ants and the appearance in 2008 on Santa Cruz and San Cristóbal of the Mediterranean fruit fly (*Ceratitis capitata*), one of the most destructive species in the world. Similarly, the impact of the invasive black fly (*Simulium bipunctatus,* introduced on San Cristóbal in 1989) is unknown,

[9] Fire ants, which arrived in Galápagos in 1981 and on Champion in 2006, attack chicks and invertebrates that are an important source of food for the mockingbirds. Phorid flies reproduce by laying their eggs in the ant's thorax. The first instar larvae eat the ant's brain, causing the ant to wander aimlessly for 10–14 days before dying. Local organizations in Galápagos, principally GNPS, spend more than $200,000 per year to control invasive ants.

[10] Eradicating other invasive species has been equally problematic. For example, between 2001 and 2007, GNPS participated in a program aimed at eradicating 36 introduced species. The program cost more than $1 million, but eradicated only four species.

[11] According to some sources, guava (also called guayaba) came to Galápagos in the 1850s when Manuel J. Cobos (Chapter 2) brought three of the plants to the archipelago. Today guava, which is arguably the most invasive plant in Galápagos, covers more than 25,000 hectares of south Isabela.

but similar species carry nematodes such as *Onchocerca volvulus,* which causes river blindness in humans.

At last count, more than 1,400 introduced species had been documented in Galápagos, and the list is growing. A study in 2006 found insects aboard 72% of the planes that landed at Baltra; a similar study three years later reported that 93 flights to Galápagos brought 74 insects to the archipelago. The flood of introduced species continues.

Preserving Galápagos

In 2006, the director of the Charles Darwin Research Station (CDRS) summarized the station's work this way: "We focus on two things: threats and threatened things." Doing so requires promoting conservation and preservation. Conserving Galápagos involves education, legislation, and eradication programs designed to change human behavior and restore the islands' altered landscapes (Chapter 3). Preserving Galápagos involves trying to manage or eliminate several threats, the most important of which is invasive species.

Documenting the archipelago's biodiversity

In 2007, the Charles Darwin Foundation began assembling an inventory of all known species in Galápagos.[12] Although the list originally included almost 14,000 entries, the elimination of duplicates and redundant names later reduced the species count to near 7,500. This list provides an important baseline to which other studies of the archipelago's biodiversity can be compared.

Blocking entry of invasive species

The most important step in controlling invasive species is preventing their arrival. Most invasive species in Galápagos are detected only after they become established, after which they are difficult to eradicate. To help prevent the arrival and establishment of invaders, Ecuador implemented SICGAL[13] to inspect (and, if necessary, quarantine) incoming luggage, planes, boats, and cargo in Galápagos.[14] These inspections have blocked thousands of insects from reaching the islands, as well as animals such as toads, reptiles, and monkeys carried by visitors. On flights to Galápagos, flight attendants spray the overhead bins with organic pesticides that kill insects such as flies, ants, and mosquitos. During its first year, SICGAL—which is funded by 5% of the GNP entrance fee—intercepted 800 prohibited products on their way to Galápagos. In 2006, SICGAL confiscated 1,022 products infected with invasive species at Galápagos airports; this

[12] This species checklist is available at http://www.darwinfoundation.org/datazone/checklists/.

[13] SICGAL, a division of the Ecuadorian Agricultural Health Service, is an acronym for Sistema de Inspección y Cuarentena Para Galápagos. It was created in 1998, and fully began operations in 2000. There are SICGAL offices on all of the inhabited islands; a temporary office also opens occasionally on the Gus Angermeyer Ecotourism Pier in Puerto Ayora.

[14] Subsequent surveys revealed that although tourists outnumber local residents by two to one, local residents bring more than 70% of the introduced species to Galápagos.

amounted to 1 interception per every 8,230 entries of plants and invertebrates. Only two years later, SICGAL confiscated more than 2,700 products carrying introduced species. Today, quarantine procedures include random visual inspections of approximately 60% of the cargo entering Galápagos. Although more than 3,000 crates of food were confiscated in 2012, docks have few facilities for pest controls or fumigation.

Many residents of Galápagos have introduced plants for agricultural, decorative, or medicinal purposes. For example, in Puerto Villamil, there are now more than 260 introduced species, of which 39 also grow in the wild. Five of these species are suspected to be invasive, especially *Leucaena leucocephala* (which IUCN lists as one of the 100 worst invasive species in the world). To minimize the introduction of invasive plants, Vivero—the CDRS Technical Office at San Cristóbal (a similar office operates on Santa Cruz)—annually provides thousands of native and endemic plants to local homes and businesses to help reduce the introduction of invasive species (more than half of the introduced species of plants in Galápagos are in gardens). Vivero also provides research, presentations, and educational materials for the community.

But these programs face daunting challenges. Small insects are difficult to find in packages and on the hundreds of large planes and boats that come to Galápagos, and many of the insects go undetected. Indeed, more than 70% of the flights landing at the Baltra airport carry insects such as *Culex quinquefasciatus,* a mosquito introduced in 1989 that can harbor West Nile virus, avian malaria, and other pathogens (female mosquitos, which feed on birds, are the probable vectors of various pathogens of Galápagos penguins and flightless cormorants).[15] Similarly, food and other materials imported to meet the growing needs of the islands' residents and visitors arrive in more than a *million* packages (weighing more than 7,800,000 kg) from mainland Ecuador each year. Although these packages, especially those containing food, are the primary way that introduced species reach Galápagos, only about 2,000 packages are confiscated by Galápagos authorities each year. Many species get through these checkpoints. In 2008, a single boat brought 11 new species to Galápagos from around the world.

Eradicating goats

GNPS has initiated several programs to eradicate goats in the archipelago (Figure 11.2). The first of these programs was relatively simple, and involved the removal in 1961 of five goats from South Plaza. A decade later, the last of 3,005 goats was removed from Santa Fé; this was the first large-scale, systematic eradication of an invasive species in Galápagos. In 1975, all of the 14 goats were removed from Rábida. Several eradication programs have failed (e.g., rats on Pinzón in 1988, and on Lobos Island in 2002), and others were inefficient because of a lack of resources. For example, it took 30 years to remove more than 18,800 pigs from Santiago; eradicating goats on Pinta also took three decades.

[15] *Aedes aegypti,* the mosquito that transmits dengue fever, was found in Puerto Ayora in 2002.

FIGURE 11.4 Residents of Galápagos continue to hunt feral animals. This sign near Isabela's Wall of Tears (Figure 2.11) warns visitors not to stray from the popular Visitor Site.

Modern eradications of goats have been more efficient. In addition to traditional ground-based hunting (Figure 11.4), other methods have included airborne hunters (i.e., shooting from helicopters) and "Judas" goats. These goats, which are collared with a radio transmitter, exploit goats' herding instincts; when they are released, the "Judas" goats quickly find other goats. Rangers then track down the "Judas" goats and kill all of their companions. "Judas" goats are also characterized by hormonally induced estrus, thus making them so attractive to other goats that they are called "Mata Hari," a reference to the irresistible double agent in World War I. These new techniques were effective; between 2006 and 2009, hunters removed 1,561 goats from Floreana, and between 2006 and 2010 they removed 9,207 goats from San Cristóbal. The last of approximately 41,000 goats were killed on Pinta in 2003, and the last goat was gone from Santiago in 2005.

Project Isabela—the world's largest eradication and restoration program—removed goats and several other invasive species from Pinta, Santiago, and northern Isabela. Such projects are expensive; Project Isabela cost $43,000 per month, most of which—despite the project's name—was spent on eradication efforts on Santiago. Here is what it cost to eradicate goats from Santiago, Pinta, and northern Isabela:

	Santiago	Northern Isabela	Pinta
Area (ha)	58,465	458,812	5,940
Year completed	2006	2006	2000
Number of goats killed	79,579	~135,000	~41,000
Cost	$5,483,373	$3,009,529	$83,891
Cost per hectare	$94	$7	$14
Cost per goat	$61[16]	$48	$1,291
Type of hunting	Ground,[17] aerial, Judas goats	Almost all aerial[18]	Ground

[16] Most goats cost $10–100 each to kill, but the final few goats cost $10,000 each. The last 1,000 goats cost $2 million to kill over 1.5 years. The GNPS then spent $467,000 to confirm the eradication.

[17] More than 90% of the ground hunters were Ecuadorians; their average salary was $550 per month.

[18] Only 2,637 feral goats were killed by ground hunters, and 3,439 were killed as a result of associating with Judas goats.

FIGURE 11.5 Helicopters have been used to distribute poison to kill invasive rats on several islands in Galápagos. For example, in 2011, GNPS helicopters dispersed poison-laced pellets to eradicate rats on Rábida (4.9 km^2). The 40 tons of pellets contained 2 pounds of poison. In 2012, a similar helicopter-based eradication program began on Pinzón and South Plaza.

Project Isabela ended in 2006 and was a stunning success, prompting CDRS to announce that it had "done what the world believed impossible. Pigs, goats and donkeys on Santiago and northern Isabela are now history."

Eradicating rats and other invasive species

GNPS has used poisons to eliminate black rats from Rábida, Santa Cruz, Sombrero Chino, Baltra, and Bartolomé (Figure 11.5); similar eradication programs elsewhere (e.g., Pinzón, San Cristóbal; Figure 11.6) are reducing the sizes of their populations.[19] GNPS has also eradicated feral goats and donkeys from northern Isabela; goats, donkeys, and pigs from Santiago; goats from Pinta; fire ants from Santa Fé and Marchena; cats from Baltra, tilapia from Laguna El Junco on San Cristóbal;[20] and pigeons (which transmit more than 40 diseases)

[19] The poisons entice rats, but repel sea lions, birds, tortoises, iguanas, and other wildlife.

[20] Tilapia (*Oreochromis niloticus*) was first detected in Laguna El Junco in 2006; within two years, all 39,958 of the fish had been eradicated with rotenone, a naturally occurring poison of fish.

FIGURE 11.6 Poison-lined tubes are used at San Cristóbal's Laguna El Junco (Figure 4.7) to control invasive rats.

from Isabela, Santa Cruz, and San Cristóbal. On many islands, GNPS uses traps to help control introduced insects, especially wasps. You will see these traps along several trails (Figure 11.7). "Judas" goats continue to be used on Santiago, Isabela, and other islands to detect reintroductions of other goats.

Eradication of pigs began in the 1970s with the use of poisons and hunting dogs. The last of Santiago's more than 18,000 feral pigs were killed by ground hunters in 2000. By 2009, the last 380 donkeys were gone from Floreana, and by 2010, the last of 498 donkeys had been killed on San Cristóbal. In 2004, the last of Santiago's 340 donkeys were killed by aerial hunters, and in 2006 Santiago was free of all large, introduced mammals. Pigeons, which were introduced

FIGURE 11.7 Baited traps control invasive insects on several islands in Galápagos.

FIGURE 11.8 Residents of inhabited islands of Galápagos can have their pet dogs and cats sterilized for free. These programs protect native and endemic species by minimizing the number of feral animals in Galápagos.

to Galápagos in 1972, were eradicated from the archipelago in 2005. GNPS and other governmental agencies sponsor programs that sterilize pet cats and dogs on inhabited islands (Figure 11.8).

GNPS is now experimenting with biological control of invasive species. The insect pest cottony cushion scale (*Icerya purchasi*) was detected on San Cristóbal in 1982 and, within a few years, threatened more than 60 native and endemic plants (including *Cordia, Piscidia, Scalesia,* and *Darwiniothamnus*). In 2002, in the first use of biological control in Galápagos, an Australian ladybird beetle (*Rodolia cardinalis*) was released on four islands to help control cottony cushion scale. Soon thereafter, the ladybird beetles were released on five other islands. In 2010, CDRS reported that the beetles had reduced the impact of cottony cushion scale, with few side effects.

When invasive species have been eliminated, the populations of native and endemic species have often rebounded. For example, before pigs and other invasive species were removed from Santiago in 1979, the survival rate of sea turtle hatchlings on the island was 1.88%. By 2007, after invasive mammals had been removed from the island, the survival rate increased to 86.4%. Similarly, when Pinta became goat-free in 2003, populations of endemic plants and animals began to recover; and when goats and donkeys were eradicated from northern Isabela, endemic plants (e.g., *Tournefortia rufo-sericea* and *Zanthoxylum fagara*) began growing there again, and Galápagos rails (*Laterallus spilonotus*) returned.

Stopping overfishing

In 1998, GNPS banned industrial fishing, and in 2005 banned long-line fishing, in the Galápagos Marine Reserve. Many boats involved in illegal fishing have been confiscated in recent years; visitors to the islands can see the results of one such confiscation in San Cristóbal. There, in 2004, GNPS confiscated the *Don Abraham* for illegal fishing. That boat was later procured "on behalf of the women" of San Cristóbal and now houses "Fisherman Souvenir Cafeteria." This shop, along with a historical marker and a historical mailbox, is at the southwestern corner of where Avenida Manuel J. Cobos intersects Avenida Charles Darwin in Puerto Baquerizo Moreno.

Despite a variety of well-intentioned laws and regulations, the Galápagos Marine Reserve has proven difficult to protect. The reserve is vast (133,000 square kilometers, an area about 1300 square kilometers smaller than Louisiana) and is patrolled by the GNPS Marine Control and Surveillance Unit and Ecuadorian Navy. As of 2012, the GNPS fleet included three oceangoing boats, two medium-range boats, six speedboats, and the *SeaWolf* hydroplane.[21] As a result of increased efforts and concerns about the health of the reserve, the number of cited infractions (e.g., fishing without a license, fishing in protected areas, and harvesting banned species) has risen to more than 400 per year. Many of these infractions involve sharks, whose dorsal fins are severed to make shark-fin soup, a prized Asian delicacy (the soup can cost up to $150 per bowl). GNPS banned shark fishing in 1989, but since then has confiscated hundreds or thousands of carcasses each year. For example, in 2011, GNPS and the Ecuadorian Navy confiscated the *P/P Fer Mary I,* which had 379 sharks (many with their fins cut off) in its hold. On average, the GNPS finds about one boat per month fishing illegally in the reserve, but many others presumably go undetected.

Why does illegal fishing continue? The high demand for lobsters, sea cucumbers, shark fins, and other organisms has produced several "gold rush" scenarios in Galápagos, and fishermen have often reaped huge profits. Moreover, the penalties for illegal fishing have often been trivial. For example, in 2001, GNPS confiscated *María Canela II,* which had 1,044 shark fins and more than 80 shark carcasses aboard. The ship and its crew later escaped and were not apprehended.[22] In 2002, the Colombian tuna boat *El Dorado* was caught

[21] These boats operate from bases on Santa Cruz and San Cristóbal, as well as from remote bases on Cartago Bay (Isabela) and Bolivar Channel.

[22] In 2011, *María Canela II* was caught carrying more than 1,100 kg of cocaine near Galápagos.

in the Galápagos Marine Reserve with more than 50 dead dolphins aboard. The judge fined the captain $0.04 per pound of dolphin and ordered the captain to serve two weeks in jail, which the captain was allowed to spend on his boat (with shore leave). In 2010, *Papate* was caught within GNP with 183 dead sharks, but when the accused captain and crew left Galápagos before their trial, they were not prosecuted. In 2011, after *Reina del Cisne* was caught with 81 sharks in its hold, the presiding judge ordered the release of the boat and its crew. That judge, claiming that his court was incompetent to address environmental crimes, annulled all of the previous proceedings. When boats are confiscated, they are usually returned within a few weeks or months. As *The Economist* has noted, "Officially, conservation has been the rule in the islands since the 1930s. But governments have often paid little more than lip service."

GNPS officials are trying to prevent overfishing, but political pressures from fishermen often compromise policies. For example, in 1993, Galápagos banned lobster fishing until 2000, but political pressures forced the reopening of the fishery the next year (1994). The islands' lobster fishery was open for only a month in 2011 with several restrictions (e.g., the catch could not exceed 10 metric tons of red lobster and four tons of green lobster at Isabela). Authorities claimed that the fishery would then be closed permanently, but political tensions threatened this closure the next year.

Protecting and restoring habitat

Environmental damage caused long ago (e.g., by buccaneers and pirates; see Chapter 2) has been exacerbated by recent visitors and the invasive species these visitors have left behind. GNPS has used a variety of strategies to minimize habitat destruction and to rebuild healthy, balanced, sustainable ecosystems to their pre-human conditions. For example, on San Cristóbal and Santiago, GNPS officials have built fences to protect endangered plants from feral animals. With these protections, the native and endemic plants soon recolonized some of the areas when the invasive animals were removed or controlled. To protect ecologically sensitive areas from damage by tourists, GNPS in 2012 began forbidding tour boats from visiting several Visitor Sites more than once every two weeks.[23] However, these improvements have often been negated by pressures from developers and tourists. For example, the flood of tourists to Galápagos in recent years (Figure 1) has increased the need for hotels, restaurants, taxis, gas stations, stores, governmental agencies, roads, and other infrastructure, all of which have altered the landscape of Galápagos. A prominent example occurred recently on Isabela, where developers wanted to build a large hotel at a site where marine iguanas move from the ocean to a nearby lagoon. Several groups warned that such a hotel would affect the population of marine iguanas

[23] These changes were part of GNPS's new System of Managing Visitors (SIMAVIS), which integrates visitation rates, Visitor Site management, and boats' itineraries throughout GNP.

FIGURE 11.9
On Isabela, signs warn motorists and others of areas where marine iguanas cross roads. A popular hotel built near this site is called "Iguana Crossing."

in the area, but the hotel—aptly named "Iguana Crossing" (Figure 11.9)—was constructed and is now open for business. These businesses, which often are granted exemptions from protective regulations, have also increased the production of pollution, trash, and other indirect damage to the islands.[24] Plastic garbage (and illegal fishing with nets) threatens flightless cormorants, sea lions, and other marine organisms (Figure 11.10). In Puerto Ayora, traffic noises mask the echolocation calls of Galápagos bats (*Lasiurus brachyotis*), and thereby may impair their ability to locate prey.

Educating residents and tourists

Although most tourists view Galápagos as a once-in-a-lifetime experience, local residents consider the islands to be home, and their lifestyles are influenced by

[24] For example, although there is a clear association between diet and beak size in wild medium ground finches (*G. fortis;* Chapter 5), this association is seldom apparent in populations such as those in Puerto Ayora that interact with humans.

FIGURE 11.10 Trash thrown overboard from boats damages wildlife in Galápagos. A GNPS guide removed the band from this sea lion.

a variety of traditions. Most of these residents are *insulares* who moved to the islands in search of economic and social opportunities. In contrast, *isleños* are residents who, if they were not born in Galápagos, have lifestyles deeply rooted in the archipelago. Whereas *insulares* may expect life in Galápagos to resemble that on the mainland (i.e., to have relatively few restrictions on activities such as fishing and interactions with animals), *isleños* are less likely to have such expectations, and are more likely to accept the restrictions of life in Galápagos.[25]

Although tourism dominates the Galápagos economy, fishing remains a popular career (you will see monuments to fishermen throughout the islands; Appendix 1), and illegal fishing operations regularly set up clandestine camps in the western islands to process their harvests. These camps often destroy the local habitat (usually coastal mangrove, which is home to the endangered mangrove finch) and introduce invasive species. Other residents living in the highlands (e.g., southern Isabela) continue to eat tortoises, despite the fact that consumption of this meat has been banned since 1933.

To help residents and tourists appreciate the importance of GNP, GNPS, and other organizations (e.g., Galápagos Conservancy; see Appendix 8) have built environmental education centers on Isabela, Santa Cruz, and San Cristóbal that sponsor research and educational programs, opened technical centers that provide residents and businesses with native and endemic plants (see above), hired fishermen to help clean and monitor GNPS, and sponsor programs to neuter dogs, cats, and other pets on the islands (Figure 11.8). The Governing Council of Galápagos works closely with GNPS, CDRS, and local governments to collect and share information, as well as to educate local residents and tourists about protecting Galápagos. These programs are based on the premise that the long-term protection of Galápagos requires a sustainable economy that is compatible

[25] The population of Galápagos is growing at a rate of more than 6% per year. To accommodate the needs of its expanding population, Puerto Ayora in 2012 was developing El Mirador, a residential community (consisting of 1,133 lots) that would almost double Puerto Ayora's geographic footprint. However, this project was put on hold because of environmental concerns.

with conservation, and an educational system that prepares and encourages its citizens to promote conservation throughout the archipelago.

Minimizing threats outside GNP

Several native species of Galápagos spend part of their time outside GNP. For example, the iconic waved albatross spends much of its time foraging for food along the coast of Perú. Long-line fishing there kills many albatrosses per year (Chapter 5), and has thereby reduced the birds' average lifespan. GNPS officials are now trying to monitor the impacts of these external threats and, where possible, work with other governments to minimize their impact on Galápagos wildlife.

Much of what happens outside GNP (e.g., in towns, at airports) affects the plants, animals, and landscapes of GNP. To minimize problems arising from these activities, the newly formed Consejo de Gobierno del Régimen Especiál de Galápagos coordinates what happens outside of GNP (e.g., the collection and recycling of wastes, the use of water, the driving patterns of taxis and buses) with the work of GNP.

Easing political tensions with fishermen

Some of the measures implemented to preserve GNP have reduced the incomes of fishermen in the archipelago. In 1993, to garner attention to their plight, furious fishermen began staging protests at CDRS; they claimed that the "Darwin Station people want to keep Galápagos as their own plantation" and demanded that they be allowed to increase their harvests. The following year, machete-wielding fishermen protesting a governmental ban on sea cucumber fishing blocked the entrance to CDRS, harassed scientists, and threatened to kill tortoises. Boats coming to collect illegal harvests of sea cucumbers brought prostitutes and drugs from the mainland, and prostitutes paid with sea cucumbers later traded them for cash. In 1994, fishermen killed more than 80 tortoises on Isabela, attacked rangers (four of whom had to be hospitalized), and threatened to block Visitor Sites and airports serving Galápagos. The fishermen killed 72 tortoises on Isabela and released goats on Pinta Island, and in 1995 seized GNPS headquarters and CDRS while shouting "Mate Solitario Jorge!" ("Kill Lonesome George!"). After they threatened to kill the CDRS's animals if there were no changes to the fishing laws, the laws were changed. But the following year, protesters slit the throats of several tortoises on Santa Cruz.

In 1997, GNPS ranger Julio Lopez was shot and seriously wounded when he discovered an illegal fishing camp that was processing almost 30,000 sea cucumbers. Almost 300 citizens marched in protest of this violence, but fishermen again seized GNPS headquarters. When GNPS restricted the harvests of sea cucumbers and spiny lobsters in 2000, more than 800 protesting fishermen rioted for six days and set fires at park headquarters, and Juan Chavez—the local director of GNPS at Puerto Villamil—had his house plundered and was forced to hide in a mangrove swamp, where he was rescued by the Ecuadorian Navy. After 11 days of riots in 2002, disgruntled fishermen in 2004 rioted for 14 days and again seized the offices of GNPS. As one protester noted, "The government follows the [CDRS] and not the people's necessities; they do not care about us; they only want money from tourism, but [there is] no tourism in Puerto Villamil [Isabela]. . . . We cannot work because of the prohibitions of the National Park. . . . We came before the scientists. . . . We are the animals least taken care of on the islands."

More recently, angry fishermen have occupied GNPS offices, assaulted GNPS officials, reintroduced invasive species on several islands, and repeatedly killed scores of animals on islands (e.g., 53 sea lions were clubbed to death in 2008 on Pinta). To help remedy the problem, GNPS now hires fishermen to monitor and collect trash in GNP, and has implemented a Participatory Management Plan that solicits input from fishermen, tour operators, guides, CDRS, and GNPS. Nevertheless, political tensions continue.

Promoting administrative stability within GNPS

In recent years, the administration of GNPS has been unstable. The director of GNPS faces a daunting challenge, for he or she must balance the ecological integrity of the park with the demands of tourists and local residents who often oppose the environmental restrictions necessary to protect the park. This balance has been difficult to achieve, and GNPS officials who have tried to enforce regulations have often been fired; this helps explain why there were 12 directors of GNPS between 2003 and 2006.[26] In 2006, GNPS director Raquel Molina increased the number of patrols within GNP and the number of prosecutions against violators. When she was fired in 2008, her supervisor claimed that she had failed to protect the park, but a memorandum from the U.S. consul to Ecuador blamed Molina's firing on her "unwillingness to turn a blind eye to corruption and unlawful activities in the park." Although GNPS now has a merit-based system for appointing a director, political pressures continue to influence the administration of GNPS.

Reducing the importation of fossil fuels

Although the *Jessica* oil spill in 2001 highlighted the perils of importing fossil fuels to Galápagos (Figure 2.13), the archipelago's use of fossil fuels has increased by more than 8% per year since the spill. Galápagos has set an ambitious goal of energy independence from fossil fuels by 2017. When this book went to press in 2012, that goal remained distant, but progress was being made. For example, in 2007 the San Cristóbal Wind Park—the first such park in

[26] These political appointments mirrored Ecuador's political instability. Between 1998 and 2010, Ecuador had eight presidents, all of whom were elected to four-year terms.

FIGURE 11.11 There are tortoise breeding programs on Santa Cruz, San Cristóbal, and Isabela. Shown here are a six-year-old tortoise (right) and Genesis (left), the only survivor of the first group of tortoises hatched at the center in 2005. The smaller tortoise is approximately 25 centimeters (10 in) long.

Ecuador—began decreasing the island's dependence on fossil fuels. The turbines of the park, which are visible from Laguna El Junco, provided 31% of San Cristóbal's electricity in their first year of operation.[27] San Cristóbal also has plans for a large solar energy facility, and Floreana has generators powered by piñon oil (which is imported from the mainland). In 2011, to conserve fuel and provide incentives for fuel-efficient engines, Ecuador removed its subsidy on ships using more than 4,000 gallons per month of diesel fuel.[28] This change added $100 to $300 to the cost of typical cruises through the islands.

Assisting threatened species

Most visitors to Galápagos visit at least one of the archipelago's three breeding centers: "Fausto Llerena" on Santa Cruz (established in 1965 and renovated in 1999), "Cerro Colorado" on San Cristóbal (established in 2004), and "Arnoldo Tupiza" on Isabela (established in 1995). These centers have focused on breeding and repatriating land iguanas and tortoises (Figure 11.11).

> *Tortoises:* The first repatriation of tortoises began in 1965, when GNPS rangers brought the first tortoise eggs from Pinzón to CDRS. Five years later, 20 tortoises hatched from these eggs were repatriated to the island. (Before this repatriation, black rats had killed *all* of the tortoise hatchlings on Pinzón for several decades.) In 1973, 17 five-year-old tortoises raised in captivity at CDRS were returned to Española; 42 more were returned in 1975, 81 more in 1976, and 74 more in 1984. In 2001, GNPS repatriated

[27] In its first year of operation, the San Cristóbal Wind Park provided 2,543 MWh of power, thereby avoiding the release of 2,034 tons of CO_2 (1 MWh = megawatt hour = 1,000 kilowatt hours = 10^6 watts of power used for an hour).

[28] This subsidy had promoted excessive consumption; between 2003 and 2010, fuel consumption in Galápagos increased by more than 8% per year.

its 1,000th tortoise to Española, and by 2012 it had repatriated more than 5,000 tortoises bred in captivity to several populations throughout Galápagos.[29] These tortoises are among only a few species ever successfully reintroduced to the wild exclusively from a captive-breeding program. In 2010, Project Pinta returned 39 non-native hybrid adult tortoises to Pinta to help restore the island's ecological balance (Chapter 6; the tortoises were sterilized to prevent cross-breeding). These were the first adult tortoises to be repatriated in the archipelago, and the first tortoises on Pinta since Lonesome George was removed in 1972. This reintroduction of tortoises on Pinta was featured on NBC's *Today Show.*

The Sex Life of a Galápagos Icon

Lonesome George, whom *Science* magazine called "an ideal poster boy for conservation," was the last known tortoise from Pinta Island. After his discovery in 1971, George was moved to the CDRS on Santa Cruz, where park officials began trying to help him reproduce. From 1972 to 1992, George lived with two female tortoises of unknown origin, but produced no offspring. CDRS officials then brought in two female tortoises from Isabela's Volcán Wolf that were morphologically more similar to George than his previous pen mates. Despite a new pen and diet designed to boost his libido, George sired no offspring.

In 2001, genetic analysis indicated that two females from Española were more closely related to Lonesome George than were the Volcán Wolf females. When CDRS paired George with these tortoises, one of the females nested in 2008, but the eggs were not viable. The following year, both females laid eggs, but none hatched.

George died in 2012; he had produced no offspring at GNPS. Soon after George died, Ecuador's President Rafael Correa mourned George's death in an address to the nation, noting "one day, science and technology will be able to reproduce him, to clone him." A plaque outside George's pen provided an important reminder for visitors to Galápagos: "Whatever happens to this single animal, let him always remind us that the fate of all living things on Earth is in human hands."

Land iguanas: One of the earliest and most important attempts to conserve the biodiversity of Galápagos was a fortuitous one. In 1932, zoologist Cy Perkins and tycoon Allan Hancock (Chapter 2) of the *Velero III*—sensing that land iguanas on Baltra were not faring well—moved 66 land iguanas from Baltra to North Seymour. The following year, they moved another 26 land iguanas from Baltra to the same island. These iguanas and their offspring were later used to repatriate Baltra when land iguanas there disappeared during the U.S. occupation of the island in the

[29] These repatriations include more than 2,500 tortoises to Santa Cruz, 1,500 to Isabela, 600 to Pinzón, and more than 1,700 to Española. In 2012, the CDRS tortoise nursery was showcasing baby tortoises from Pinzón, Santiago, and Española.

1940s (Chapter 2; see also "Did American Troops Kill All of Baltra's Land Iguanas?" in Chapter 6).

When feral dogs threatened several populations of land iguanas in 1975, officials moved iguanas from the threatened populations to islets of Venecia just northwest of Santa Cruz. When the iguanas could find no areas for nesting, GNPS officials moved 100 square meters of soil from Santa Cruz to the islets to build an artificial nesting-site for the iguanas. The populations thrived and, after the feral dogs were eliminated, the iguanas were repatriated.

To date, CDRS has restored endangered populations of land iguanas on Baltra, Santa Cruz (Cerro Dragón), Isabela (Cartago Bay), and elsewhere in the archipelago. The breeding of Baltra land iguanas at CDRS ended in 2008 after the successful repatriation of more than 800 iguanas.

GNPS officials have gone to great lengths to protect species threatened with extinction. For example, when an eruption of Cerro Azul in 1998 threatened resident tortoises, GNPS brought 18 adults to the breeding center at Isabela. Two years later, these adults had produced more than 200 baby tortoises. More recently, in 2011, GNPS officials began building homes for penguins on three islands of the archipelago to ensure that they have suitable nesting areas. That same year, GNPS temporarily moved hawks from Rábida, Bartolomé, and Sombrero Chino to aviaries on Santiago while eradicating rats with poisons on Rábida.

Finding, identifying, propagating, and repatriating other threatened species

Increased vigilance has uncovered several species once believed to be extinct. For example, Santiago's *Scalesia atractyloides* was assumed extinct because of grazing by feral goats, but in 1995 GNPS rangers found five individuals growing on cliffs inaccessible to goats. Similarly, *Linum cratericola* (Floreana flax) had not been seen since 1981, but in 1997 four plants were discovered on Floreana at one of the two original sites of growth (a road-stone quarry from which it had originally disappeared because of habitat destruction). In 1997, after having not been seen since 1906, the endemic mouse *Nesoryzomys swarthi* was rediscovered on Santiago. These and similar species remain endangered, but GNPS officials are working to prevent their extinction.

Genetic analyses are helping identify and find threatened species. For example, in early 2012, genetic analyses of 1,669 tortoises indicated that a Floreana giant tortoise (*Chelonoidis elephantopus*) thought to be extinct for more than 150 years may be alive at Isabela's Volcán Wolf. (Whalers and pirates often left tortoises gathered from other islands at Volcán Wolf before departing Galápagos.) More than 5% of the tortoises had DNA indicating a Floreana tortoise as a parent. Many of these tortoises were less than 15 years old, suggesting that the Floreana tortoise could still be alive. While GNPS searches for this tortoise, repatriation programs continue.

Recycling waste

Galápagos imports large amounts of food, water, and other materials, and the islands' dependence on imports continues to grow. Indeed, between 2006 and 2011, the amount of freight arriving in Galápagos increased by more than 18% per year. These imports, combined with the increasing numbers of visitors and residents, has increased the production of garbage in the archipelago by more than 10% per year during the past decade. In 2006, GNPS and local residents of Santa Cruz created the Fabricio Valverde Environmental Center to recycle wastes produced on the island. During its first year of operation, the amount of recycled materials increased from 17 to 35 tons per month. Today, more than 40% of all household waste on Santa Cruz is processed by the island's recycling plant; much of this waste is made into compost. Additional recycling centers are planned for San Cristóbal and Isabela.

THE FUTURE OF GALÁPAGOS

Galápagos is no longer the isolated, relatively unknown archipelago that Charles Darwin visited in 1835. Today, Galápagos is a thriving tourist destination visited by more than 170,000 visitors per year (Figure 1).[30] This popularity has changed the islands and has created many challenges, most of which arise from the fact that the Galápagos economy is often at odds with the archipelago's sustainability and long-term conservation. Indeed, the Galápagos economy demands an increasing flow of goods and services from elsewhere, but its conservation depends on the relative isolation that has characterized Galápagos for all of its prehuman history. Can this conflict be resolved or, if not, managed?

In some ways, Galápagos is in better shape than it was 100 years ago; several invasive species have been eliminated, environmental impacts are now monitored, and GNPS and CDRS are trying to protect several endangered species. However, these successes have been tempered by setbacks. For example, since the early 1990s, vengeful fishermen—using invasive species as a political tool—have repeatedly reintroduced goats on several islands as a means of influencing policies. These reintroductions are damaging and expensive to remedy; for example, the six goats reintroduced onto Santiago in 2009 cost GNPS $32,393 to eradicate. Illegal fishing remains relatively common, and penalties are often insignificant. Also, several invasive species have proven exceedingly difficult to

[30] To accommodate these tourists, the number of restaurants in Galápagos increased from 20 to 141 between 1982 and 2006, and the number of hotels increased from 18 to 65 in the same period.

eradicate. For example, GNPS has been trying to eradicate guava and quinine since 1971, but these species remain abundant and continue to displace native and endemic plants.

Preserving Galápagos will require vast resources; removing invasive plants from farmland can cost more than $2,000 per hectare (plus maintenance costs exceeding $500 per hectare per year),[31] and a field study of just one island can cost more than $90,000. Individual donors and governmental agencies have donated millions of dollars for preservation, but these donations have not met the growing needs of the islands. Large amounts of money for GNPS and CDRS must continue to come from tourists, who are now a foundation of the Galápagos economy.[32] All visitors (except those aboard private yachts) must enter from mainland Ecuador and pay a park entrance fee. Although tourism increases the chances of exotic organisms reaching Galápagos, 40% of the entrance fee goes to GNPS.[33] These fees generate millions of dollars for GNPS. Thus, although people have been (and continue to be) the greatest threat to the biodiversity of Galápagos, people are also the only thing that can save the islands. Galápagos could be saved by eradicating all invasive species, closing all land and water ports, and abandoning the archipelago, but this is not going to happen. High-quality, low-impact ecotourism is the only viable commercial activity capable of both protecting the islands and providing sustainable livelihoods for the islands' residents. Your visit can be fun and informative, and can also help conserve the islands for future generations (Figure 11.12).

Explorers from Dampier to Darwin—and thousands more since—have seen Galápagos and shared tales of its history, landscapes, and wildlife. We have written books, such as this one, that describe the plants, animals, and evolutionary stories that must be seen to be believed. We are impressed by the dazzling colors of Sally lightfoot crabs, and we are stunned when we see the prehistoric lumbering of a land iguana or giant tortoise, or when we realize that we are snorkeling midocean with a penguin. We know these kinds of experiences are common in Galápagos, and we save for years, or for lifetimes, to go see them for ourselves.

In *Galápagos* (1960), biologist Irenäus Eibl-Eibesfeldt noted, "Slowly but surely we men are covering our planet with asphalt and concrete and we can

[31] Adding to the challenge is the fact that much of the agricultural area on inhabited islands has been abandoned, thereby creating a fertile seeding area for the invasion of the surrounding GNP.

[32] Whereas more than half of the economy of the archipelago was based on agriculture, fishing, and ranching in the 1960s, today these sectors account for less than 5% of the islands' economy and provide fewer than 15% of the jobs in the islands. Tourism has significantly increased the gross domestic product of Galápagos, but per-capita income in the archipelago has lagged far behind. Most of the money spent by tourists to visit Galápagos continues to be spent outside of the archipelago.

[33] The Special Law for Galápagos (1998) divides the National Park Entrance Fee as follows: GNPS: 40%; Galápagos municipalities: 20%; Provincial government: 10%; INGALA: 10%; Ministry of the Environment: 5%; Galápagos Marine Reserve: 5%; SICGAL: 5%; Ecuadorian Navy: 5%.

see how, in a few decades, natural beauty which has lasted for millions of years, has been destroyed forever. . . . Let us, then, do our best to see that at least the . . . Galápagos Islands, that are so rich in natural marvels, are kept undisturbed for ourselves and for those who come after us."[34] Let us hope that the money, energy, and hopes of the islands' tourists, residents, and other people concerned about the future of Galápagos make it into the right hands and can be used to preserve this magical place.

We promise to tell your story and to share your conservation message.

—Memorial to Lonesome George at the Charles Darwin Research Station (Puerto Ayora, Santa Cruz)

---

[34] As this book went to press, there were encouraging signs that Ecuador's judicial system was becoming more vigilant about enforcing its environmental-protection laws. For example, German national Dirk Bender was arrested in 2012 at the Baltra airport while trying to smuggle four land iguanas out of Galápagos. (In 2011, Bender had been arrested for a similar offense in Fiji.) Although other tourists had been caught trying to smuggle animals out of Galápagos, Bender was the first defendant to be denied bail and kept in jail (and, therefore, unable to flee the islands) for an environmental crime. Bender faced up to four years in prison, and conservationists hope his case represents a turning point in how Galápagos enforces its environmental-protection laws.

fundación
Charles Darwin
foundation

Charles Darwin Foundation-*Working for Galápagos since 1959*

Understanding Galápagos: What You'll See and What It Means will appreciate why Galápagos is world-renowned as a storehouse of unique terrestrial and marine biological diversity and a natural laboratory for biological evoluvtion and speciation. However, during the more than five decades the Charles Darwin Foundation (CDF) has been operating in Galápagos, the archipelago has become a microcosm of changes taking place the world over.

CDF and our partners have been very successful in finding solutions to the inherited problems associated with the human footprint from centuries past and mitigating modern threats. CDF was created to provide scientific information to the Ecuadorian Government to safeguard the Galápagos environment and biodiversity of Galápagos. Our information is used by decision-makers, especially those involved in conservation, bio-security, sustainable development, and planning. We work with government and other partners from many sectors: Fisheries, quarantine, conservation, tourism, planning, education, and agriculture.

We are also very proud to provide opportunities to the islands' residents, be that in the form of employment, scholarships or volunteer positions, and providing tools so local people can play an active role in protecting their home. Many are now contributing to science, policy and the archipelago from key positions in Galápagos, Ecuador, or other parts of the world.

The CDF is the only organization offering onsite research, knowledge, information and technical assistance to ensure the conservation of Galápagos ecosystems and biodiversity. Our success depends on the generosity of individuals and organizations and it is the trust and ongoing commitment of our supporters that enables our legacy to endure.

The authors of this book have pledged a part of their royalties to CDF. I hope that you will join them today by making a gift and helping to ensure that we help bring Galápagos, unique in the entire world, closer to a sustainable future.

Thank you for choosing to join with our efforts.

With best wishes,

S. Lorenz

Swen Lorenz
Executive Director

Puerto Ayora, Santa Cruz
Galapagos, Ecuador
PO Box 17-1-3891 Quito - Ecuador
(593-5) 2526 146/147
cdrs@fcdarwin.org.ec
www.darwinfoundation.org

The Charles Darwin Foundation operates the Charles Darwin Research Station in Puerto Ayora, Santa Cruz Island, Galapagos Islands, Ecuador. The Charles Darwin Foundation is an *Association Internationale Sans But Lucratif* (AISBL), registered in Belgium under the number 371359 and subject to Belgian law. The address in Belgium is Avenue Louise 50, 1050 Brussels.

FIGURE 11.12
The Charles Darwin Foundation protects the unique ecosystems of Galápagos. The success of the Charles Darwin Foundation depends on individuals and organizations concerned about the future of the archipelago.

CHAPTER 12

A Checklist for What You Will See in Galápagos

Wildlife in Galápagos is diverse and remarkably approachable. A great way to keep track of the species you see is with the island-by-island checklist presented here. This list includes all of the native and endemic plants and animals described in this book. Chapter 4 describes the particular sites and trails on islands where you will see many of these organisms, and Chapters 5–10 describe their evolution and behaviors. As you tour the islands, use this list to compile your personal inventory of the plants and animals of Galápagos. There is also room for you to record other species that you see, as well as space for notes about your encounters with these organisms.

Although the plants and animals you will see during your tour of the islands will depend on when and where you go (Chapter 3), you *will* see many of the native and endemic plants and animals listed here and discussed in this book. You will probably not visit all of the islands during your visit, and land visits to several islands (marked in the checklist with an asterisk *) are prohibited. However, your boat may get close enough to some of these islands for you to see their plants and animals (e.g., if you circumnavigate Daphne Major, or if you are aboard a dive boat near Pinta, Pinzón, Marchena, Darwin, or Wolf Islands). When you finish your hike, compare your notes with those of others in your group. An updated inventory of the species living in Galápagos is the *Galápagos Species Checklist* at the CDRS.

Do not hesitate to ask your GNPS guide to identify the plants and animals that you do not recognize. Also be on the lookout for new creatures; according to the CDRS, more than 50% of all species in Galápagos are probably unknown, and you may discover one of these new species.

Enjoy looking for and watching the plants and animals of Galápagos.

	Baltra	Bartolomé	Daphne Major	Darwin*	Española	Fernandina	Floreana	Genovesa	Isabela	Marchena*	North Seymour	Pinta*	Pinzón*	Rábida	San Cristóbal	Santa Cruz	Santa Fé	Santiago	South Plaza	Wolf*	Notes
BIRDS																					
Land Birds																					
Medium ground finch (*Geospiza fortis*)																					
Large ground finch (*Geospiza magnirostris*)																					
Small ground finch (*Geospiza fuliginosa*)																					
Sharp-beaked ground finch (*Geospiza difficilis*)																					
Common cactus finch (*Geospiza scandens*)																					
Large cactus finch (*Geospiza conirostris*)																					
Vegetarian finch (*Platyspiza crassirostris*)																					
Medium tree finch (*Camarhynchus pauper*)																					
Small tree finch (*Camarhynchus parvulus*)																					
Large tree finch (*Camarhynchus psittacula*)																					
Woodpecker finch (*Cactospiza pallidus*)																					
Mangrove finch (*Cactospiza heliobates*)																					
Green warbler finch (*Certhidea olivacea*)																					
Grey warbler finch (*Certhidea fusca*)																					
Galápagos hawk (*Buteo galapagoensis*)																					
Short-eared owl (*Asio flammeus galapagoensis*)																					
Galápagos mockingbird (*Mimus parvulus*)																					
San Cristóbal mockingbird (*Mimus melanotis*)																					
Floreana mockingbird (*Mimus trifasciatus*)																					
Hood mockingbird (*Mimus macdonaldi*)																					
Galápagos dove (*Zenaida galapagoensis*)																					
Galápagos rail (*Laterallus spilonotus*)																					
Galápagos flycatcher (*Myriarchus magnirostris*)																					
Vermilion flycatcher (*Pyrocephalus rubinus*)																					

	Baltra	Bartolomé	Daphne Major	Darwin*	Española	Fernandina	Floreana	Genovesa	Isabela	Marchena*	North Seymour	Pinta*	Pinzón*	Rábida	San Cristóbal	Santa Cruz	Santa Fé	Santiago	South Plaza	Wolf*	Notes
Sea and Shorebirds																					
Blue-footed booby (*Sula nebouxii excisa*)																					
Nazca booby (*Sula granti*)																					
Red-footed booby (*Sula sula websteri*)																					
Galápagos penguin (*Spheniscus mendiculus*)																					
Galápagos petrel (*Pterodroma phaeopygia*)																					
Galápagos shearwater (*Puffinus subalaris*)																					
Red-billed tropicbird (*Phaeton aethereus*)																					
Brown pelican (*Pelecanus occidentalis urinator*)																					
Magnificent frigatebird (*Fregata magnificens*)																					
Great frigatebird (*Fregata minor ridgwayi*)																					
Flightless cormorant (*Phalacrocorax harrisi*)																					
Lava gull (*Larus fuliginosus*)																					
Lava heron (*Butorides sundevalli*)																					
Yellow-crowned night heron (*Nyctanassa violacea*)																					
White-cheeked pintail (*Anas bahamensis galapagensis*)																					
Greater flamingo (*Phoenicopterus ruber*)																					
American oystercatcher (*Haematopus palliatus*)																					
Waved albatross (*Phoebastria irrorata*)																					
Other:																					
Other:																					
Other:																					
Other:																					
Other:																					
Other:																					

(*Continued*)

	Baltra	Bartolomé	Daphne Major	Darwin*	Española	Fernandina	Floreana	Genovesa	Isabela	Marchena*	North Seymour	Pinta*	Pinzón*	Rábida	San Cristóbal	Santa Cruz	Santa Fé	Santiago	South Plaza	Wolf*	Notes
REPTILES																					
Galápagos giant tortoise (*Chelonoidis nigra*)																					
Galápagos green turtle (*Chelonia mydas agassizii*)																					
Lava lizard (*Microlophus* sp.)																					
Common land iguana (*Conolophus subcristatus*)																					
Santa Fé land iguana (*Conolophus pallidus*)																					
Galápagos pink land iguana (*Conolophus marthae*)																					
Marine iguana (*Amblyrhynchus cristatus*)																					
Other:																					
MAMMALS																					
Galápagos sea lion (*Zalophus wollebaeki*)																					
Galápagos fur sea lion (*Arctocephalus galapagoensis*)																					
Common dolphin (*Delphinus delphis*)																					
Santiago rice rat, or Santiago Galápagos mouse (*Nesoryzomys swarthi*)																					
Galápagos rice rat (*Aegialomys galapagoensis*)																					
Large Fernandina rice rat (*Nesoryzmys narboroughi*)																					
Fernandina Galápagos mouse, or Fernandina rice rat (*Nesoryzmys fernandinae*)																					
Galápagos bat (*Lasiurus brachyotis*)																					
Hoary bat (*Lasiurus cinereus*)																					
Black rat, or ship rat (*Rattus rattus*)																					
Norway rat, or brown rat (*Rattus norvegicus*)																					
Other:																					
Other:																					
Other:																					

	Baltra	Bartolomé	Daphne Major	Darwin*	Española	Fernandina	Floreana	Genovesa	Isabela	Marchena*	North Seymour	Pinta*	Pinzón*	Rábida	San Cristóbal	Santa Cruz	Santa Fé	Santiago	South Plaza	Wolf*	Notes
FISH																					
Goldrim surgeonfish (*Acanthurus nigricans*)																					
Yellowfin surgeonfish (*Acanthurus xanthopterus*)																					
Yellowtail (or razor) surgeonfish (*Prionurus laticlavius*)																					
Moorish idol (*Zanclus cornutu*)																					
King angelfish (*Holacanthus passer*)																					
Barberfish (*Johnrandallia nigrirostris*)																					
Threebanded butterflyfish (*Chaetodon humeralis*)																					
Giant damselfish (*Microspathodon dorsalis*)																					
Bumphead damselfish (*Microspathodon bairdii*)																					
Galápagos ringtail damselfish (*Stegastes beebei*)																					
Panamic sergeant major (*Abudefduf troschelii*)																					
Yellowtail damselfish (*Stegastes arcifrons*)																					
Scissortail (or whitespot, or brown) chromis (*Chromis atrilobata*)																					
Pacific creolefish (*Paranthias colonus*)																					
Bumphead parrotfish (*Scarus perrico*)																					
Bicolor parrotfish (*Scarus rubroviolaceus*)																					
Blue-chin parrotfish (*Scarus ghobban*)																					
Mexican hogfish (*Bodianus diplotaenia*)																					
Galápagos (or goldspot) sheephead (*Semicossyphus darwini*)																					
Harlequin wrasse (*Bodianus eclancheri*)																					
Cortez rainbow wrasse (*Thalassoma lucasanum*)																					
Chameleon wrasse (*Halichoeres dispilus*)																					
Panamic (or Crimson) soldierfish (*Myripristis leiognathos*)																					
Blacktip cardinalfish (*Apogon atradorsatus*)																					

(*Continued*)

Species	Baltra	Bartolomé	Daphne Major	Darwin*	Española	Fernandina	Floreana	Genovesa	Isabela	Marchena*	North Seymour	Pinta*	Pinzón*	Rábida	San Cristóbal	Santa Cruz	Santa Fé	Santiago	South Plaza	Wolf*	Notes
Pink cardinalfish (*Apogon pacificus*)																					
Galápagos barnacle blenny (*Acanthemblemaria castroi*)																					
Sabretooth blenny (*Plagiotremus azalea*)																					
Panamic fanged blenny (*Ophioblennius steindachneri*)																					
Galápagos four-eyed blenny (*Dialommus fuscus*)																					
Bravo clinid (*Labrisomus dendriticus*)																					
Galápagos triplefin blenny (*Lepidonectes corallicola*)																					
Marbled goby (*Eleotrica cableae*)																					
Blue-banded goby (*Lythrypnus gilberti*)																					
Red-lipped batfish (*Ogcocephalus darwini*)																					
Tropical flounder (*Bothus mancus*)																					
Pacific leopard flounder (*Bothus leopardinus*)																					
Rainbow scorpionfish (*Scorpaenodes xyris*)																					
Stone scorpionfish (*Scorpaena plumieri mystes*)																					
Player scorpionfish (*Scorpaena histrio*)																					
Giant hawkfish (*Cirrhitus rivulatus*)																					
Coral hawkfish (*Cirrhitichthys oxycephalus*)																					
Guineafowl puffer (*Arothron meleagris*)																					
Bullseye (or concentric) puffer (*Sphoeroides annulatus*)																					
Orangeside (or yellow-bellied) triggerfish (*Sufflamen verres*)																					
Blunthead triggerfish (*Pseudobalistes naufragium*)																					
Mexican (or yellow-tailed) goatfish (*Mulloidichthys dentatus*)																					
Steel pompano (*Trachinotus stilbe*)																					
Bigeye trevalle (*Caranx sexfasciatus*)																					
Indo-Pacific bonito (*Sarda orientalis*)																					
Yellowfin tuna (*Thunnus albacares*)																					
Longfin halfbeak (*Hemiramphus saltator*)																					

	Baltra	Bartolomé	Daphne Major	Darwin*	Española	Fernandina	Floreana	Genovesa	Isabela	Marchena*	North Seymour	Pinta*	Pinzón*	Rábida	San Cristóbal	Santa Cruz	Santa Fé	Santiago	South Plaza	Wolf*	Notes
Reef cornetfish (*Fistularia commersonii*)																					
Trumpetfish (*Aulostomus chinensis*)																					
Fantail pipefish (*Doryrhamphus excisus*)																					
Pacific seahorse (*Hippocampus ingens*)																					
Galápagos garden eel (*Taenioconger klausewitzi*)																					
Tiger snake eel (*Myrichthys tigrinus*)																					
Fine-spotted moray (*Gymnothorax dovii*)																					
Jewel (or lentil) moray (*Muraena lentiginosa*)																					
Galápagos shark (*Carcharhinus galapagensis*)																					
White-tipped reef shark (*Triaenodon obesus*)																					
Whale shark (*Rhincodon typus*)																					
Hammerhead shark (*Sphyrna lewini*)																					
Diamond (or whiptail) stingray (*Dasyatis brevis*)																					
Golden cowray (*Rhinoptera steindachneri*)																					
Spotted eagle ray (*Aetobatus narinari*)																					
Manta ray (*Manta hamiltoni*)																					
Other:																					
Other:																					
Other:																					
Other:																					
INVERTEBRATES																					
Orange cup coral (*Tubastrea coccinea*)																					
Green sea urchin (*Lytechinus semituberculatus*)																					
Pencil-spined sea urchin, or slate-pencil urchin (*Eucidaris thouarsii*)																					

(*Continued*)

	Baltra	Bartolomé	Daphne Major	Darwin*	Española	Fernandina	Floreana	Genovesa	Isabela	Marchena*	North Seymour	Pinta*	Pinzón*	Rábida	San Cristóbal	Santa Cruz	Santa Fé	Santiago	South Plaza	Wolf*	Notes
Crowned sea urchin (or hatpin sea urchin; *Centrostephanus coronatus*)																					
Galápagos sea cucumber, or "pepino" (*Isostichopus fuscus*)																					
Chocolate-chip sea star (*Nidorellia armata*)																					
Panamic cushion star (*Pentaceraster cumingi*)																					
Variable sea star (*Linckia columbiae*)																					
Sally lightfoot crab (*Grapsus grapsus*)																					
Ghost crab (*Ocypode gaudichaudii*)																					
Fiddler crab (*Uca helleri*)																					
Galápagos hermit crab (*Calcinus explorator*)																					
Galápagos giant centipede (*Scolopendra galapagoensis*)																					
Galápagos scorpion (*Hadruroides maculatus galapagoensis*)																					
Galápagos black widow spider (*Latrodectus apicalis*)																					
Zig-zag spider (*Neoscona cooksoni*)																					
Galápagos blue butterfly (*Leptodes parrhasioides*)																					
Galápagos sulfur butterfly (*Phoebis sennae marcellina*)																					
Monarch butterfly (*Danaus plexippus*)																					
Queen butterfly (*Danaus gilippus*)																					
Green hawkmoth (*Eumorpha labruscae yupanquii*)																					
Galápagos carpenter bee (*Xylocopa darwini*)																					
Galápagos carpenter ants (*Camponotus* spp.)																					
Little red fire ant (*Wasmannia auropunctata*)																					
Galápagos field cricket (*Gryllus abditus*)																					
Large painted locust (*Schistocerca melanocera*)																					
Small painted locust (*Schistocerca literosa*)																					
Galápagos flightless grasshopper (*Halemus robustus*)																					

	Baltra	Bartolomé	Daphne Major	Darwin*	Española	Fernandina	Floreana	Genovesa	Isabela	Marchena*	North Seymour	Pinta*	Pinzón*	Rábida	San Cristóbal	Santa Cruz	Santa Fé	Santiago	South Plaza	Wolf*	Notes
Spotless ladybug (*Cycloneda galapagoensis*)																					
Vedelia beetle, or cardinal ladybug (*Rodolia cardinalis*)																					
Galápagos green-eyed horsefly (*Tabanus vittiger*)																					
Other:																					
Other:																					
Other:																					
PLANTS																					
Black mangrove (*Avicennia germinans*)																					
Red mangrove (*Rhizophora mangle*)																					
White mangrove (*Laguncularia racemosa*)																					
Button mangrove (*Conocarpus erectus*)																					
Beach morning glory (*Ipomoea pes-caprae*)																					
Lava morning glory (*Ipomoea habeliana*)																					
Leatherleaf (*Maytenus octogona*)																					
Saltbush (*Cryptocarpus pyriformis*)																					
Saltwort (*Batis maritima*)																					
Scorpion weed (*Heliotropium curassavicum*)																					
Salt sage (*Atriplex peruviana*)																					
Galápagos clubleaf (*Nolana galapagensis*)																					
Galápagos carpetweed (*Sesuvium edmonstonei*)																					
Common carpetweed (*Sesuvium portulacastrum*)																					
Galápagos shore petunia (*Exedeconus miersii*)																					
Lava cactus (*Brachycereus nesioticus*)																					
Candelabra cactus (*Jasminocereus thouarsii*)																					
Prickly pear cactus (*Opuntia* spp.)																					

(Continued)

	Baltra	Bartolomé	Daphne Major	Darwin*	Española	Fernandina	Floreana	Genovesa	Isabela	Marchena*	North Seymour	Pinta*	Pinzón*	Rábida	San Cristóbal	Santa Cruz	Santa Fé	Santiago	South Plaza	Wolf*	Notes
Palo santo (*Bursera graveolens*)																					
Dwarf palo santo (*Bursera malacophylla*)																					
Matazarno (*Piscidia carthagenensis*)																					
Pega pega (*Pisonia floribunda*)																					
Guayabillo (*Psidium galapageium*)																					
Poison apple (*Hippomane mancinella*)																					
Muyuyo, or yellow cordia (*Cordia lutea*)																					
Parkinsonia (*Parkinsonia aculeate*)																					
Galápagos acacia (*Acacia rorudiana*)																					
Mesquite (*Prosopis juliflora*)																					
Galápagos croton (*Croton scouleri*)																					
Darwin's cotton (*Gossypium darwinii*)																					
Lance-leafed cordia (*Cordia leucophylctis*)																					
Galápagos lantana (*Lantana peduncularis*)																					
White-haired tournefortia (*Tournefortia pubescens*)																					
Red-haired tournefortia (*Tournefortia rufo-sericea*)																					
Spiny bush (*Scutia spicata*)																					
Bitterbush (*Castela galapageia*)																					
Desert thorn (*Lycium minimum*)																					
Pearl berry (*Vallesia glabra*)																					
Thread-leafed chaff flower (*Alternanthera* spp.)																					
Thin-leafed Darwin's shrub (*Darwiniothamnus tenuifolius*)																					
Cutleaf daisy (*Lecocarpus pinnatifidus*)																					
Needle-leafed daisy (*Macraea laricifolia*)																					
Galápagos passion flower (*Passiflora foetida*)																					
Grey matplant (*Tiquilia* spp.)																					
Spurge (*Chamaesyce* spp.)																					
Galápagos milkwort (*Polygala* spp.)																					
Mollugo (*Mollugo* spp.)																					

	Baltra	Bartolomé	Daphne Major	Darwin*	Española	Fernandina	Floreana	Genovesa	Isabela	Marchena*	North Seymour	Pinta*	Pinzón*	Rábida	San Cristóbal	Santa Cruz	Santa Fé	Santiago	South Plaza	Wolf*	**Notes**
Galápagos sarcostemma (*Sarcostemma angustissimum*)																					
Galápagos tomato (*Solanum cheesmanii*)																					
Goat's head (*Tribulus cistoides*)																					
Stickleaf (*Mentzelia aspera*)																					
Galápagos purslane (*Portulaca howellii*)																					
Arrow-leafed morning glory (*Ipomoea linearifolia*)																					
Oily pectis (*Pectis tenuifolia*)																					
Galápagos bean (*Phaseolus mollis*)																					
Scalesia (*Scalesia* spp.)																					
Soapberry (*Sapindus saponaria*)																					
Miconia (*Miconia robinsoniana*)																					
Cat's claw (*Zanthoxylum fagara*)																					
Galápagos mistletoe (*Phoradendron henslowii*)																					
Galápagos dead nettle (*Pilea baurii*)																					
Galápagos peperomia (*Peperomia galapagensis*)																					
Galápagos orchid (*Epidendrum spicatum*)																					
Galápagos tillandsia (*Tillandsia insularis*)																					
Galápagos justicia (*Justicia galapagana*)																					
Galápagos jaegeria (*Jaegeria gracilis*)																					
Galápagos three-awn grass (*Aristida subspicata*)																					
Indefatigable passion flower (*Passiflora colinvauxii*)																					
Galápagos sedge (*Cyperus anderssonii*)																					
Galápagos tree fern (*Cyathea weatherbyana*)																					
Club moss (*Lycopodium* spp.)																					
Other:																					
Other:																					
Other:																					

APPENDIX 1

GPS Coordinates of Various Sites in Galápagos

This appendix will help you locate many of the sites mentioned in this book. GPS coordinates for specific sites are accurate to ±10 meters.

Galápagos Archipelago		01°40N–01°25S	89°16W–92°06W
Center of the Archipelago		00°32.132S	90°31.156W
Baltra	Abandoned airstrip from World War II	00°28.322S	90°17.096W
Baltra	Aeolian Bay	00°26.238S	90°17.062W
Baltra	Airport (GPS)	00°26.588S	90°16.413W
Baltra	Airport, area to board bus	00°26.737S	90°16.160W
Baltra	Airport, baggage claim	00°26.761S	90°16.159W
Baltra	Black Turtle Cove, eastern entrance	00°29.740S	90°19.588W
Baltra	Black Turtle Cove, inner cove	00°30.200S	90°19.833W
Baltra	Itabaca Channel, boarding site	00°29.004S	90°16.643W
Baltra	Mosquera	00°24.175S	90°16.415W
Baltra	Fueling station (from World War II)	00°26.187S	90°17.107W
Bartolomé	Base of boardwalk	00°16.836S	90°33.125W
Bartolomé	First overlook	00°16.937S	90°33.001W
Bartolomé	Second overlook	00°16.975S	90°33.005W
Bartolomé	Third overlook	00°17.033S	90°32.991W
Bartolomé	Fourth overlook	00°17.047S	90°33.035W
Bartolomé	Landing site for climb to summit	00°16.833S	90°33.122W
Bartolomé	North beach, landing site	00°17.109S	90°33.328W
Bartolomé	North beach, monument	00°17.066S	90°33.304W
Bartolomé	Pinnacle Rock	00°16.868S	90°33.253W
Bartolomé	South beach	00°17.176S	90°33.283W
Bartolomé	Summit, near beacon	00°17.006S	90°33.044W
Bartolomé	*Tiquilia* forest	00°16.896S	90°33.052W
Daphne Major		00°25.225S	90°22.202W
Daphne Minor		00°23.445S	90°21.526W
Darwin		01°39.236N	92°00.193W

Española	Gardner Bay, beach	01°21.080S	89°39.727W
Española	Gardner Bay, anchorage	01°21.077S	89°39.650W
Española	Gardner Island, docking site	01°20.635S	89°38.901W
Española	Punta Cevallos	01°23.399S	89°37.054W
Española	Punta Suárez, albatross colony	01°22.352S	89°44.623W
Española	Punta Suárez, viewpoint to see blowhole	01°22.442S	89°44.544W
Española	Punta Suárez, dry-landing site	01°22.143S	89°44.680W
Española	Punta Suárez, monument and beacon	01°22.146S	89°44.701W
Española	Punta Suárez, albatross runway	01°22.415S	89°44.464W
Fernandina	Bolivar Channel	00°16.113S	91°26.514W
Fernandina	Cabo Douglas	00°18.169S	91°39.230W
Fernandina	Cabo Hammond	00°29.564S	91°35.355W
Fernandina	Punta Espinosa, anchorage	00°16.115S	91°26.510W
Fernandina	Punta Espinosa, flightless cormorants	00°15.972S	91°26.741W
Fernandina	Punta Espinosa, forest of lava cactus	00°15.904S	91°26.801W
Fernandina	Punta Espinosa, lagoon-like canal	00°15.999S	91°26.999W
Fernandina	Punta Espinosa, landing site	00°15.975S	91°26.731W
Fernandina	Punta Espinosa, alternate landing site	00°15.970S	91°26.726W
Fernandina	Punta Espinosa, monument	00°15.924S	91°26.775W
Fernandina	Punta Espinosa, pond in lava field	00°16.203S	91°27.000W
Fernandina	Punta Espinosa, rocky point	00°15.796S	91°26.744W
Fernandina	Punta Espinosa, whale skeleton	00°15.931S	91°26.800W
Fernandina	Punta Mangle	00°27.054S	91°23.136W
Floreana	Asilo de la Paz	01°18.517S	90°17.135W
Floreana	Asilo de la Paz, sign near Wittmer Farm	01°18.783S	90°27.280W
Floreana	Asilo de la Paz, trail from parking area	01°18.782S	90°27.274W
Floreana	Black Beach	01°16.574S	90°29.293W
Floreana	Black Beach, former residence of Wittmers	01°16.628S	90°29.306W
Floreana	Carving by Heinz Wittmer in highlands	01°18.843S	90°27.235W
Floreana	Caldwell	01°18.104S	90°20.161W
Floreana	Cemetery	01°16.677S	90°28.879W
Floreana	Cerro Pajas	01°16.174S	90°26.205W
Floreana	Champion Island	01°07.326S	90°22.570W
Floreana	Devil's Crown	01°12.569S	90°25.258W
Floreana	Enderby Island	01°13.519S	90°21.331W
Floreana	Gardner Island	01°19.520S	90°18.432W
Floreana	Grave of Friedrich Ritter	01°17.042S	90°28.167W
Floreana	Lagoon, north side	01°13.597S	90°25.704W
Floreana	Lagoon, overlook	01°13.618S	90°25.575W
Floreana	Mirador de la Baronesa, apex	01°13.879S	90°26.383W
Floreana	Mirador de la Baronesa, landing site	01°13.841S	90°26.424W
Floreana	Mirador de la Baronesa, beacon	01°13.876S	90°26.391W

Floreana	Original settlement	01°18.364S	90°27.066W
Floreana	Pirate caves	01°18.833S	90°27.232W
Floreana	Plaque honoring Rolf Wittmer	01°16.447S	90°29.336W
Floreana	Post Office Bay, anchorage	01°14.073S	90°27.206W
Floreana	Post Office Bay, landing site	01°14.185S	90°26.941W
Floreana	Post Office Bay, lava tube	01°14.272S	90°26.808W
Floreana	Post Office Bay, Post Office barrel	01°14.200S	90°26.927W
Floreana	Post Office Bay, remains of Casa Matriz	01°14.216S	90°26.907W
Floreana	Puerto Velasco Ibarra, anchorage	01°16.440S	90°29.497W
Floreana	Puerto Velasco Ibarra, dock and lighthouse	01°16.440S	90°29.342W
Floreana	Punta Cormorán, anchorage	01°13.329S	90°25.708W
Floreana	Punta Cormorán, lagoon, west side	01°13.601S	90°25.630W
Floreana	Punta Cormorán, landing site	01°13.516S	90°25.669W
Floreana	Punta Cormorán, trailhead	01°13.543S	90°25.694W
Floreana	Punta Cormorán, western beach	01°13.698S	90°25.425W
Floreana	Punta Cormorán, mirador	01°13.617S	90°25.574W
Floreana	Remains of house where the baroness and Nuggeröd lived temporarily	01°13.845S	90°26.573W
Floreana	Road to highlands, near Black Beach	01°16.427S	90°28.412W
Floreana	SICGAL office	01°16.460S	90°29.334W
Floreana	Tortoise pen	01°18.832S	90°27.334W
Floreana	Water spring	01°18.881S	90°27.237W
Floreana	White Beach	01°13.656S	90°25.420W
Floreana	Wittmer family gravesite	01°16.679S	90°28.881W
Floreana	Wittmer Pension and Museum	01°16.581S	90°29.274W
Genovesa	Darwin Bay, anchorage	00°18.927N	89°56.843W
Genovesa	Darwin Bay, beach	00°19.109N	89°56.943W
Genovesa	Darwin Bay, end of trail	00°19.076N	89°57.289W
Genovesa	Darwin Bay, monument	00°19.115N	89°56.933W
Genovesa	Darwin Bay, plaque honoring divers lost at sea	00°10.088N	89°57.180W
Genovesa	Prince Philip's Steps, base	00°18.203N	89°57.107W
Genovesa	Prince Philip's Steps, lava field with storm petrels	00°17.985N	89°56.949W
Genovesa	Prince Philip's Steps, top	00°18.189N	89°57.112W
Isabela	General Villamil Airport (SEII)	00°56.690S	90°57.266W
Isabela	Banks Bay	00°02.398S	91°27.119W
Isabela	Beagle Crater, north rim	00°16.192S	91°21.062W
Isabela	Bust of José de Villamil at public pier	00°57.539S	90°57.928W
Isabela	Cabo Berkeley	00°00.406S	91°35.392W
Isabela	Cabo Marshall	00°01.714S	91°12.468W
Isabela	Cabo Woodford	00°45.299S	90°46.509W

Isabela	Caleta Iguana, dock	00°58.481S	91°26.455W
Isabela	Cartago Bay	00°36.000S	90°55.599W
Isabela	Cerro Orchilla, base	00°57.813S	91°00.452W
Isabela	Cerro Orchilla, peak	00°57.785S	91°00.461W
Isabela	Church with memorial to penal colony	00°57.400S	90°57.949W
Isabela	Concha de Perla, end of boardwalk	00°57.729S	90°57.448W
Isabela	Concha de Perla, entrance	00°57.716S	90°57.536W
Isabela	Darwin Lake, first overlook	00°15.509S	91°22.190W
Isabela	Darwin Lake, highest overlook from north rim	00°15.217S	91°22.238W
Isabela	El Embarcadero, landing site	00°57.777S	90°57.543W
Isabela	Elizabeth Bay	00°36.108S	91°04.548W
Isabela	Flamingo lagoon near breeding center	00°56.712S	90°58.501W
Isabela	GNPS Technical Office	00°57.416S	90°58.048W
Isabela	Grave of Antonio Gil	00°57.346S	90°59.109W
Isabela	Grove of manzanilla	00°57.478S	90°59.754W
Isabela	Iguana Crossing Hotel	00°57.368S	90°58.262W
Isabela	Isla Cowley	00°23.211S	90°57.464W
Isabela	Islote Tintoreras, landing site	00°58.233S	90°57.632W
Isabela	Islote Tintoreras, marine iguana nursery	00°58.298S	90°57.667W
Isabela	Islote Tintoreras, sea lion breeding area	00°58.331S	90°57.594W
Isabela	Islote Tintoreras, shark channel	00°58.307S	90°57.660W
Isabela	Laguna Salinas	00°57.374S	90°58.054W
Isabela	Mirador El Mango	00°53.043S	91°00.821W
Isabela	Muro de las Lágrimas	00°57.858S	91°00.809W
Isabela	Muro de las Lágrimas, parking area	00°57.887S	91°00.730W
Isabela	Muro de las Lágrimas, foundation of building	00°57.870S	91°00.791W
Isabela	Penguin colony	00°57.649S	90°59.402W
Isabela	Perry Bay	00°36.398S	90°55.587W
Isabela	Playa de Amor, parking area	00°57.621S	90°59.346W
Isabela	Poza Baltazar	00°57.133S	90°58.302W
Isabela	Poza Puerta del Jelí	00°57.291S	90°58.283W
Isabela	Puerto Villamil, anchorage site	00°57.917S	90°57.753W
Isabela	Punta Albemarle	00°09.384N	91°20.099W
Isabela	Punta Cristóbal	00°52.550S	91°30.378W
Isabela	Punta Garcia	00°18.109S	91°05.524W
Isabela	Punta Moreno	00°41.221S	91°19.336W
Isabela	Punta Vicente Roca	00°03.190S	91 °33.374W
Isabela	Santo Tomás	00°48.386S	91°05.144W
Isabela	SICGAL office	00°57.690S	90°57.567W
Isabela	Tagus Cove, anchorage	00°15.677S	91°22.231W

Isabela	Tagus Cove, atop steps	00°15.566S	91°22.157W
Isabela	Tagus Cove, first overlook	00°15.510S	91°22.190W
Isabela	Tagus Cove, landing site	00°15.610S	91°22.168W
Isabela	Tagus Cove, monument	00°15.598S	91°22.161W
Isabela	Tagus Cove, peak	00°15.023S	91°22.263W
Isabela	Trailhead to Minas de Azufre	00°50.224S	91°05.448W
Isabela	Tortoise breeding center	00°56.837S	90°58.439W
Isabela	Tortoise breeding center, entrance from road	00°56.868S	90°58.470W
Isabela	Tortuga Island	01°01.171S	90°51.447W
Isabela	Trail to tortoise breeding center	00°57.365S	90°58.289W
Isabela	Túnel del Estero, entrance	00°57.639S	90°59.383W
Isabela	Túnel del Estero, parking area	00°57.621S	90°59.346W
Isabela	Urbina Bay	00°23.439S	91°13.380W
Isabela	Volcán Alcedo	00°26.190S	91°05.195W
Isabela	Volcán Azul	00°55.355S	91°23.060W
Isabela	Volcán Darwin	00°11.485S	91°18.412W
Isabela	Volcán Ecuador	00°01.307S	91°34.408W
Isabela	Volcán Sierra Negra	00°49.505S	91°10.170W
Isabela	Volcán Sierra Negra, parking area	00°50.208S	91°05.424W
Isabela	Volcán Sierra Negra, rim	00°48.392S	91°05.272W
Isabela	Volcán Wolf	00°01.071N	91°21.174W
Isabela	Wetlands	00°57.105S	90°58.279W
Kicker Rock	Kicker Rock (Leon Dormido)	00°46.371S	89°31.116W
Las Plazas	Las Plazas, anchorage	00°34.935S	90°09.827W
Las Plazas	South Plaza, colony of bachelor sea lions	00°35.101S	90°09.644W
Las Plazas	South Plaza, beacon	00°35.049S	90°09.825W
Las Plazas	South Plaza, naval marker	00°34.960S	90°09.888W
Las Plazas	South Plaza, landing site	00°34.948S	90°09.882W
Las Plazas	South Plaza, monument	00°34.960S	90°09.889W
Las Plazas	North Plaza	00°34.497S	90°09.423W
Marchena		00°19.509N	90°28.269W
North Seymour	Frigatebird colony	00°23.833S	90°17.483W
North Seymour	Landing site, base of steps	00°24.044S	90°17.459W
North Seymour	Landing site, atop steps	00°24.036S	90°17.349W
North Seymour	Nesting site	00°24.021S	90°17.364W
North Seymour	*Sesuvium* near landing site	00°24.020S	90°17.367W
North Seymour	Site to see breaking waves	00°23.796S	90°17.560W
Pinta		00°35.319N	90°45.141W
Pinzón		00°36.392S	90°39.563W
Rábida	Anchorage	00°23.935S	90°42.667W
Rábida	Landing site	00°23.977S	90°42.401W

Rábida	Overlook at eastern end of beach	00°23.950S	90°42.340W
Rábida	*Opuntia* forest along cliff	00°24.013S	90°42.356W
Roca Redonda		00°16.426N	91°38.027W
Roca Sin Nombre		00°40.157S	90°35.068W
San Cristóbal	Airport (SCY)	00°54.375S	89°37.035W
San Cristóbal	Association of Naturalist Guides	00°54.173S	89°36.752W
San Cristóbal	Intersection of Avenida Charles Darwin and 12 de Febrero	00°54.147S	89°36.730W
San Cristóbal	CDRS Technical Office (Vivero)	00°54.141S	89°36.490W
San Cristóbal	Centro de Interpretación	00°53.620S	89°36.580W
San Cristóbal	Cerro Brujo	00°45.368S	89°27.355W
San Cristóbal	Intersection of Avenidas Charles Darwin and Herman Melville	00°54.027S	89°36.642W
San Cristóbal	Intersection of Avenidas Charles Darwin and Manuel Cobos	00°54.175S	89°36.785W
San Cristóbal	Intersection of Avenidas Charles Darwin and Teodoro Wolf	00°54.080S	89°36.688W
San Cristóbal	Charles Darwin International Conference Center	00°54.120S	89°36.461W
San Cristóbal	Confiscated boat *Don Abraham*	00°54.186S	89°36.742W
San Cristóbal	Darwin monument, Frigatebird Hill	00°53.282S	89°36.518W
San Cristóbal	Darwin monument, town plaza	00°54.148S	89°36.743W
San Cristóbal	Darwin monument, near naval station	00°54.138S	89°36.839W
San Cristóbal	*Darwin Shop—Since 1835...*	00°54.206S	89°36.798W
San Cristóbal	Darwin's landing site	00°53.288S	89°36.500W
San Cristóbal	Darwin's landing site, original steps	00°53.291S	89°36.519W
San Cristóbal	El Cafetal, entrance	00°55.466S	89°32.951W
San Cristóbal	El Cafetal, processing plant	00°55.525S	89°32.395W
San Cristóbal	El Progreso, largest tree in Galápagos	00°54.488S	89°33.459W
San Cristóbal	El Progresso, Charles Darwin bust	00°54.488S	89°33.342W
San Cristóbal	El Progreso, machinery from Cobos's plantation	00°54.474S	89°33.533W
San Cristóbal	El Progreso, ruins of Cobos's home	00°54.439S	89°33.485W
San Cristóbal	El Progreso, ruins of housing for Cobos's servants	00°54.440S	89°33.473W
San Cristóbal	El Progreso, tomb of Manuel Cobos	00°54.511S	89°33.490W
San Cristóbal	Fish market	00°53.935S	89°36.577W
San Cristóbal	Fishermen monument	00°53.948S	89°36.563W
San Cristóbal	Frigatebird Hill, apex	00°53.349S	89°36.466W
San Cristóbal	Frigatebird Hill, first overlook on upper trail	00°53.322S	89°36.421W
San Cristóbal	Frigatebird Hill, lower overlook	00°53.359S	89°36.482W
San Cristóbal	Galapaguera de Cerro Colorado	00°54.808S	89°26.184W
San Cristóbal	Interpretation Center, boardwalk	00°53.641S	89°36.566W

San Cristóbal	Interpretation Center, entrance	00°53.618S	89°36.578W
San Cristóbal	Interpretation Center, entrance from Predial trail	00°53.650S	89°36.761W
San Cristóbal	Interpretation Center, exit from road	00°53.651S	89°36.579W
San Cristóbal	Isla Lobos, anchorage	00°51.480S	89°33.957W
San Cristóbal	Isla Lobos, landing site	00°51.373S	89°33.963W
San Cristóbal	*Jessica* wreckage	00°53.695S	89°37.322W
San Cristóbal	*Jessica*, buoy near wreckage	00°53.704S	89°37.259W
San Cristóbal	*Karahua* wreckage	00°53.651S	89°37.144W
San Cristóbal	La Honda	00°56.252S	89°29.265W
San Cristóbal	La Predial, landing site	00°53.603S	89°36.696W
San Cristóbal	La Predial, old buildings	00°53.594S	89°36.707W
San Cristóbal	La Predial, pier	00°53.602S	89°36.695W
San Cristóbal	Laguna El Junco, junco plants	00°53.733S	89°28.728W
San Cristóbal	Laguna El Junco, northern side	00°53.642S	89°28.848W
San Cristóbal	Laguna El Junco, roadside stop	00°53.546S	89°29.051W
San Cristóbal	Laguna El Junco, southwestern side	00°53.770S	89°28.868W
San Cristóbal	Laguna El Junco, western (main) side	00°53.715S	89°28.893W
San Cristóbal	Love Beach	00°53.436S	89°36.725W
San Cristóbal	Love Beach, large entrance to trail	00°53.572S	89°36.673W
San Cristóbal	Love Beach, western entrance to trail	00°53.573S	89°36.704W
San Cristóbal	Mirador de la Soledad, peak	00°53.202S	89°32.470W
San Cristóbal	Mirador de la Soledad, view toward Isla Lobos	00°53.202S	89°32.476W
San Cristóbal	*Mockingbird Café*	00°54.089S	89°36.644W
San Cristóbal	Mount Pitt	00°43.090S	89°14.539W
San Cristóbal	Old Darwin statue	00°54.136S	89°36.839W
San Cristóbal	Pan de Azucar	00°43.234S	89°21.204W
San Cristóbal	Parque Ecológico Artesanal San Francisco	00°54.216S	89°36.602W
San Cristóbal	Playa de Oro	00°53.978S	89°36.563W
San Cristóbal	Playa Mann	00°53.743S	89°36.561W
San Cristóbal	Puerto Baquerizo Moreno, public dock	00°53.997S	89°36.653W
San Cristóbal	Puerto Grande, beach	00°45.310S	89°27.299W
San Cristóbal	Punta Carola	00°53.381S	89°36.816W
San Cristóbal	Punta Pitt	00°41.295S	89°15.320W
San Cristóbal	Roca Ballena	00°56.571S	89°35.323W
San Cristóbal	Shipyard	00°54.197S	89°36.785W
San Cristóbal	SICGAL office on Charles Darwin Avenue	00°54.045S	89°36.655W
San Cristóbal	Statue of tortoise reaching for *Opuntia*	00°54.211S	89°36.761W
San Cristóbal	Stephen's Bay, southwest shore	00°49.246S	89°31.255W
San Cristóbal	Tijeretas Bay, Darwin statue	00°53.284S	89°36.502W

San Cristóbal	Tijeretas Bay, Darwin's landing site	00°53.289S	89°36.502W
San Cristóbal	Galápagos Academic Institute for the Arts and Sciences, Universidad San Francisco de Quito, Galápagos campus	00°53.737S	89°36.542W
San Cristóbal	Wreck Bay, anchorage	00°53.935S	89°36.794W
San Cristóbal	Wreck Bay, docking site	00°53.929S	89°36.790W
Santa Cruz	Angermeyer Point	00°45.040S	90°18.553W
Santa Cruz	Angermeyer Point, Waterfront Inn (Gus's Cave)	00°45.036S	90°18.574W
Santa Cruz	Artisanal market	00°44.689S	90°18.696W
Santa Cruz	El Chato, entrance	00°38.284S	90°25.710W
Santa Cruz	Las Grietas, trailhead	00°45.165S	90°18.463W
Santa Cruz	Bellavista	00°41.669S	90°19.520W
Santa Cruz	Bellavista, Angermeyer family burial plot	00°41.635S	90°19.897W
Santa Cruz	CDRS, dock	00°44.511S	90°18.442W
Santa Cruz	CDRS, entrance	00°44.530S	90°18.253W
Santa Cruz	CDRS, office	00°44.492S	90°18.392W
Santa Cruz	CDRS, breeding center	00°44.503S	90°18.365W
Santa Cruz	CDRS, California Academy of Science plaque	00°44.458S	90°18.150W
Santa Cruz	CDRS, first building built at station	00°44.449S	90°18.260W
Santa Cruz	CDRS, land iguana pens	00°44.450S	90°18.153W
Santa Cruz	CDRS, pen of mixed tortoises	00°44.462S	90°18.147W
Santa Cruz	CDRS, Shade House	00°44.542S	90°18.231W
Santa Cruz	CDRS, store	00°44.493S	90°18.181W
Santa Cruz	CDRS, Super Diego	00°44.454S	90°18.161W
Santa Cruz	CDRS, tortoise nursery	00°44.476S	90°18.200W
Santa Cruz	CDRS, tortoise pen	00°44.479S	90°18.176W
Santa Cruz	Cerro Crocket	00°38.389S	90°10.274W
Santa Cruz	Darwin monument near Darwin arch	00°44.538S	90°18.605W
Santa Cruz	Dragon Hill, beach	00°31.746S	90°29.362W
Santa Cruz	Dragon Hill, cove	00°31.559S	90°29.288W
Santa Cruz	Dragon Hill, lagoon	00°31.627S	90°29.266W
Santa Cruz	Dragon Hill, landing site	00°31.550S	90°29.290W
Santa Cruz	Dragon Hill, lower peak	00°31.909S	90°29.208W
Santa Cruz	Dragon Hill, summit	00°31.913S	90°29.280W
Santa Cruz	Eden Island	00°33.298S	90°32.172W
Santa Cruz	El Carmen	00°40.556S	90°22.870W
Santa Cruz	El Mirador de los Túneles	00°43.924S	90°19.712W
Santa Cruz	Escuela Fiscal Mixta, mural of creation story	00°44.724S	90°18.821W
Santa Cruz	Genesis billboard at Seventh Day Adventist Church	00°44.716S	90°18.691W
Santa Cruz	GNPS, Inspection Building	00°44.491S	90°18.393W

Santa Cruz	GNPS, store	00°44.518S	90°18.497W
Santa Cruz	GNPS, control station	00°39.292S	90°24.221W
Santa Cruz	Gordon Rocks	00°34.285S	90°08.359W
Santa Cruz	Guy Fawkes	00°30.504S	90°31.368W
Santa Cruz	*Hotel Galápagos*	00°44.578S	90°18.540W
Santa Cruz	Intersection of Avenidas Charles Darwin and 12 de Febrero	00°44.747S	90°18.706W
Santa Cruz	La Feria market	00°44.336S	90°18.925W
Santa Cruz	La Garrapatero	00°41.393S	90°13.147W
Santa Cruz	Las Bachas, anchorage	00°29.509S	90°20.438W
Santa Cruz	Las Bachas, lagoon	00°29.649S	90°20.640W
Santa Cruz	Las Bachas, landing site	00°29.644S	90°20.443W
Santa Cruz	Las Bachas, nesting sites of turtles	00°29.647S	90°20.446W
Santa Cruz	Las Bachas, remains of largest pier	00°29.608S	90°20.547W
Santa Cruz	Las Bachas, remains of smaller pier	00°29.607S	90°20.589W
Santa Cruz	Las Grietas	00°45.256S	90°18.559W
Santa Cruz	Lava tube, entrance	00°40.102S	90°25.471W
Santa Cruz	Lava tube, parking area	00°40.077S	90°25.476W
Santa Cruz	Los Gemelos, edge of western crater	00°37.540S	90°23.129W
Santa Cruz	Los Gemelos, parking area	00°37.557S	90°23.102W
Santa Cruz	Los Gemelos, roadside sign	00°37.557S	90°23.102W
Santa Cruz	Man and the Biosphere plaque	00°44.837S	90°18.807W
Santa Cruz	Monument to fishermen	00°44.601S	90°18.724W
Santa Cruz	Monument to GNPS Park Rangers	00°44.484S	90°18.401W
Santa Cruz	Playa de los Alemanos	00°45.013S	90°18.455W
Santa Cruz	Public dock (Gus Angermeyer Ecotourism Pier)	00°44.868S	90°18.751W
Santa Cruz	Public plaza	00°44.843S	90°18.778W
Santa Cruz	Puerto Ayora, panga landing	00°44.866S	90°18.759W
Santa Cruz	Puerto Ayora, Centro Comunitario de Education Ambiental	00°44.751S	90°19.133W
Santa Cruz	Puerto Ayora, commemorative plaque	00°44.488S	90°18.403W
Santa Cruz	Puerto Ayora, Darwin Arch	00°44.541S	90°18.603W
Santa Cruz	Puerto Ayora, fish market	00°44.614S	90°18.730W
Santa Cruz	Puerto Ayora, house made of Baltra pine	00°44.785S	90°18.787W
Santa Cruz	Puerto Ayora, Laguna de las Ninfas	00°44.860S	90°18.982W
Santa Cruz	Puerto Ayora, Playa de la Estación	00°44.556S	90°18.344W
Santa Cruz	Puerto Ayora, Playa de la Estación, exit from road	00°44.535S	90°18.334W
Santa Cruz	Puerto Ayora, tourism office	00°44.706S	90°18.793W
Santa Cruz	Rancho Círculo	00°40.224S	90°23.455W

Santa Cruz	Rancho Primicias	00°40.182S	90°25.805W
Santa Cruz	Rancho Primicias, exit from Avenida Baltra	00°38.888S	90°24.823W
Santa Cruz	Residencial Angermeyer (now Hotel Silberstein)	00°44.566S	90°18.673W
Santa Cruz	Santa Rosa	00°39.098S	90°24.393W
Santa Cruz	SICGAL office and mural of history of Galápagos	00°44.889S	90°18.831W
Santa Cruz	Sin Nombre	00°40.109S	90°35.636W
Santa Cruz	*The Rock*	00°44.789S	90°18.751W
Santa Cruz	Tortuga Bay, southern end of iguana trail	00°45.558S	90°20.278W
Santa Cruz	Tortuga Bay, check point	00°44.782S	90°19.172W
Santa Cruz	Tortuga Bay, surfing beach	00°45.424S	90°20.591W
Santa Cruz	Tortuga Bay, swimming beach	00°45.479S	90°20.258W
Santa Cruz	Túnel de Miguel Angel Arias, entrance	00°38.346S	90°25.979W
Santa Cruz	UNESCO plaque	00°40.182S	90°18.403W
Santa Cruz	Van Straelen Interpretation Center	00°44.488S	90°18.292W
Santa Cruz	Whale Bay, anchorage site	00°36.162S	90°32.284W
Santa Cruz	Whale Bay, landing site	00°36.084S	90°32.226W
Santa Cruz	Windmill, site of original source of fresh water	00°44.580S	90°18.728W
Santa Cruz	Itabaca Channel	00°28.993S	90°16.641W
Santa Fé	Apex of long trail	00°48.466S	90°02.715W
Santa Fe	Barrington Bay, beach landing	00°48.232S	90°02.471W
Santa Fé	Barrington Bay, alternate landing site	00°48.238S	90°02.473W
Santa Fé	Barrington Bay, anchorage	00°48.271S	90°02.251W
Santa Fé	Barrington Bay, docking site	00°48.280S	90°02.283W
Santa Fé	Barrington Bay, landing site	00°48.233S	90°02.471W
Santa Fé	Cactus Arch	00°48.088S	90°02.491W
Santa Fé	Monument	00°48.219S	90°02.465W
Santiago	Albany Rock	00°10.270S	90°50.434W
Santiago	Bainbridge Rocks	00°39.155S	90°33.497W
Santiago	Buccaneer Cove	00°09.476S	90°49.446W
Santiago	Cerro Cowan	00°10.538S	90°50.120W
Santiago	Cerro Pelado	00°12.222S	90°47.031W
Santiago	Cousins Rock	00°14.755S	90°34.264W
Santiago	Grottos	00°14.766S	90°52.116W
Santiago	James Bay, anchorage	00°14.463S	90°51.586W
Santiago	James Bay, colony of fur sea lions	00°14.860S	90°52.147W
Santiago	James Bay, Darwin's Toilet	00°14.774S	90°52.110W
Santiago	James Bay, Egas's house and shed	00°14.545S	90°51.632W

Santiago	James Bay, goat pens	00°14.482S	90°51.731W
Santiago	James Bay, landing site	00°14.495S	90°51.670W
Santiago	James Bay, monument	00°14.469S	90°51.715W
Santiago	James Bay, soccer field	00°14.542S	90°51.649W
Santiago	James Bay, water barrel	00°14.503S	90°51.683W
Santiago	Lava field	00°17.490S	90°33.912W
Santiago	Lava field, end of lava river	00°17.762S	90°34.161W
Santiago	Lava field, landing site	00°17.394S	90°33.961W
Santiago	Lava field, old and new lava	00°17.671S	90°34.208W
Santiago	Lava field, intersection of older/ newer lava	00°17.766S	90°34.161W
Santiago	Lava fields, landing site	00°17.406S	90°33.929W
Santiago	Sulivan Bay	00°17.325S	90°33.787W
Santiago	Sulivan Bay, landing site	00°17.436S	90°33.887W
Sombrero Chino	Alternate landing site	00°22.164S	90°35.119W
Sombrero Chino	Anchorage	00°22.060S	90°35.119W
Sombrero Chino	Landing site	00°22.087S	90°35.062W
Wolf		01°23.031N	91°48.410W

APPENDIX 2

Getting around Galápagos

There are two ways to see Galápagos: (1) hotel-based tours, and (2) boat-based tours. Both have distinct advantages and disadvantages.

Hotel-Based Tours

During these tours, you will live in hotels on Santa Cruz, San Cristóbal, Isabela, or Floreana, from which you can take day trips to some other islands. These tours usually are cheaper, more flexible (e.g., you can visit other islands on days of your choosing), provide nicer accommodations than most boats, and give you more time to explore the island on which you are staying. However, your access to other islands is limited (e.g., you will not get to Genovesa, Española, or Fernandina). These tours reduce your odds of getting seasick, but your travel between islands (on speedboats) can be rough (Chapter 3).

Boat-Based Tours

Boat-based tours are usually more expensive than hotel-based tours. Itineraries are set by GNPS and are seldom flexible. You will be at sea for 7 to 10 days, during which you will visit several different islands not accessible by day trips from hotels. Approximately 60% of foreign visitors to Galápagos stay only on boats.

Tour boat operators spend much of their time moving tourists to the various Visitor Sites on their itinerary. Many of these sites are near each other; because boats can get from one of these sites to another in less than an hour or so, these navigations are usually done before or after lunch. Other more-distant sites, especially those to northern or western islands such as Genovesa and Fernandina, require longer navigations. These navigations usually start after dinner so that visitors awaken at the new island.

The amount of time you will spend going from one site to another depends on a variety of factors, including your boat's maximum speed, ocean currents, winds, the ocean's choppiness, the technical difficulty of the navigation, and the distance between sites.

Speed. Tour boats have different cruising speeds. For example, *Yate Darwin* can cruise at 9 knots (10.4 mph = 16.7 km h^{-1}), whereas *Galápagos Explorer* can cruise at almost 16 knots (18.4 mph = 29.6 km h^{-1}). However, captains usually navigate at speeds significantly slower than their boat's maximum speed to conserve fuel and produce a smoother ride.[1]

Currents. Currents are strongest in August and September, and weakest in February and March.

Winds. Winds are strongest in August and September, and weakest in March.

Waves. Oceans are usually roughest in August and September, and calmest from February to April (Chapter 3).

Technical difficulty of the navigation. Oceanic charts warn captains of "CAUTION" at five locations in Galápagos: (1) the southeastern side of Isabela near Puerto Villamil and Isla Tortuga, where boats are typically hit by waves from the side, (2) the northwestern tip of Isabela just past Cabo Berkeley, where boats are hit by waves from the front, (3) Isabela's Elizabeth Bay just east of Punta Moreno, (4) Bolivar Channel, and (5) the northeastern side of Isabela near Cabo Marshall. These cautions usually result from combinations of strong currents, high winds, and/or shallow waters. The shallow or narrow entrances to Darwin Bay (Genovesa), Puerto Villamil (Isabela), and Wreck Bay (San Cristóbal) also require caution.[2]

Distances between Islands

GPS coordinates for more than 350 sites are included in Appendix 1. You can use any of the several excellent programs available on the Internet and elsewhere to determine the distances between the sites of your choice.

Some navigational charts and websites will list distances in nautical miles (M, NM, or nmi), which is the distance equal to a minute of arc of longitude at the equator. (If you view the equator as a circle, a degree is 1/360 of that circle. You could then divide a degree into 60 minutes. One minute of arc is a nautical mile.) Here are some handy conversion factors: 1 NM = 1,852 meters = 1.852 kilometers = 1.152 miles = 6,076 feet. A ship moving at a speed of 1 knot (kt) is moving 1 nautical mile per hour (i.e., 1 kt = 1 NM h^{-1}).

[1] When navigating, a typical 16-passenger boat consumes about 15 gallons (56.8 L) of fuel per hour; when docked, consumption drops to 3 gallons (11.4 L) per hour. These boats usually hold about 1,400 gallons (5,300 L) of fuel. About 1,000 gallons (3,700 L) are required to go from Galápagos to Guayaquil.

[2] Typical 16-passenger tour boats extend approximately 1.8 meters below the surface of the water and need water that is at least 5 meters deep to sail.

APPENDIX 3

Darwin in Galápagos

Following is a list of the more permanent uses of Darwin's name and image in the archipelago; not included are the many tour agencies, hotels, boats, schools, and other businesses that have adopted his name.[1] GPS coordinates for most of these sites are listed in Appendix 1.

Statues and Busts	Puerto Ayora	Near the blue "Darwin Arch" on the way to CDRS
	Puerto Baquerizo Moreno	Frigatebird Hill
		Naval Station
		Plaza along Malecón Charles Darwin (Figure A3.1)
	El Progreso	Escuela Carlos Darwin
Buildings and Public Structures	Puerto Baquerizo Moreno	Charles Darwin International Convention Center
	Puerto Ayora	Darwin Arch on Avenida de Charles Darwin
Street Names	Puerto Ayora	Avenida de Charles Darwin
	Puerto Baquerizo Moreno	Avenida de Charles Darwin
Organizations	Puerto Ayora	Charles Darwin Research Station
Sites	Genovesa	Darwin Bay
	Darwin	Darwin Arch
	Isabela	Volcán Darwin
	Isabela	Darwin Lake[2]

(Continued)

[1] There are reminders of Charles Darwin and Robert FitzRoy almost everywhere the *Beagle* went, including Australia, Tierra del Fuego, Falkland Islands, New Zealand, and Great Britain. Some of the information in this appendix is modified from *Darwin in Galápagos* (Grant and Estes, 2009).

[2] Darwin wrote about this bay, but never explored it. However, he did explore the lake in Isabela's Beagle Crater. Darwin did not visit all of the sites in this appendix, nor did he collect all of the organisms that today bear his name.

FIGURE A3.1 This bust of a young Charles Darwin is at the public plaza along Avenida de Charles Darwin in Puerto Baquerizo Moreno on San Cristóbal.

Life	Molluscs	*Bulimulus darwini*	
	Marine Invertebrates	*Cavernulina darwini*	
		Pacifigorgia darwinii	
	Arthropods	*Capsus darwini*	
		Darwin's darklings	
		Phalacrus darwini	
		Nitela darwini	
		Darwinysius marginalis	
		Darwinysius wenmanensis	
		Amblyomma darwini	
		Darwinneon crypticus	
		Xylocopa darwinii	Carpenter bee
	Fish	*Semicossyphus darwini*	Pacific red sheephead
		Ogcocephalus darwini	Red-lipped batfish
	Birds	*Geospiza* spp.	Darwin's finches
	Reptiles	*Geochelone nigra* var. *darwini*	James Island tortoise
	Mammals	*Nesoryzomys darwini*	Darwin's rice rat[3]
	Plants	*Croton scouleri* var. *darwini*	Galápagos croton
		Scalesia darwinii	
		Darwiniothamnus spp.	Darwin's shrub
		Tiquilia darwini	Gray matplant
		Lecocarpus darwinii	Darwin's daisy
		Gossypium darwinii	Darwin's cotton
		Pleuropetalum darwinii	

[3] Extinct from Santa Cruz.

APPENDIX 4

Rules of the Galápagos National Park

When you board your tour boat or meet your guide, you will learn GNPS's rules for protecting the islands' wildlife and landscapes:

Do not disturb or remove any plant, animal, or their remains (e.g., shells, pieces of wood, bones) or other natural objects.

Do not take any live material (e.g., fruits, seeds, insects) to the islands, or from island to island.

Do not take any food to any of the uninhabited islands. Smoking is forbidden at GNPS Visitor Sites.

Do not feed, touch, or handle the animals. Stay at least 2 meters away from the animals.

Do not startle or chase any animal. When you take photographs, turn off your camera's flash.

Stay within the areas designated as Visitor Sites. Stay on the trails, which are marked by black-and-white stakes (Figure 3.4) made of matazarno (*Piscidia carthagenensis*), a plant native to Galápagos.

Do not leave any litter on the islands, or throw any litter off your boat. Nothing may be taken from any of the islands.

Do not buy souvenirs or other objects made from plants or animals of the islands (e.g., coral, sea shells, sea lion teeth, tortoise shells, volcanic rock, endemic woods).

Do not write on, scratch, or otherwise deface rocks.

Do not visit the islands unless you are accompanied by a licensed GNPS guide. When walking on trails, stay near your guide.

Restrict your visit to officially approved Visitor Sites.

Fishing from boats is prohibited, except for boats authorized by GNPS for fishing.

Jet skis, water skiing, submarines, and aerial tourism are forbidden.

Camping is allowed in specific areas, but only if you have a permit to do so from GNPS.

Show your conservationist attitude. For example, if you see trash, pick it up. Report any park violations or concerns to your guide, boat captain, or GNPS.

APPENDIX 5

Diving in Galápagos

Galápagos includes more than 80 Marine Visitor Sites, most of which feature snorkeling and scuba diving. If you are going to scuba dive, visibility is usually 12 to 18 meters and the water is cold (13–22°C); you will need to wear a wetsuit (5–7 mm). Currents and surges can be strong, the rocks to which you will cling at some sites are sharp volcanic tuff, and the nearest help is often far away. Except for a few sites near Puerto Ayora and Puerto Baquerizo Moreno, diving in Galápagos is for experienced divers.[1] The closest hyperbaric chambers are in Puerto Ayora and Guayaquil.

Galápagos includes several world-class sites for scuba diving. To see these sites, you will either (1) stay on an inhabited island and make day trips to dive sites, or (2) visit several parts of the archipelago during a 7-to-10-day trip aboard a live-aboard dive boat. Like tours of Terrestrial Visitor Sites, the live-aboard boats follow an itinerary approved by GNPS that cannot be altered without permission from GNPS. Night diving is not allowed in Galápagos.

Here are several of the most famous dive sites in Galápagos:

Darwin Island

Diving at this island, the northernmost and most remote island of the archipelago, features colorful corals and warm-water species found nowhere else in Galápagos. The most famous dive site at Darwin Island is near a giant arch named Darwin's Arch, which is along the southeastern side of the island. Here, you will usually find creole fish, rainbow chubs, snappers, mackerels, Moorish idols, moray eels, hogfish, and bottle-nosed dolphins, who may escort you to your dive site. Hammerhead sharks are abundant year-round; you will often see them being cleaned by small groups of king angelfish. Giant manta rays and whale sharks—the biggest fish in the world—are also common here between June and November. Currents here are strong, and you will probably hold onto

[1] For additional details about diving and other dive sites in Galápagos, we recommend *The Diving Guide: Galápagos Islands* (Rosenberg and Sarbone, 2004).

rocks and watch the wildlife go by. Darwin and Wolf Islands are among the best dive sites in the world, although they are often rough, and currents change quickly.

Wolf Island

Diving at this northern island showcases schools of hammerhead sharks and Galápagos sharks, as well as moray eels, whales, tuna, barracudas, green spotted moray eels, and cleaning stations where reef fish are cleaned by barberfish and king angelfish. The warm water at Wolf Island is home to many fish not found in the southern islands of Galápagos, including zebra and white-mouth morays, rainbow runners, trumpetfish, and cornetfish. Cruising sharks are cleaned by butterfly fish among the boulders on the island's eastern side. Experienced divers like to visit the northeastern tip of the island, where they swim up through a tube and are pitched out into a strong current near the surface. Water around Wolf Island, like that around Darwin Island, is often rough, and there are several strong down-currents. Wolf and Darwin Islands are visited only by live-aboard dive boats.

Roca Redonda

This monolithic islet north of Isabela features barracuda, hammerhead sharks, Galápagos sharks, white-tipped sharks, groupers, manta rays, bacalao (sea bass), sea lions, and many species of reef fish. Giant sunfish, dolphins, and whales are also common here, and the Harlequin wrasse is a favorite target of most photographers. There are strong currents, eddies, and unpredictable down-surges. Along the southeastern side of Roca Redonda (at a depth of 10–18 meters), you will see fumaroles emitting bubbles of hydrogen sulfide. Water and rocks near these fumaroles are warm.

Gordon Rocks

These islets—each a part of the top of an eroded, 100-meter-wide crater—east of Santa Cruz are home to four dive sites showcasing numerous species of golden rays, spotted eagle rays, Galápagos sharks, hammerheads, manta rays, king angelfish, turtles, stingrays, reef fish, and pelagic fish such as tuna and sailfish. The wall at this site is among the best in the central islands of Galápagos. Currents here are usually strong and unpredictable.

North Seymour

The five dive sites around this island—including Punta Carrion, Daphne Minor, and Mosquera—feature hammerhead sharks, sea lions, garden eels, snappers, fur sea lions, eagle and manta rays, tuna, and the endemic barnacle blenny. In

February and March you may see pregnant white-tipped reef sharks ready to deliver their young. Dolphins and humpback whales visit occasionally. Dives here are usually suitable for all divers.

Devil's Crown (Onslow Island)

Both inside and just outside the "crown" (the top of an eroded crater) you will find wrasses, yellow-tailed surgeonfish, and many other colorful reef-fish and hammerhead sharks. Most divers will enjoy seeing juvenile giant damselfish with their iridescent blue spots, and red-backed juvenile ring-tailed damsels. Large schools of creole fish are cleaned by barberfish, and along the eastern side of the crown (at a depth of about 20 m) lives a colony of endemic garden eels. The center of the crown is less than 5 meters deep and features sand and coral. Currents through the crown are usually east-to-west.

Cousins Rock

This triangle-shaped islet northeast of Santiago is accessible by day on dive boats from Santa Cruz, and features spotted eagle rays, barracuda, manta rays, rainbow wrasses, Pacific seahorses, nudibranchs, frogfish, hammerhead sharks, white-tipped sharks, reef fish, octopuses, forests of black juvenile corals, and sponges along its walls. These walls are steep on the northern and western sides of the island, and sloping on the eastern side. Cousins Rock is famous for its many ledges and overhangs. Currents here are usually moderate and from the southeast.

Punta Vicente Roca

This site, just northwest of Isabela, is called "The Ice Box" because of its cold water. Dives here feature camouflaged frogfishes, electric rays, red-lipped batfishes, barracuda, salemas, brown-striped snapper, mola mola, Pacific seahorses, bullhead sharks, and endemic camotillo. There is also a wall and two shallow underwater caves that feature invertebrates such as corals, anemones, and plantlike gorgonians. Just outside one of the caves, you may see marine iguanas eating algae at the waterline.

Floreana

Surrounding Floreana are numerous excellent dive sites (e.g., Champion, Enderby, and Gardner islands) that feature, tunas, seahorses, turtles, red-lipped batfish, snappers, sharks (hammerhead, Galápagos, and white-tipped), rays, and groupers. Champion is a drift-dive along a wall; among the black corals you will find resting sea turtles. Along the southern end of Gardner Islet are caves, and along the northern end are tunnels. At half-moon-shaped Enderby Island, dives are usually along the ridge where currents hit the reef; you will see lots of black coral here. Turtles often visit Enderby to have their parasites removed by cleaner fish (Chapter 8).

Cape Marshall

Just northeast of Isabela is Cape Marshall, which is one of the best places to see sharks and pillow lava in Galápagos. Here, you will dive along a near-vertical wall of lava rocks where lava from Volcán Wolf flowed into the ocean. The protected reef here is visited by manta rays, hammerhead sharks, turtles, and giant schools of fish, including the endemic brown-striped salema. Other common species include anemones, white zebra morays, eagle rays, blue marlin, and orcas.

Tagus Cove

Despite its sandy bottom, this cold site along northwest Isabela features frogfish, seahorses, scorpionfish, hawkfish, anemones, horn sharks, rainbow tonguefish (buried in the sand), and black coral. (Black coral has colorful tissue, and gets its name from its distinctive black or dark brown skeleton). High concentrations of plankton here often decrease visibility but also support the area's diverse life. Attentive divers may see giant sea stars (more than a meter wide), red-lipped batfish hopping across the ocean floor, and the male pipe blenny, which emerges from tubeworms and tries to attract females by showing its colored dorsal coat. This site, which is relatively calm, is one the best in northern Galápagos for inexperienced divers.

APPENDIX 6

Photography in Galápagos

There are few places in the world as photogenic as Galápagos; every day, visitors take thousands of photos of the islands' spectacular wildlife, stunning vistas, and beautiful shorelines, volcanoes, and lava. At some places, you will find it hard to stop taking pictures. Because most people visit the islands only once, it is worth some effort to get photos that you admire and that will help you preserve your memories of the archipelago.

Before you go

Get the right camera. There are many different types of cameras, and your choice of cameras will be influenced by the camera's cost, size, image quality, and versatility. You will enjoy the most versatility with a single-lens reflex (SLR) autofocus camera with interchangeable lenses, but these cameras and lenses are bulkier and more expensive than smaller point-and-shoot cameras. For specific needs, you can also buy underwater cameras and those with GPS. If you plan to buy a camera for your trip to Galápagos, do some research to ensure that your camera meets your needs.

Get the right lenses. If you use an SLR that can accommodate different lenses, you will have to decide which lenses you will take on your trip. Most books claim that photographers do not need long lenses because the animals of Galápagos can be easily approached, but this does not mean that the closest animal will always be the best subject (e.g., they may be turned away from you, be doing something unremarkable, or be in poor light). If you can take extra lenses, take a medium zoom lens (70–300 mm), a short zoom lens (28–105 mm), and a macro lens. If you increase the length of your lenses by adding a teleconverter, (e.g., 1.4× or 2.0×), remember that doing so will lengthen your exposures and dim the image, often to the point of defeating the camera's autofocus system.

Get the right equipment. Other equipment to take includes a wide camera strap, filters (see below), lens cleaners, lens caps, batteries, battery chargers, media cards, and a plastic bag large enough to keep your camera dry in rain or garúa. Having a tripod and a shutter-release cable will stabilize your camera and improve virtually every photo you take (especially those in lower light), but you will have to weigh that benefit against the increased bulk and trouble of carrying and using the tripod.

Test all of your equipment. If you get to Galápagos and something does not work, you will be disappointed, and you will miss the great photographic opportunities that lure people to the islands. Test your equipment before you go.

Know how to use your equipment. Point-and-shoot cameras are remarkably versatile, and you will need this versatility to take your best pictures in Galápagos. Here is a summary of what the automatic settings on most point-and-shoot cameras do:

Mode	What It Does
Landscape	Uses the smallest aperture to ensure the greatest depth of field. Use this setting when you want as much of the scene as possible to be in focus. With this setting, your camera may use slower shutter speeds, so be sure to hold your camera steady.
Portrait	Uses the largest aperture, which reduces the depth of field and creates a blurred background.
Action	Uses a fast shutter speed to minimize blurring. With this setting, your camera may use a wider aperture to compensate for the faster shutter, thereby decreasing the depth of field.
Nighttime	Uses the slow-sync mode of the flash to brighten the background of the image. Do not use this mode if you want the background to be dark.
Close-up (Macro)	Use this setting when you want to fill the frame with your subject from a short distance. Your depth of field with this mode will usually be small.

Although automatic exposures can often produce good images, you lose control of your image when you use the camera's automatic settings. To minimize problems, know how to bracket images as a way of altering the exposure; if you photograph a Nazca booby on black lava, your image will likely be overexposed by your camera's autoexposure system (in this example, you will need to underexpose the image or increase the ISO to avoid a washed-out image). Know how to turn off your flash (Appendix 4).

Read—or at least glance over—a book about nature photography. For example, know to avoid placing the subject in the center of your photographs, know the importance of the background and lighting for the quality of your photos, and know how to control the depth of field of your image.

While you are there

Do not get so involved with photography that you miss the experience of Galápagos. At some point, put your camera aside and enjoy *being* in Galápagos.

When in the panga, wear the camera strap around your neck, not over your shoulder. This will help prevent your camera from going overboard during rough rides.

Start each day by photographing an index card on which you have written where you will be that day. If you take another picture of that card at the end of the day, you will know when and where you took your photos. If you do not do this, you may forget where some of your photos were taken.

If possible, back up your images on a laptop, iPad, digital storage device, or similar gadget as soon as conveniently possible.

When you take pictures

Photographs are personal creations, and there is no one "best" way to compose an image. Moreover, for every "rule" of composition there is an image that proves the exception. Nevertheless, there are several principles of composition that will usually improve the impact of your images. You will be surprised how the framing of a subject, zooming in for a closer view, and changing the camera angle can improve a photo.

Take a lot of pictures. To paraphrase a famous photographer, "If you want to take a good picture, take a lot of pictures."

Place the subject off-center in your photographs. For maximum impact, resist the urge to place your subject in the center of the frame. Instead, mentally divide the frame into horizontal and vertical thirds like a tic-tac-toe game, and place the main subject at an intersection of the lines. To remind you of this principle, many cameras include the tic-tac-toe grid on their viewfinders.

Turn off your flash when you photograph animals on trails. When you do this in dim light, your camera will compensate by slowing its shutter speed. This means that you will need to concentrate on holding the camera steady. You can also increase your camera's sensitivity to light by raising the ISO.

Get close to show details. Using a zoom to show salt on a marine iguana, or a carpenter bee on a flower, will often increase the impact of your photograph.

Focus on your subject's eyes. This sense of eye contact increases the drama of your photograph. If you can wait until the animal looks at you, you will have a more dramatic photo.

Draw the viewer's eye across the frame. Compose the image so that viewers' eyes are led from one edge of the frame to another. Leave a margin of background in the direction that the animal or person is looking.

Do as much of your photography as possible just after sunrise, and just before sunset. At these times, there are usually shadows that highlight the subject. Moreover, sunlight at these times goes through more of the atmosphere, which results in warmer colors. The bright skies of midday minimize shadows and often confuse a camera's exposure sensors; this is why correct exposures at these times of the day are challenging, especially without filters (see below).

Avoid long exposures. Because most of your photos in Galápagos will be in bright light, long exposure times will seldom be a problem. However, you will probably take some photos in lower light, thereby increasing exposure times. Shoot for exposures shorter than 1/125 second. If the exposure is longer than about 1/40 second, steady the camera by holding it against a post, railing, or tree.

Level the horizon and get low. Photos of animals and landscapes will usually be most dramatic if they are shot from low positions.

Use a large aperture to highlight your subject. If you use a large aperture (e.g., *f*2.8), the background will be out of focus, and viewers' eyes will be drawn to the subject. Remember, the smaller the aperture number, the larger the aperture.

Include people in photographs of landscapes. This will add perspective and enable viewers to place themselves in the scene.

Check and clean your equipment at the end of each day. You will be surprised how quickly they are coated with dirt, sand, and salt.

Buy postcards of sights that you especially like. You will probably not be able to duplicate the quality of the images on these cards.

Do not take photographs through a window if you can avoid doing so. These are usually poor photos that you will discard.

Ask for permission before photographing people. Some people do not want to be photographed, and others may be offended. Similarly, be sensitive when taking photos inside museums and churches. Turn off your flash unless you have been given permission to use it.

Consider varying the depth of field to accomplish the purpose of your image. Remember: a small *f*-number creates a small depth of field that isolates a subject, whereas a large *f*-number creates a large depth of field. You can check the depth of field on many cameras by pressing the depth-of-field preview button.

Consider using filters to improve your photos. For example, on sunny days, a circular polarizing filter will increase color saturation and decrease glare, and a graduated neutral-density filter will decrease differences in contrast. However, these filters will lengthen exposure times.

For wide-angle pictures, include something interesting in the foreground to add perspective to your image.

Remember that your subject is not the only focus of your picture. Before taking a picture, consider the lighting, background, and other things you may want to include in (or exclude from) your photograph.

When you get home

Recharge or recycle your spent batteries.

Download and categorize your photographs as soon as you can. The sooner you do this, the more likely you are to recall important details about the images.

Clean your equipment.

Enjoy your photographs!

APPENDIX 7

What to Pack

For your flight to Galápagos, there is a weight limit of 44 pounds (20 kg) of luggage per person, plus a carry-on weighing no more than 15 pounds (7 kg). Your luggage and carry-on may be inspected to minimize the introduction of invasive species.

- ❑ Passport, airplane tickets, hotel reservations, and supporting documents
- ❑ Photocopies of important documents, kept in a separate place from originals
- ❑ Swimsuit(s)
- ❑ Pants, pajamas, skirts
- ❑ T-shirts, long-sleeved shirts, blouses
- ❑ Shorts, socks, underwear
- ❑ Windbreaker, lightweight jacket, rain jacket, and/or sweatshirt
- ❑ Sun hat or visor
- ❑ Scarf or bandana
- ❑ Motion sickness pills and/or Scopolamine[1]
- ❑ Sturdy water sandals
- ❑ Toiletry items (e.g., razors, deodorant, comb, shampoo, clippers, toothpaste and toothbrush, floss, deodorant)
- ❑ Wetsuit or shorty
- ❑ Swim mask, snorkel, fins[2]
- ❑ Medicines (e.g., prescription drugs, aspirin, contraceptives, antibiotics)
- ❑ Money (especially denominations under $50), credit cards, traveler's checks
- ❑ Sunblock and lip balm with sunblock
- ❑ Insect repellant

[1] The use of Scopolamine requires a prescription from a physician. If you will be entering Galápagos from Quito (altitude = 9,300 ft), consider taking medications for altitude sickness, such as acetazolamide (Diamox).
[2] You may be able to rent these items on your boat or in town.

- ❑ Sunglasses and lanyard
- ❑ Ziploc bags
- ❑ Earplugs (for protection against snoring roommates)
- ❑ Small backpack for hikes
- ❑ Small flashlight (for lava tunnels)
- ❑ Camera(s), lenses, filters, batteries and chargers, lens cleaner, media cards, and associated photography equipment (Appendix 6)
- ❑ Travel alarm clock
- ❑ Binoculars
- ❑ Water bottle
- ❑ Addresses of friends and relatives
- ❑ Reference books, notepads, pen/pencil
- ❑ Personal first-aid kit

APPENDIX 8

Friends of Galápagos

Protecting and conserving Galápagos—truly a global treasure—requires serious ongoing efforts and commitment. Our passionate, dedicated staff members and volunteers take on that challenge each and every day.

—Fernando Espinoza, Executive Director,
Charles Darwin Foundation

You may leave Galápagos inspired by what you have experienced and wondering what you can do to help preserve this unique ecosystem. There are a variety of organizations dedicated to saving Galápagos by raising awareness of the problems plaguing the islands, providing technical assistance, and raising money. Although tourism and the government of Ecuador finance much of the work done by GNPS, this funding is inadequate to address the many complex challenges facing the park (Chapter 11). CDRS is supported entirely by private funds.

If you would like to help preserve Galápagos, contact any of the following partners of Galápagos:

ECUADOR
Charles Darwin Foundation
www.darwinfoundation.org

UNITED STATES
Galápagos Conservancy
www.galapagos.org

CANADA
Charles Darwin Foundation of Canada
http://www.charlesdarwinfoundationofcanada.ca

UNITED KINGDOM
Galápagos Conservancy Trust
www.savegalapagos.org

FINLAND & SCANDINAVIA
Nordic Friends of Galápgaos
www.galapagosnordic.org

THE NETHERLANDS & BELGIUM
Stichting Vrienden van de Galápagos Eilanden
www.galapagos.nl

JAPAN
The Japanese Association for Galápagos
www.j-galapagos.org

NEW ZEALAND
Friends of Galápagos New Zealand
www.galapagos.org.nz

SWITZERLAND
Freunde der Galápagos Inseln
www.galapagos-ch.org

GERMANY
Zoologische Gesellschaft, Frankfurt
www.zgf.de

REFERENCES

Preface

Oxford, Peter, & Watkins, Graham. 2009. *Galápagos: Both Sides of the Coin.* Morganville, NJ: Imagine Books.

Perry, R. (Ed.). 1984. *Key Environments: Galápagos.* New York: Pergamon Press.

Rachowiecki, Rob. 1989. *Ecuador and the Galápagos Islands: A Travel Survival Kit* (Second Edition). Berkeley, CA: Lonely Planet.

Thornton, Ian W. B. 1971. *Darwin's Islands: A Natural History of the Galápagos.* Garden City, NY: Natural History Press (American Museum of Natural History).

Chapter 1—Evolution and the Galápagos Archipelago

Darwin, Charles. 1845. *Journal of Researches in the National History of Geology of the Countries Visited during the Voyage of HMS.* Beagle *Round the World.* London: John Murray.

Darwin, Charles. 1859. *On the Origin of Species by Natural Selection, or the Preservation of Favoured Races in the Struggle for Life.* London: John Murray.

Grant, Peter R., & Grant, B. Rosemary. 1998. *Evolution on Islands.* Oxford: Oxford University Press.

Weiner, Johnathan. 1994. *The Beak of the Finch: A Story of Evolution in Our Own Time.* New York: Knopf.

Chapter 2—History of the Galápagos Islands

Angermeyer, Johanna. 1990. *My Father's Island: A Galápagos Quest.* New York: Viking.

Banning, George Hugh. 1933. *Hancock Expedition to the Galápagos Islands, 1933.* San Diego, CA: Zoological Society of San Diego, Bulletin No. 10 (May).

Barlow, N. (Ed.). 1934. *Charles Darwin's Diary of the Voyage of H.M.S. "Beagle."* Cambridge: Cambridge University Press.

Barrett, John, & Yánes, Francisco J. 1911. Galápagos. *Bulletin of the Pan American Union,* XXXII, 222–39.

Beebe, William. 1924. *Galápagos: World's End.* New York: G. P. Putnam's Sons.

Beebe, William. 1926. *The Arcturus Adventure.* New York: G. P. Putnam's Sons.

Black, Juan. 1973. *Galápagos, Archipiélago del Ecuador.* Quito: Fundacion Charles Darwin.

Bowman, Robert I. (Ed.). 1966. *The Galápagos: Proceedings of the Symposia of the Galápagos International Scientific Project.* Berkeley: University of California Press.

British Admiralty. 1841. *Galápagos Islands Surveyed by Capt. Robt. FitzRoy and the Officers of H.M.S.* Beagle, *1836.* London: British Admiralty Chart #1375.

Colnett, James. 2010 (1798). *A Voyage to the South Atlantic and Round Cape Horn into the Pacific Ocean for the Purpose of Extending the Spermaceti Whale Fisheries.* London: W. Bennett.

Cowley, William Ambrose. 1688. *Cowley's Voyage Round the World.* British Library Sloane MS 1050.

Dampier, William. 1697. *A New Voyage Round the World.* London: James Knapton.

De Berlanga, T. 1884. A letter to his majesty from Fray Tomás de Berlanga, describing his voyage from Panama to Puerto Viejo, and the hardships he encountered in this navigation. Vol. 41, 538–44 in *Colección de Documentos Inéditos Relativos al Descubrimiento, Conquista y Organización de las Antiguas Posesiones Españolas de América y Oceania.* Madrid: Manuel G. Hernández.

De Roy, Tui (Ed.). 2009. *Galápagos: Preserving Darwin's Legacy.* Buffalo, NY: Firefly Books.

Eibl-Eibesfeldt, Irenäus. 1959. Survey on the Galápagos Islands. *UNESCO Mission Reports* 8, 1–31.

Eibl-Eibesfeldt, Irenäus. 1960. *Galápagos.* London: MacGibbon & Kee.

Fraser, C. McLean. 1934. *The Voyage of the* Velero III, *1933–1934.* Unpublished diary. In *Papers of Waldo LaSalle Schmitt,* Box 86, Folder 9. Washington, DC: Smithsonian Institution.

Gordillo, Jacinto. 2011. *Stories of 44 Years in the Galápagos Islands* (Second Edition). Quito: Ediciones Abya-Yala.

Gottlieb, L. D. 1975. The uses of place: Darwin and Melville on the Galápagos. *BioScience* 25(3), 172–75.

Grant, K. Thalia, & Estes, Gregory B. 2009. *Darwin in Galápagos: Footsteps to a New World*. Princeton, NJ: Princeton University Press.

Grant, R. 1975. Four Galápagos islands. *Geographical Journal* 141, 76–91.

Harris, Paul. 1996. Galápagos 1946—Last Flight Out: A Brief History of the U.S.A.F. Shutdown of the Island's Military Complex. *51st Squadron Fighter Newsletter* 3(1).

Heyerdahl, T., & Skjolsvold, A. 1956. Archaeological evidence of pre-Spanish visits to the Galápagos Islands. *American Antiquities* 22, 1–71.

Hickman, John. 2009. *Galápagos: The Enchanted Islands Through Writers' Eyes*. London: Eland.

Hydrographic Office of the U.S. Navy. 1928. *South America Pilot, Volume III* (Third Edition). Washington, DC: U.S. Government Printing Office.

Idrovo, H. 2005. *Galápagos: Footsteps in Paradise*. Quito: Ediciones Libri Mundi.

James, Matthew J. 2010. Collecting evolution: The vindication of Charles Darwin by the 1905–1906 Galápagos expedition of the California Academy of Sciences. *Proceedings of the California Academy of Sciences, Series 4,* 61 (Suppl. II, No. 12), 197–210.

Keynes, Richard Darwin (Ed.). 1988. *Charles Darwin's* Beagle *Diary*. Cambridge: Cambridge University Press.

Larson, Edward J. 2001. *Evolution's Workshop: God and Science on the Galápagos Islands*. New York: Basic Books.

McEwen, Alec. 1988. The English place-names of the Galápagos. *Geographical Journal* 154(2), 234–42.

Melville, Herman. 1854. *The Encantadas, or Enchanted Isles*. Published serially in the March, April, and May issues of *Putnam's Monthly Magazine* under the pseudonym Salvator R. Tarnmoor.

Nicholls, Henry. 2006. *Lonesome George: The Life and Loves of a Conservation Icon*. London: Macmillan.

Ortega, Antonio Constante. 2003. *Basalto: Etapa de Terror y Lagrimas Durante la Colónia Penal en Isabela*. Gobernia Municipal de Santa Cruz: Santa Cruz, Galápagos.

Otterman, Lillian. 1993. *Clinker Islands: A Complete History of the Galápagos Archipelago*. Bradenton, FL: McGuinn & McGuire.

Parks, E. Taylor, & Rippy, J. Fred. 1940. The Galápagos Islands, a neglected phase of American strategic diplomacy. *Pacific Historical Review* IX(1), 37–45.

Porter, David. 1815. *Journal of the Cruise Made to the Pacific Ocean*. Philadelphia: Bradford & Inskeep.

Ritter, Friedrich. 1931. Adam and Eve in the Galápagos. *Atlantic Monthly,* CXLVIII (October), 409–18.

Ritter, Friedrich. 1931. Eve calls it a day. *Atlantic Monthly,* CXLVIII (December), 733–43.

Ritter, Friedrich. 1931. Satan walks in the garden. *Atlantic Monthly,* CXLVIII (November), 565–75.

Robinson, William A. 1936. *Voyage to Galápagos*. New York: Harcourt, Brace and Co.

Slevin, Joseph R. 1959. *The Galápagos Islands: A History of their Exploration,* Occ. Paper No. 25. San Francisco, CA: California Academy of Sciences.

Sulloway, Frank J. 1982. Darwin's conversion: The *Beagle* voyage and its aftermath. *Journal of the History of Biology* 15(3), 325–96.

Sulloway, Frank J. 2009. Tantalizing tortoises and the Darwin-Galápagos legend. *Journal of the History of Biology* 42, 3–31. DOI 10.1007/s10739-008-9173-9.

Treherne, J. 2002. *The Galápagos Affair.* London: Pimlico.

Wittmer, Margret. 1961 [1959]. *Floreana [Postlagernd Floreana]*. London: Michael Joseph Ltd.

Woram, John. 2005. *Charles Darwin Slept Here: Tales of Human History at World's End*. Rockville Centre, NY: Rockville Press.

Worman, John. 2011. Las Encantadas: Human and Cartographic History of the Galápagos Islands. Retrieved on July 3, 2012, at www.galapagos.to

Chapter 3—Geology and Climate

Christie, D. M., et al. 1992. Drowned islands downstream from the Galápagos hotspot imply extended speciation times. *Nature* 335(6357), 246–48. DOI:10.1038/355246a0.

Geist, D. 1996. On the emergence and submergence of the Galápagos islands, Ecuador. *Noticias de Galápagos* 56, 5.

Snell, H. M., Stone, A., & Snell, H. L. 1996. A summary of geographic characteristics of the Galápagos Islands. *Journal of Biogeography* 23, 619–24.

Chapter 4—The Islands: Where You Will Go and What You Will See

Constant, Pierre. 2006. *Galápagos: A Natural History Guide,* 7th edition. Hong Kong: Odyssey.

Fitter, Julian, Fitter, Daniel, & Hosking, David. 2000. *Safari Guides: Galápagos*. London: HarperCollins.

Jackson, M. H. 1993. *Galápagos, A Natural History.* Calgary, Alberta: University of Calgary Press.

Kricher, John. 2002. *Galápagos.* Washington, DC: Smithsonian Institution Press.

Woram, John M. 1989. Galápagos island names. *Noticias de Galápagos* 48, 22–32.

Chapter 5—Birds

Abzhanov, A. 2010. Darwin's Galápagos finches in modern biology. *Philosophical Transactions of the Royal Society B: Biological Sciences* 365(1543), 1001.

Abzhanov, A., et al. 2006. The calmodulin pathway and evolution of elongated beak morphology in Darwin's finches. *Nature* 442(7102), 563–67.

Abzhanov, A., Protas, M., Grant, B. R., Grant, R., & Tabin, C. J. 2004. *BMP4* and morphological variation of beaks in Darwin's finches. *Science* 305(5689), 1462.

Anderson, D. J. 1991. Apparent predator-limited distribution of Galápagos Red-footed Boobies *Sula sula. Ibis* 133(1), 26–29.

Anderson, D. J., et al. 2002. Population size and trends of the waved albatross *Phoebastria irrorata. Marine Ornithology* 30, 63–69.

Anderson, D. J., et al. 2003. At-sea distribution of waved albatrosses and the Galápagos Marine Reserve. *Biological Conservation* 110(3), 367–73.

Anderson, D. J., et al. 2008. Population status of the critically endangered waved albatross *Phoebastria irrorata,* 1999 to 2007. *Endangered Species Research* 5(2–3), 185–92.

Arbogast, B. S., et al. 2006. The origin and diversification of Galápagos mockingbirds. *Evolution* 60(2), 370–82.

Awkerman, J. A., Huyvaert, K. P., Mangel, J., Shigueto, J. A., & Anderson, D. J. 2006. Incidental and intentional catch threatens Galápagos waved albatross. *Biological Conservation* 133(4), 483–89.

Baker, A. J., Pereira, S. L., Haddrath, O. P., & Edge, K. A. 2006. Multiple gene evidence for expansion of extant penguins out of Antarctica due to global cooling. *Proceedings of the Royal Society B: Biological Sciences* 273(1582), 11–17.

Beck, R. H. 1904. Bird life among the Galápagos Islands. *Condor* 6(1), 5–11.

Bertelli, S., & Giannini, N. P. 2005. A phylogeny of extant penguins (Aves: Sphenisciformes) combining morphology and mitochondrial sequences. *Cladistics* 21(3), 209–39.

Boag, T. 1983. The heritability of external morphology in Darwin's ground finches (*Geospiza*) on Isla Daphne Major, Galápagos. *Evolution* 37(5), 877–94.

Boag, T., and Grant, R. 1981. Intense natural selection in a population of Darwin's finches (Geospizinae) in the Galápagos. *Science* 214(4516), 82–85.

Boersma, D. 1998. Population trends of the Galápagos penguin: Impacts of El Niño and La Niña. *Condor* 100(2), 245–53.

Boersma, D. 2008. Penguins as marine sentinels. *Bioscience* 58(7), 597–607.

Bollmer, J. L., et al. 2003. Variation in morphology and mating system among island populations of Galápagos hawks. *Condor* 105(3), 428–38.

Bowman, Robert I. 1961. *Morphological Differentiation and Adaptation in the Galápagos Finches.* Berkeley: University of California Press.

Bugoni, L., Griffiths, K., & Furness, R. W. 2011. Sex-biased incidental mortality of albatrosses and petrels in longline fisheries: Differential distributions at sea or differential access to baits mediated by sexual size dimorphism? *Journal of Ornithology,* 152(2), 261–268.

Burns, K. J., Hackett, S. J., & Klein, N. K. 2002. Phylogenetic relationships and morphological diversity in Darwin's finches and their relatives. *Evolution* 56(6), 1240–52.

Castro, Isabel, & Phillips, Antonia. 1996. *A Guide to the Birds of the Galápagos Islands.* Princeton, NJ: Princeton University Press.

Costa, D. P. 1991. Reproductive and foraging energetics of high latitude penguins, albatrosses and pinnipeds: Implications for life history patterns. *American Zoologist* 31(1), 111–130.

Curry, R. L. 1988a. Group structure, within-group conflict and reproductive tactics in cooperatively breeding Galápagos mockingbirds, *Nesomimus parvulus. Animal Behaviour* 36(6), 1708–28.

Curry, R. L. 1988b. Influence of kinship on helping behavior in Galápagos mockingbirds. *Behavioral Ecology and Sociobiology* 22(2), 141–52.

Curry, R. L. 1989. Geographic variation in social organization of Galápagos mockingbirds: Ecological correlates of group territoriality and cooperative breeding. *Behavioral Ecology and Sociobiology* 25(2), 147–60.

Darwin, Charles D. 1963 [1836]. *Darwin's Ornithological Notes.* Edited with an Introduction, Notes, and Appendix by Nora Barlow. Bulletin of the British Museum (Natural History) Historical Series, 2(7).

Dearborn, D. C., Anders, A. D., & Parker, G. 2001. Sexual dimorphism, extrapair fertilizations, and operational sex ratio in great frigatebirds (*Fregata minor*). *Behavioral Ecology* 12(6), 746–752.

Deem, S. L., et al. 2010. Exposure to *Toxoplasma gondii* in Galápagos penguins (*Spheniscus mendiculus*) and flightless cormorants (*Phalacrocorax harrisi*) in the Galápagos Islands, Ecuador. *Journal of Wildlife Diseases* 46(3), 1005–1011.

DeLay, L. S., Faaborg, J., Naranjo, J., Paz, S. M., de Vries, T., & Parker, G. 1996. Paternal care in the cooperatively polyandrous Galápagos hawk. *Condor* 98, 300–11.

Dentressangle, F., Boeck, L., & Torres, R. 2008. Maternal investment in eggs is affected by male feet colour and breeding conditions in the blue-footed booby, *Sula nebouxii*. *Behavioral Ecology and Sociobiology* 62(12), 1899–1908.

Donohue, Kathleen (Ed.). 2011. *Darwin's Finches: Readings in the Evolution of a Scientific Paradigm*. Chicago: University of Chicago Press.

Drummond, H., & Garcia Chavelas, C. 1989. Food shortage influences sibling aggression in the blue-footed booby. *Animal Behaviour* 37, 806–19.

Eibl-Eibesfeldt, Irenäus. 1966. Darwin's finches. *Animals* 9(3), 137–39.

Faaborg, J., et al. 1995. Confirmation of cooperative polyandry in the Galápagos hawk (*Buteo galapagoensis*). *Behavioral Ecology and Sociobiology* 36(2), 83–90.

Fessl, B., Loaiza, A. D., Tebbich, S., & Young, H. G. 2011. Feeding and nesting requirements of the critically endangered Mangrove Finch *Camarhynchus heliobates*. *Journal of Ornithology*, 152(2), 453–460.

Fessl, B., Sinclair, B. J., & Kleindorfer, S. 2006. The life-cycle of *Philornis downsi* (Diptera: Muscidae) parasitizing Darwin's finches and its impacts on nestling survival. *Parasitology* 133 (Pt 6), 739–47.

Gould, John. 1841. *The Zoology of the Voyage of H.M.S.* Beagle: *Part III, Birds*. London: Smith, Elder.

Grant, B. R., & Grant, P. R. 1996. Cultural inheritance of song and its role in the evolution of Darwin's finches. *Evolution* 50(6), 2471–87.

Grant, B. R., & Grant, P. R. 2002. Lack of premating isolation at the base of a phylogenetic tree. *American Naturalist* 160(1), 1–19.

Grant, B. R., & Grant, P. R. 2010. Songs of Darwin's finches diverge when a new species enters the community. *Proceedings of the National Academy of Sciences of the United States of America* 107(47), 20156–63.

Grant, P. R., & Grant, B. R. 1993. Evolution of Darwin's finches caused by a rare climatic event. *Proceedings of the Royal Society B: Biological Sciences* 251(1331), 111–17.

Grant, P. R., & Grant, B. R. 1997a. Hybridization, sexual imprinting, and mate choice. *American Naturalist* 149(1), 1–28.

Grant, P. R., & Grant, B. R. 1997b. Mating patterns of Darwin's finch hybrids determined by song and morphology. *Biological Journal of the Linnean Society* 60(3), 317–43.

Grant, P. R., & Grant, B. R. 2002. Unpredictable evolution in a 30-year study of Darwin's finches. *Science* 296(5568), 707–11.

Grant, P. R., & Grant, B. R. 2006. Evolution of character displacement in Darwin's finches. *Science* 313(5784), 224–226.

Grant, P. R., et al. 2004. Convergent evolution of Darwin's finches caused by introgressive hybridization and selection. *Evolution* 58(7), 1588–99.

Grant, P. R., Grant, B. R., & Abzhanov, A. 2006. A developing paradigm for the development of bird beaks. *Biological Journal of the Linnean Society* 88(1), 17–22.

Grant, P. R., Grant, B. R., Keller, L. F., Markert, J. A., & Petren, K. 2003. Inbreeding and interbreeding in Darwin's finches. *Evolution* 57(12), 2911–16.

Grant, P. R., Grant, B. R., & Petren, K. 2000. The allopatric phase of speciation: The sharp-beaked ground finch (*Geospiza difficilis*) on the Galápagos islands. *Biological Journal of the Linnean Society* 69(3), 287–317.

Grant, P. R., Grant, B. R., & Petren, K. 2005. Hybridization in the recent past. *American Naturalist* 166(1), 56–67.

Grant, P. R., Grant, B. R., Petren, K., & Keller, L. F. 2005. Extinction behind our backs: The possible fate of one of the Darwin's finch species on Isla Floreana, Galápagos. *Biological Conservation* 122(3), 499–503.

Grant, Peter R. 1986. *Ecology and Evolution of Darwin's Finches*. Princeton, NJ: Princeton University Press.

Grant, Peter R., Curry, R. L., & Grant, B. R. 2000. A remnant population of the Floreana mockingbird on Champion Island, Galápagos. *Biological Conservation* 92, 285–90.

Grant, Peter R., & Grant, B. Rosemary. 2008. *How and Why Species Multiply: The Radiation of Darwin's Finches*. Princeton, NJ: Princeton University Press.

Hailer, F., et al. 2010. Long-term isolation of a highly mobile seabird on the Galápagos. *Proceedings of the Royal Society B: Biological Sciences* 278(1707), 817.

Harris, M. P. 1982. *Collins Field Guide to the Birds of Galápagos, Revised Edition*. New York: Viking Penguin.

Hatchwell, B. J., & Komdeur, J. 2000. Ecological constraints, life history traits and the evolution of cooperative breeding. *Animal Behaviour* 59(6), 1079–86.

Heinzel, Hermann, & Barnaby Hall. 2000. *Galápagos Diary: A Complete Guide to the Archipelago's Birdlife*. Berkeley: University of California Press.

Hoeck, E. A., Bollmer, J. L., Parker, G., & Keller, L. F. 2010. Differentiation with drift: A spatio-temporal genetic analysis of Galápagos mockingbird populations (*Mimus* spp.). *Philosophical Transactions of the Royal Society B: Biological Sciences* 365(1543), 1127–1138.

Hosken, D. J., & Balloux, F. 2002. Thirty years of evolution in Darwin's finches. *Trends in Ecology & Evolution* 17(10), 447–48.

Humphries, C. A., Arevalo, V. D., Fischer, K. N., & Anderson, D. J. 2006. Contributions of marginal offspring to reproductive success of Nazca booby (*Sula granti*) parents: Tests of multiple hypotheses. *Oecologia* 147(2), 379–90.

Huyvaert, K. P., Anderson, D. J., Parker, G., & Lank, D. B. 2006. Mate opportunity hypothesis and extra-pair paternity in waved albatrosses (*Phoebastria irrorata*). *The Auk* 123(2), 524–36.

Huyvaert, K. P., & Parker, G. 2010. Extra-pair paternity in waved albatrosses: Genetic relationships among females, social mates and genetic sires. *Behaviour* 147(12), 1591–1613.

Juola, F. A., & Dearborn, D. C. 2007. Does the differential cost of sons and daughters lead to sex ratio adjustment in great frigatebirds *Fregata minor? Journal of Avian Biology* 38(1), 94–104.

Kinnaird, M. F., & Grant, P. R. 1982. Cooperative breeding by the Galápagos mockingbird, *Nesomimus parvulus*. *Behavioral Ecology and Sociobiology* 10(1), 65–73.

Kleindorfer, S. 2007. Nesting success in Darwin's small tree finch, *Camarhynchus parvulus:* Evidence of female preference for older males and more concealed nests. *Animal Behaviour* 74(4), 795–804.

Kleindorfer, S. 2007. The ecology of clutch size variation in Darwin's Small Ground Finch *Geospiza fuliginosa:* Comparison between lowland and highland habitats. *Ibis* 149(4), 730–41.

Kleindorfer, S., & Dudaniec, R. Y. 2009. Love thy neighbour? Social nesting pattern, host mass and nest size affect ectoparasite intensity in Darwin's tree finches. *Behavioral Ecology and Sociobiology* 63(5), 731–39.

Kleindorfer, S., Sulloway, F. J., & O'Connor, J. 2009. Mixed species nesting associations in Darwin's tree finches: Nesting pattern predicts predation outcome. *Biological Journal of the Linnean Society* 98(2), 313–24.

Lack, David. 1947. *Darwin's Finches*. Cambridge: Cambridge University Press.

Levin, I. I., Outlaw, D. C., Vargas, F. H., & Parker, G. 2009. Plasmodium blood parasite found in endangered Galápagos penguins (*Spheniscus mendiculus*). *Biological Conservation* 142(12), 3191–95.

Lindström, K. M., Foufopoulos, J., Pärn, H., & Wikelski, M. 2004. Immunological investments reflect parasite abundance in island populations of Darwin's finches. *Proceedings of the Royal Society of London. Series B: Biological Sciences* 271(1547), 1513–1519.

Lougheed, L. W., & Anderson, D. J. 1999. Parent blue-footed boobies suppress siblicidal behavior of offspring. *Behavioral Ecology and Sociobiology* 45(1), 11–18.

Lowe, R. 1936. The finches of the Galápagos in relation to Darwin's conception of species. *Ibis* 78(310), 320–21.

Madsen, V., Dabelsteen, T., Osorio, D., & Osorno, J. L. 2007. Morphology and ornamentation in male magnificent frigatebirds: Variation with age class and mating status. *American Naturalist* 169, 93–111.

Miller, G. D., Hofkin, B. V., Snell, H., Hahn, A., & Miller, R. D. 2001. Avian malaria and Marek's disease: Potential threats to Galápagos penguins *Spheniscus mendiculus*. *Marine Ornithology* 29(1), 43–46.

Mills, K. L. 1998. Multispecies seabird feeding flocks in the Galápagos Islands. *Condor* 100(2), 277–85.

Morgan, S. M., Ashley-Ross, M. A., & Anderson, D. J. 2003. Foot-mediated incubation: Nazca booby (*Sula granti*) feet as surrogate brood patches. *Physiological and Biochemical Zoology,* 76(3), 360–66.

Müller, M. S., et al. 2011. Maltreated nestlings exhibit correlated maltreatment as adults: Evidence of a "cycle of violence" in Nazca boobies (*Sula granti*). *The Auk* 128(4), 615–19.

Nelson, Bryan. 1968. *Galápagos: Islands of Birds*. New York: William Murrow.

O'Connor, J. A., Dudaniec, R. Y., & Kleindorfer, S. 2010. Parasite infestation and predation in Darwin's small ground finch: Contrasting two elevational habitats between islands. *Journal of Tropical Ecology* 26(3), 285–92.

O'Connor, J. A., Sulloway, F. J., & Kleindorfer, S. 2010. Avian population survey in the Floreana highlands: Is Darwin's Medium Tree Finch declining in remnant patches of Scalesia forest? *Bird Conservation International* 20(4), 343–53.

O'Connor, J., Sulloway, F., Robertson, J., & Kleindorfer, S. 2010. *Philornis downsi* parasitism is the primary cause of nestling mortality in the critically endangered Darwin's medium tree finch (*Camarhynchus pauper*). *Biodiversity and Conservation* 19(3), 853–66.

Petren, K., Grant, B. R., & Grant, P. R. 1999. A phylogeny of Darwin's finches based on microsatellite DNA length variation. *Proceedings of the Royal Society of London. Series B: Biological Sciences* 266(1417), 321–329.

Petren, K., Grant, P. R., Grant, B. R., & Keller, L. F. 2005. Comparative landscape genetics and the adaptive radiation of Darwin's finches: The role of peripheral isolation. *Molecular Ecology* 14(10), 2943–57.

Sato, A., et al. 1999. Phylogeny of Darwin's finches as revealed by mtDNA sequences. *Proceedings of the National Academy of Sciences* 96(9), 5101–5106.

Schluter, D., Price, T. D., & Grant, P. R. 1985. Ecological character displacement in Darwin's finches. *Science* 227(4690), 1056–59.

Sulloway, Frank J. 1982. Darwin and his finches: The evolution of a legend. *Journal of the History of Biology* 15, 20–36.

Tebbich, S., & Bshary, R. 2004. Cognitive abilities related to tool use in the woodpecker finch, *Cactospiza pallida*. *Animal Behaviour* 67(4), 689–97.

Tebbich, S., Taborsky, M., Fessl, B., & Blomqvist, D. 2001. Do woodpecker finches acquire tool-use by social learning? *Proceedings of the Royal Society of London. Series B: Biological Sciences* 268(1482), 2189–2193.

Tonnis, B., Grant, P. R., Grant, B. R., & Petren, K. 2005. Habitat selection and ecological speciation in Galápagos warbler finches (*Certhidea olivacea* and *Certhidea fusca*). *Proceedings of the Royal Society B: Biological Sciences* 272(1565), 819–827.

Torres, R., & Velando, A. 2005. Male preference for female foot colour in the socially monogamous blue-footed booby, *Sula nebouxii*. *Animal Behaviour* 69(1), 59–65.

Vargas, F. H., Harrison, S., Rea, S., & Macdonald, D. W. 2006. Biological effects of El Niño on the Galápagos penguin. *Biological Conservation* 127(1), 107–14.

Velando, A., Beamonte-Barrientos, R., & Torres, R. 2006. Pigment-based skin colour in the blue-footed booby: An honest signal of current condition used by females to adjust reproductive investment. *Oecologia* 149(3), 535–42.

Velando, A., Torres, R., & Espinosa, I. 2005. Male coloration and chick condition in blue-footed booby: A cross-fostering experiment. *Behavioral Ecology and Sociobiology* 58(2), 175–80.

Vincek, V., et al. 1997. How large was the founding population of Darwin's finches? *Proceedings of the Royal Society of London. Series B: Biological Sciences* 264(1378), 111–118.

Wilson, R. P., et al. 2008. What grounds some birds for life? Movement and diving in the sexually dimorphic Galápagos cormorant. *Ecological Monographs* 78(4), 633–52.

Chapter 6—Reptiles

Andersen, J. B. 2008. Iguanas eavesdrop on mockingbirds. *Journal of Experimental Biology* 211(5), vi.

Benavides, E., Baum, R., Snell, H. M., Snell, H. L., & Sites, J.W., Jr. 2009. Island biogeography of Galápagos lava lizards (Tropiduridae: Microlophus): Species diversity and colonization of the archipelago. *Evolution* 63(6), 1606–26.

Bowen, B. W., & Karl, S. A. 2007. Population genetics and phylogeography of sea turtles. *Molecular Ecology* 16(23), 4886–4907.

Caccone, A., J. P. Gibbs, V. Ketmaier, E. Suatoni, & J. R. Powell. 1999. Origin and evolutionary relationships of giant Galápagos tortoises. *Proceedings of the National Academy of Sciences of the USA* 96, 13223–228.

Censky, E. J., Hodge, K., & Dudley, J. 1998. Over-water dispersal of lizards due to hurricanes. *Nature* 395(6702), 556.

Chambers, Paul. 2006. *A Sheltered Life: The Unexpected History of the Giant Tortoise.* Oxford: Oxford University Press.

Ciofi, C., et al. 2009. Genetics and conservation on islands: The Galapagos giant tortoise as a case study, pp. 269–293. Chapter 12 in G. Bertorelle, et al., *Population Genetics for Animal Conservation.* Cambridge: Cambridge University Press.

Fitter, Julian, Daniel Fitter, & David Hosking. 2007. *Wildlife of the Galápagos.* London: HarperCollins.

Fuentes, M., Hamann, M., & Limpus, C. J. 2010. Past, current and future thermal profiles of green turtle nesting grounds: Implications from climate change. *Journal of Experimental Marine Biology and Ecology* 383(1), 56–64.

Garrick, R. C., et al. 2012. Genetic rediscovery of an "extinct" Galápagos giant tortoise species. *Current Biology* 22(1), R10–R11.

Gentile, G., & Snell, H. 2009. *Conolophus marthae* sp. nov. (Squamata, Iguanidae), a new species of land iguana from the Galápagos archipelago. *Zootaxa* 2201, 1–10.

Gentile, G., et al. 2009. An overlooked pink species of land iguana in the Galápagos. *Proceedings of the National Academy of Sciences* 106(2), 507.

Gibbs, J. P., Marquez, C., & Sterling, E. J. 2008. The role of endangered species reintroduction in ecosystem restoration: Tortoise-cactus interactions on Espanola Island, Galapagos. *Restoration Ecology* 16(1), 88–93.

Gibbs, J. P., Sterling, E. J., & Zabala, F. J. 2010. Giant tortoises as ecological engineers: A long-term quasi-experiment in the Galápagos Islands. *Biotropica* 42(2), 208–14.

Green, D. 1984. Long-distance movements of Galapagos green turtles. *Journal of Herpetology* 18(2), 121–30.

Günther, A. 1877. *The Gigantic Land-Tortoises (Living and Extinct) in the Collection of the British Museum*. London: Taylor & Francis.

Hawkes, L. A., Broderick, A. C., Godfrey, M. H., & Godley, B. J. 2009. Climate change and marine turtles. *Endangered Species Research* 7(2), 137–54.

Isaksson, C., et al. 2011. Aggression, but not testosterone, is associated to oxidative status in a free-living vertebrate. *Behaviour 148,* 713–31.

Jordan, M. A., & Snell, H. L. 2008. Historical fragmentation of islands and genetic drift in populations of Galapagos lava lizards (*Microlophus albemarlensis* complex). *Molecular Ecology* 17(5), 1224–37.

Kizirian, D., Trager, A., Donnelly, M. A., & Wright, J. W. 2004. Evolution of Galapagos Island lava lizards (Iguania: Tropiduridae: Microlophus). *Molecular Phylogenetics and Evolution* 32(3), 761–69.

Nemésio, A. 2009. On the live holotype of the Galápagos pink land iguana, *Conolophus marthae* Gentile & Snell, 2009 (Squamata: Iguanidae): Is it an acceptable exception? *Zootaxa* 2201, 21–25.

Parham, J. F., & Zug, G. R. 1996. *Chelonia agassizii*—valid or not? *Marine Turtle Newsletter* 72, 2–5.

Phillips, R. B., et al. 2011. Eradicating feral cats to protect Galapagos land iguanas: Methods and strategies. *Pacific Conservation Biology* 11(4), 257–67.

Pritchard, Peter C. H. 1996. *The Galápagos Tortoises: Nomenclatural and Survival Status*. Lunenburg, MA: Chelonian Research Foundation.

Rassmann, K., Trillmich, F., & Tautz, D. 1997. Hybridization between the Galápagos land and marine iguana (*Conolophus subcristatus* and *Amblyrhynchus cristatus*) on Plaza Sur. *Journal of Zoology* 242(4), 729–39.

Romero, L. M., & Wikelski, M. 2001. Corticosterone levels predict survival probabilities of Galápagos marine iguanas during El Niño events. *Proceedings of the National Academy of Sciences* 98(13), 7366–7370.

Romero, L. M., & Wikelski, M. 2002. Exposure to tourism reduces stress-induced corticosterone levels in Galapagos marine iguanas. *Biological Conservation* 108(3), 371–74.

Romero, L. M., & Wikelski, M. 2010. Stress physiology as a predictor of survival in Galapagos marine iguanas. *Proceedings of the Royal Society B: Biological Sciences* 277(1697), 3157–62.

Schwanz, L. E., Spencer, R. J., Bowden, R. M., & Janzen, F. J. 2010. Climate and predation dominate juvenile and adult recruitment in a turtle with temperature-dependent sex determination. *Ecology* 91(10), 3016–26.

Seminoff, J. A., et al. 2008. Post-nesting migrations of Galápagos green turtles *Chelonia mydas* in relation to oceanographic conditions: Integrating satellite telemetry with remotely sensed ocean data. *Endangered Species Research* 4(1–2), 57–72.

Stone, A., Snell, H. L., & Snell, H. M. 1994. Behavioral diversity as biological diversity: Introduced cats and lava lizard wariness. *Conservation Biology* 8(2), 569–73.

Sulloway, F. J. 2009. Tantalizing tortoises and the Darwin-Galápagos legend. *Journal of the History of Biology* 42(1), 3–31.

Thomson, S. A., S. Irwin, & T. Irwin. 1996. Harriet, la Tortuga de Galápagos. *Reptilia* 2(4), 46–49.

Vágvölgyl, J. 1974. Pinta tortoise: rediscovered. *Pacific Discovery* 27(2), 21–23.

Van de Merwe, J. P., Hodge, M., Whittier, J. M., Ibrahim, K., & Lee, S. Y. 2010. Persistent organic pollutants in the green sea turtle *Chelonia mydas:* Nesting population variation, maternal transfer, and effects on development. *Marine Ecology Progress Series* 403, 269–78.

Vitousek, M. N., Adelman, J. S., Gregory, N. C., & St. Clair, J. J. H. 2007. Heterospecific alarm call recognition in a non-vocal reptile. *Biology Letters* 3(6), 632–34.

Vitousek, M. N., Romero, L. M., Tarlow, E., Cyr, N. E., & Wikelski, M. 2010. Island tameness: An altered cardiovascular stress response in Galápagos marine iguanas. *Physiology & Behavior* 99(4), 544–48.

Vitousek, M. N., Rubenstein, D. R., Nelson, K. N., & Wikelski, M. 2008. Are hotshots always hot? A longitudinal study of hormones, behavior, and reproductive success in male marine iguanas. *General and Comparative Endocrinology* 157(3), 227–32.

Wheeler, E., Hong, Y., Bedon, L. C., & Mackie, R. I. 2012. Carriage of antibiotic-resistant enteric bacteria varies among sites in Galápagos reptiles. *Journal of Wildlife Diseases* 48(1), 56–67.

Wikelski, M. 2005. Evolution of body size in Galapagos marine iguanas. *Proceedings of the Royal Society B: Biological Sciences* 272(1576), 1985–1993.

Wikelski, M., Carbone, C., Bednekoff, A., Choudhury, S., & Tebbich, S. 2001. Why is female choice not unanimous? Insights from costly mate sampling in marine iguanas. *Ethology* 107(7), 623–38.

Wikelski, M., Carbone, C., & Trillmich, F. 1996. Lekking in marine iguanas: Female grouping and male reproductive strategies. *Animal Behaviour* 52(3), 581–96.

Wikelski, M., Steiger, S. S., Gall, B., & Nelson, K. N. 2005. Sex, drugs and mating role: Testosterone-induced phenotype-switching in Galapagos marine iguanas. *Behavioral Ecology* 16(1), 260–268.

Wikelski, M., & Thom, C. 2000. Marine iguanas shrink to survive El Niño. *Nature* 403(6765), 37–38.

Wikelski, M., & Trillmich, F. 1994. Foraging strategies of the Galapagos marine iguana (*Amblyrhynchus cristatus*): Adapting behavioral rules to ontogenetic size change. *Behaviour* 128(3–4), 255–79.

Wikelski, M., & Trillmich, F. 1997. Body size and sexual size dimorphism in marine iguanas fluctuate as a result of opposing natural and sexual selection: An island comparison. *Evolution* 51(3), 922–36.

Chapter 7—Mammals

Arnold, W., & Trillmich, F. 1985. Time budget in Galápagos fur seal pups: The influence of the mother's presence and absence on pup activity and play. *Behaviour* 92(3/4), 302–21.

Dellinger, T., & Trillmich, F. 1999. Fish prey of the sympatric Galápagos fur seals and sea lions: Seasonal variation and niche separation. *Canadian Journal of Zoology* 77(8), 1204–16.

Dowler, R. C., Carroll, D. S., & Edwards, C. W. 2000. Rediscovery of rodents (Genus *Nesoryzomys*) considered extinct in the Galápagos Islands. *Oryx* 34(2), 109–18.

Harris, D. 2009. Review of negative effects of introduced rodents on small mammals on islands. *Biological Invasions* 11(7), 1611–30.

Harris, D. B., & Macdonald, D. W. 2007a. Interference competition between introduced black rats and endemic Galápagos rice rats. *Ecology* 88(9), 2330–44.

Harris, D. B., & Macdonald, D. W. 2007b. Population ecology of the endemic rodent *Nesoryzomys swarthi* in the tropical desert of the Galápagos Islands. *Journal of Mammalogy* 88(1), 208–19.

Kavanagh, James. 2001. *Galápagos Wildlife: An Introduction to Familiar Species*. Phoenix: Waterford Press.

Mueller, B., Pörschmann, U., Wolf, J. B. W., & Trillmich, F. 2011. Growth under uncertainty: The influence of marine variability on early development of Galápagos sea lions. *Marine Mammal Science* 27(2), 350–65.

Schramm, Y., et al. 2009. Phylogeography of California and Galápagos sea lions and population structure within the California sea lion. *Marine Biology* 156(7), 1375–87.

Swash, Andy, & Still, Rob. 2006. *Birds, Mammals, and Reptiles of the Galápagos Islands*, 2nd edition. New Haven, CT: Yale University Press.

Trillmich, F. 1981. Mutual mother-pup recognition in Galápagos fur seals and sea lions: Cues used and functional significance. *Behaviour* 78(1–2), 21–42.

Trillmich, F. 1986. Maternal investment and sex-allocation in the Galápagos fur seal, *Arctocephalus galapagoensis*. *Behavioral Ecology and Sociobiology* 19(3), 157–64.

Trillmich, F. 1990. The behavioral ecology of maternal effort in fur seals and sea lions. *Behaviour* 114(1–4), 3–20.

Trillmich, F., & Limberger, D. 1985. Drastic effects of El Niño on Galápagos pinnipeds. *Oecologia* 67(1), 19–22.

Trillmich, F., & Mohren, W. 1981. Effects of the lunar cycle on the Galápagos fur seal, *Arctocephalus galapagoensis*. *Oecologia* 48(1), 85–92.

Trillmich, F., & Wolf, J. B. W. 2008. Parent–offspring and sibling conflict in Galápagos fur seals and sea lions. *Behavioral Ecology and Sociobiology* 62, 363–75, DOI: 10.1007/s00265-007-0423-1.

Villegas-Amtmann, S., Atkinson, S., & Costa, D. P. 2009. Low synchrony in the breeding cycle of Galápagos sea lions revealed by seasonal progesterone concentrations. *Journal of Mammalogy* 90(5), 1232–37.

Wolf, J. B. W., Kauermann, G., & Trillmich, F. 2005. Males in the shade: Habitat use and sexual segregation in the Galápagos sea lion (*Zalophus californianus wollebaeki*). *Behavioral Ecology and Sociobiology* 59(2), 293–302.

Wolf, J. B. W., Tautz, D., & Trillmich, F. 2007. Galápagos and Californian sea lions are separate species: Genetic analysis of the genus *Zalophus* and its implications for conservation management. *Frontiers in Zoology* 4(1), 20–33.

Chapter 8—Fish

Baldwin, C. C., & McCosker, J. E. 2001. Wrasses of the Galápagos Islands, with the description of a new deepwater species of Halichoeres (Perciformes: Labridae). *Revista de Biología Tropical* 49 (Suppl. 1), 89–100.

Edgar, G. J., Banks, S., Fariña, J. M., Calvopiña, M., & Martínez, C. 2004. Regional biogeography of shallow reef fish and macro-invertebrate communities in the Galapagos archipelago. *Journal of Biogeography* 31(7), 1107–24.

Edgar, G. J., et al. 2008. Conservation of threatened species in the Galápagos Marine Reserve through identification and protection of marine key biodiversity areas. *Aquatic Conservation: Marine and Freshwater Ecosystems* 18(6), 955–68.

Grove, Jack Stein, & Lavenberg, Robert J. 1997. *The Fishes of the Galápagos Islands*. Stanford, CA: Stanford University Press.

Humann, Paul. 2003. *Reef Fish Identification: Galápagos*. Jacksonville, FL: New World.

Irving, A. D., & Witman, J. D. 2009. Positive effects of damselfish override negative effects of urchins to prevent an algal habitat switch. *Journal of Ecology* 97(2), 337–47.

Merlen, Godfrey. 1988. *A Field Guide to the Fishes of Galápagos*. London: Wilmot Books.

Chapter 9—Invertebrates

Dumont, C., Drolet, D., Deschenes, I., & Himmelman, J. 2007. Multiple factors explain the covering behaviour in the green sea urchin, *Strongylocentrotus droebachiensis*. *Animal Behavior* 73(6), 979–86.

Foucaud, J., et al. 2010. Worldwide invasion by the little fire ant: routes of introduction and eco-evolutionary pathways. *Evolutionary Applications* 3(4), 363–74.

Hickman, Cleveland. 1998. *A Field Guide to Sea Stars & Other Echinoderms of Galápagos*. Lexington, VA: Sugar Spring Press.

Hickman, Cleveland. 1999. *A Field Guide to Marine Molluscs of Galápagos*. Lexington, VA: Sugar Spring Press.

Hickman, Cleveland. 2008. *A Field Guide to Corals and Other Radiates of Galápagos*. Lexington, VA: Sugar Spring Press.

Hickman, Cleveland, & Zimmerman, Todd L. 2000. *A Field Guide to Crustaceans of Galápagos*. Lexington, VA: Sugar Spring Press.

Irving, A. D., & Witman, J. D. 2009. Positive effects of damselfish override negative effects of urchins to prevent an algal habitat switch. *Journal of Ecology* 97(2), 337–47.

Jaramillo, P., Trigo, M. M., Ramírez, E., & Mauchamp, A. 2010. Insect pollinators of *Jasminocereus thouarsii*, an endemic cactus of the Galápagos Islands. *Galápagos Research* 67, 21–25.

Parent, C. E. 2012. Biogeographical and ecological determinants of land snail diversification on islands. *American Malacological Bulletin* 30(1), 207–15.

Parent, C. E., & Crespi, B. J. 2006. Sequential colonization and diversification of Galápagos endemic land snail genus *Bulimulus* (Gastropoda, Stylommatophora). *Evolution* 60(11), 2311–28.

Parent, C. E., & Crespi, B. J. 2009. Ecological opportunity in adaptive radiation of Galápagos endemic land snails. *American Naturalist* 174(6), 898–905.

Peck. Stewart B. 2006. *The Beetles of the Galápagos Islands, Ecuador: Evolution, Ecology, and Diversity. (Insecta: Coleoptera)*. Ottawa: National Research Council of Canada.

Chapter 10—Plants

Darwin, Sarah C. 2009. The systematics and genetics of tomatoes on the Galápagos Islands (*Solanum*, Solanaceae). Doctoral thesis, University College London.

Henslow, J. S. 1837. Description of two new species of *Opuntia;* with remarks on the structure of the fruit of *Rhipsalis*. *Magazine of Zoology and Botany* 1, 466–469.

Hooker, J. D. 1847. Enumeration of the plants of the Galápagos Archipelago, with descriptions of those which are new. *Transactions of the Linnaean Society, London* 20, 163–233.

McMullen, Conley K. 1999. *Flowering Plants of the Galápagos*. Ithaca: Cornell University Press.

McMullen, Conley K., & Close, David D. 1993. Wind pollination in the Galápagos Islands. *Noticias de Galápagos* 52, 12–17.

Peralta, I.E., Spooner, D.M., and Knapp, S. 2008. Taxonomy of wild tomatoes and their relatives. *Systematic Botany Monographs* 84, 1–186.

Porter, D. 1986. Charles Darwin's vascular plant specimens from the voyage of the HMS *Beagle*. *Botanical Journal of the Linnean Society* 93(1), 1–172.

Stöcklin, Jürg. 2009. Darwin and the plants of the Galápagos Islands. *Bauhinia* 21, 33–48.

Chapter 11—The Future of Galápagos

Abedrabbo, S. 1992. Introduction of the black fly on San Cristóbal. *Galápagos Bulletin*. Charles Darwin Foundation, Washington, DC.

Bassett, Carol Ann. 2009. *Galápagos at the Crossroads: Pirates, Biologists, Tourists, and Creationists Battle for Darwin's Cradle of Evolution*. Washington, DC: National Geographic Society.

Benchley, Peter. 1999. Galápagos: Paradise in peril. *National Geographic Magazine* 195, 16–28.

Campbell, Karl, Donlan, C. Josh, Cruz, Felipe, & Carrion, Victor. 2004. Eradication of feral goats *Capra hircus* from Pinta Island, Galápagos, Ecuador. *Oryx* 38(3), 1–6.

Carrion, V., Donlan, C. Josh, Campbell, Karl, Lavoie, Christian, & Cruz, Felipe. 2007. Feral donkey (*Equus asinus*) eradications in the Galápagos. *Biodiversity and Conservation* 16, 437–45.

Carrion, V., Donlan, C. Josh, Campbell, Karl, Lavoie, Christian, & Cruz, Felipe. 2011. Archipelago-wide island restoration in the Galápagos Islands:

Reducing costs of invasive mammal eradication programs and reinvasion risk. *PLoS ONE* 6(5): e18835. DOI:10.1371/journal.pone.0018835

Charles Darwin Foundation, Galápagos National Park Service, and Governing Council of Galápagos. 2010. *Galápagos Report 2009–2010*. Puerto Ayora, Galápagos, Ecuador.

Charles Darwin Foundation & WWF. 2002. *A Biodiversity Vision for the Galápagos Islands*. Ed. R. Bensted-Smith, Charles Darwin Foundation, Puerto Ayora, Galápagos.

Cruz, Felipe, Carrion, Victor,Campbell, Karl J., Lavoie, Christian, & Donlan, C. Josh. 2009. Bio-economics of large-scale eradication of feral goats from Santiago Island, Galápagos. *Journal of Wildlife Management* 73(2), 191–200. DOI: 10.2193/2007-551

Cruz, Felipe, Donlan, C. Josh, Campbell, Karl, & Carrion, Victor. 2005. Conservation action in the Galápagos: Feral pig (*Sus scrofa*) eradication from Santiago Island. *Biological Conservation* 121, 473–78.

D'Orso, Michael. 2002. *Plundering Paradise: The Hand of Man on the Galápagos Islands*. New York: HarperCollins.

Dillard, Annie. 1975. Innocence in the Galápagos. *Harper's Magazine,* 250(May), 74, 77–78.

Donlan, C. Josh, et al. 2007. Recovery of the Galápagos rail (*Laterallus spilonotus*) following the removal of invasive mammals. *Biological Conservation* 138(3–4), 520–24.

Epler, Bruce. 2007. *Tourism, the Economy, Population Growth, and Conservation in Galápagos*. Puerto Ayora, Province of Galápagos, Ecuador: Charles Darwin Foundation.

Galápagos National Park Service. 2002. *Visitor Sites Guide of the Galápagos National Park for Naturalist Guides*. Quito, Ecuador: Testugo.

Gonzalez, José A., et al. 2008. Rethinking the Galápagos Islands as a complex social-ecological system: Implications for conservation and management. *Ecology and Society* 13(2), 13.

Gould, Stephen Jay. 1999. Galápagos Islands. *National Geographic Traveler* 16, 98–100.

Harmer, S. F. 1928. History of whaling. *Proceedings of the Linnaean Society of London* 140, 51–95.

Honey, Martha. 2008. *Ecotourism and Sustainable Development: Who Owns Paradise?* (Second Edition). Washington, DC: Island Press.

Lavoie, C., et al. 2007. *The Thematic Atlas of Project Isabela: An illustrative document describing, step-by-step, the biggest successful goat eradication project on the Galapagos Islands, 1998–2006*. Puerto Ayora, Galapagos: Charles Darwin Foundation, 60

Lemonick, Michael D. 1995. Can the Galápagos survive? *Time* 146(18), (October 30), 80–82.

Mauchamp, A. 1997. Threats from alien plant species in the Galápagos Islands. *Conservation Biology* 11(1), 260–63.

Merlen, Godfrey. 1999. *Restoring the Tortoise Dynasty: The Decline and Recovery of the Galápagos Giant Tortoise*. Puerto Ayora, Province of Galápagos, Ecuador: Charles Darwin Foundation.

Quiroga, Diego. 2009. Crafting nature: The Galápagos and the making and unmaking of a "natural laboratory." *Journal of Political Ecology* 16, 124–40.

Stewart, Paul D. 2007. *Galápagos: The Islands that Changed the World*. New Haven, CT: Yale University Press.

Stone, Richard. 1995. Fishermen threaten Galápagos. *Science* 267, 612.

Townsend, C. H. 1925a. Impending extinction of the Galápagos tortoises. *Bulletin of the New York Zoological Society* 27, 55–56.

Townsend, C. H. 1925b. The Galápagos tortoises in their relation to the whaling industry. *Zoologica* 4, 55–135.

Townsend, C. H. 1942. Can we save the giant tortoise? *Bulletin of the New York Zoological Society* 55, 109–111.

Watkins, Graham, & Cruz, Felipe. 2007. *Galápagos at Risk: A Socioeconomic Analysis of the Situation in the Archipelago*. Puerto Ayora, Province of Galápagos, Ecuador: Charles Darwin Foundation.

Wikelski, Martin, Foufopoulos, Johannes, Vargas, Hernan, & Howard Snell. 2004. Galápagos birds and diseases: Invasive pathogens as threats for island species. *Ecology and Society* 9(1), 5–14.

Appendices

Grant, K. Thalia, & Estes, Gregory B. 2009. *Darwin in Galápagos: Footsteps to a New World*. Princeton, NJ: Princeton University Press.

Instituto Oceanográfico de la Armada. 1992. *Carta Americana No. 22000. América del Sur, República del Ecuador, Archipiélago de Colón (Provincia Insular de Galápagos),* 15th Edition (Issue date: July 22, 1978).

Milicic, D., Lucic, L., and Pavkovic-Lucic, S. 2011. How many Darwins? List of animals named after Charles Darwin. *Natura Montenegrina*, Podgorica 10(4), 515–532.

National Geospatial-Intelligence Agency. 1993. *Agency Notice to Mariners #199312, Archipelago de Colon (Galápagos Islands),* 16th Edition (Issue date: January 30, 1993).

Rosenberg, Steve, & Sarbone, Eilen I. 2004. *The Diving Guide: Galápagos Islands*. Dunedin, FL: Cruising Guide Publications.

Zagier, Sergio, & Woram, John. 2008. *Islas Galápagos Historical Chart*. Buenos Aires: Zagier & Urruty.

GLOSSARY

a'a' lava Rough, generally black lava that forms when gas in magma escapes as the lava cools. A'a' lava, which is more difficult to walk on than pahoehoe lava, is common on volcanoes in the western islands of Galápagos.

abiotic Environmental features that are physical rather than biological, such as temperature, humidity, altitude, and light intensity.

adaptation An inherited trait that enhances fitness by improving an organism's chances of survival and reproduction. The feeding ecology of Darwin's finches matches their morphological diversity; this is why the radiation of these finches is adaptive.

adaptive radiation Development of many different forms from an originally homogeneous group of organisms as it fills different ecological niches; the diversification of species into separate forms that each adapt to occupy a specific environmental niche. In Galápagos, finches and *Scalesia* are excellent examples of adaptive radiation.

allele Alternate form of a gene.

allopatric Geographic separation of a population from others of the same species. Compare with *sympatric.*

alternate A type of phyllotaxis in which there is one leaf at each node.

amniote Vertebrates whose embryos develop in an amnion, which is a protective membranous sac. Mammals, birds, and reptiles are amniotes.

anapsid A lineage of reptiles characterized by skull morphology; anapsids are amniotes that lack temporal holes in their skulls.

angiosperm A flowering plant.

annual A plant that completes its life cycle and dies within one growing season

archipelago A large group of islands; Galápagos is an archipelago.

basalt A dense, dark gray, fine-grained, igneous rock made primarily of plagioclase feldspar and pyroxene. Basalt is the most common type of solidified lava. The Galápagos Islands are formed almost entirely from basalt.

Beagle British gunship on which Charles Darwin spent the years 1831–1836. During this time, Darwin sailed more than 40,000 miles, most of which were focused on South America. Darwin visited Galápagos in September–October of 1835 while aboard the *Beagle.*

biogeography Study of the geographic distribution of organisms and the changes in those distributions over time. Biogeography provides much evidence for evolution.

bilaterally symmetrical The arrangement of body parts oriented around a front-to-back axis such that an organism can be divided by a single longitudinal cut. A bilaterally symmetrical organism has mirror-image right and left sides.

biological control The use by humans of a particular organism to control a particular pest. In Galápagos, Australian ladybugs have been used to control cottony cushion scale.

biotic Environmental features relating to, or resulting from, living organisms.

buccaneer A pirate who raided ships along the Spanish coasts of America during the 17th and 18th centuries. Many buccaneers were licensed by the king or queen of England, thereby forming an unofficial British navy. The word *buccaneer* comes from the French word *boucanier,* which means "meat smoker," which was a common occupation of sailors at the time.

bycatch Fish and other animals caught (and often killed) in a fishery while intending to catch other fish. Bycatch of Peruvian fisheries has significantly reduced the population of waved albatrosses in Galápagos.

caldera A volcanic crater; a caldera forms when a volcano's magma chamber collapses after an eruption, thereby creating a void into which the caldera floor sinks.

carapace The shell of a tortoise or turtle.

CDF Charles Darwin Foundation for the Galápagos Islands. CDF is a nongovernmental, international, not-for-profit organization (registered in Belgium) created in 1959 that provides scientific research, technical information, and assistance to help preserve Galápagos. CDF was founded under the auspices of UNESCO and IUCN. More than 80% of the staff of CDF in Galápagos is Ecuadorian.

CDRS Charles Darwin Research Station; CDRS is the operational branch of the CDF.

cerro Spanish word for *hill;* in Galápagos, *cerros* are often eroded volcanic cones (e.g., Cerro Crocker on Santa Cruz).

character displacement The enhancement of differences between coexisting species by natural selection. The characters (i.e., traits) that diverge may have reproductive or ecological functions, or both. In Galápagos, finches and *Scalesia* are well-documented examples of character displacement.

cinder cone Cones formed by explosive eruptions of gas in magma. In Galápagos, cinder cones are usually further inland than tuff cones.

Cocos finch *Pinaroloxias inornata;* the only one of this group of finches not native to Galápagos. The Cocos finch, which has plumage similar to the Galápagos ground finch, was derived from birds on Galápagos.

coevolution Reciprocal adaptation occurring between species. Examples of coevolution have occurred between flowers and pollinators, predators and prey, and parasites and their hosts.

commensalism Symbiosis in which one symbiont benefits and the other neither benefits nor is harmed.

common ancestor The most recent ancestral form or species from which two species evolved.

competition The seeking of a resource that is in short supply. Competition usually limits the size of a population.

compound leaf A leaf composed of two or more leaflets.

continental drift Movement of continents and other tectonic plates in geologic time over Earth's surface.

critically endangered IUCN classification for taxa when the best available evidence indicates that it faces extremely high risk of extinction in the wild.

crypsis When an organism tends to be overlooked by predators as a result of its size, shape, colors, or behavior.

cultivated species A species intentionally brought to one area from another by humans for cultivation.

Darwin's finches The finches of Galápagos and Cocos Islands, each species of which is adapted to exploit a different food source. These finches vary from grosbeak-like birds weighing up to 35 grams, to warbler-like birds weighing only 8 grams. The term *Darwin's finches* was first proposed by Percy Lowe in 1936 and was made famous by David Lack's book *Darwin's Finches* in 1947. While in Galápagos, Darwin paid little attention to the birds that now bear his name.

deciduous A plant that drops its leaves during a specific season.

descent with modification The phrase used by Charles Darwin to refer to the process by which natural selection favors some variations, resulting in their becoming more common in the next generation. *Descent with modification* was Darwin's term for *evolution.*

diapsid A lineage of reptiles characterized by skull morphology; diapsids have two temporal holes in each side of their skulls.

dioecious Producing eggs and sperm on different individuals. About 3% of the native and endemic plants of Galápagos are dioecious.

dry landing An exit from a panga directly onto a dock or other dry site.

El Niño Southern Oscillation (ENSO) A warming of surface waters (usually 3–5°C above average) in the equatorial eastern Pacific that occurs every 3 to 8 years when the warmer, nutrient-poor waters of the Panama Current flow further south than usual. Also referred to as *El Niño*; compare with *La Niña.*

endemic Organisms that are restricted to a region. Organisms endemic to Galápagos live only in Galápagos; these organisms have evolved longest and are adapted specifically to Galápagos habitats.

ENSO See *El Niño Southern Oscillation.*

epiphyte A plant that grows on another plant without gaining any nutrients from it. In Galápagos, mosses are common epiphytes on many islands. By definition, epiphytes are not parasites.

equator An imaginary circle around the center of the Earth that is equally distant from both poles, dividing the planet into northern and southern hemispheres. The equator constitutes the parallel of latitude 0°.

evolution Change in the genetic makeup of a population over time. Evolution occurs over successive generations, not in the lifetimes of individual organisms. Charles Darwin referred to evolution as *descent with modification.*

evolutionary developmental biology (evo-devo) A discipline of biology that compares the development of different organisms to determine ancestral relationships and the developmental mechanisms that produce evolutionary change.

extirpate Eliminate; eradicate. The best examples of successful extirpations in Galápagos involve the eradication of donkeys, goats, and pigs from some of the larger islands.

exotic An organism brought to an area from another by humans; also called *introduced species*.

extinct No longer existing or living.

extinction The disappearance of a breeding population.

feral An animal that has returned to the wild condition after once being domesticated; a domesticated animal "gone wild." Feral animals have radically changed the landscapes and populations of several animals on several islands in Galápagos.

fitness The ability to survive and reproduce, as measured—most simply—by the number of offspring produced; the relative success of an organism measured by its ability to contribute genes to future generations.

flower The reproductive structure of a flowering plant (i.e., angiosperm).

fruit The seed-bearing part of a flowering plant.

fumaroles Volcanic vents that emit gases, often containing sulfur dioxide. In Galápagos, Volcán Alcedo, Volcán Sierra Negra, and Roca Redonda have accessible fumaroles.

Galápagos Volcanic islands in the Pacific Ocean approximately 930 kilometers west of Ecuador (the country to which they belong) visited by Charles Darwin in September 1835, while aboard the *Beagle*. Darwin's observations in Galápagos later helped him formulate his theory of evolution by natural selection. Of the large islands of Galápagos, only Marchena, Pinta, Genovesa, Wolf, Darwin, and Northern Isabela are in the North Pacific.

garúa The mist and fog that often hang in the higher elevations during the dry season (i.e., July to December) in Galápagos.

gene A unit of heredity. Genes are sequences of DNA that contain information needed to make a protein.

gene flow Spread of an allele through a population; the addition of genes to a population through interbreeding with members of another population, usually of the same species.

genetic drift Random change in the frequencies of alleles from one generation to the next; a change in the allelic frequencies of a small population due to chance rather than natural selection.

genotype The genetic composition of an individual.

GNP Galápagos National Park.

GNPS Galápagos National Park Service.

gular On or of the throat region, especially in birds. In Galápagos, a male frigatebird seeking a mate inflates and displays its bright red gular sac.

habitat The part of an ecosystem in which an organism lives.

herb A plant that lacks a persistent, above-ground woody stem.

heritability A measure of family resemblance as a result of shared genes.

hermaphrodite An individual with male and female reproductive parts.

hornito A small cone (usually less than 0.5 m high) formed in lava by escaping gas. Hornitos, which are often colored by their minerals, are usually found on pahoehoe lava.

hotspot An upwelling of abnormally hot rock within Earth's mantle that creates volcanic activity. The Galápagos Islands were formed by the Galápagos Hotspot, which is now beneath Fernandina Island. Hotspots are also referred to as *mantle plumes*.

hyper-saline Saltier than the ocean.

igneous Rocks formed by solidification of molten or partially molten magma. Igneous rocks form when magma cools.

immutable Unable to be changed.

inflorescence The arrangement of flowers on a floral axis.

INGALA Galápagos National Institute, the regional planning body of the archipelago. In 2009, INGALA merged with the Provincial Government of Galápagos to form the Governing Council of the Special Region of Galápagos.

insulares Residents of Galápagos who moved to the archipelago in search of economic and social opportunities.

intersexual selection Differential reproduction based on the ability to acquire a mate ("mate choice").

interspecific A term describing interactions between members of different species.

intrasexual selection Differential reproduction based on the ability to out-compete same-sex rivals for access to mates.

intraspecific A term describing interactions between members of the same species.

introduced species A species brought to Galápagos by humans, either accidentally or intentionally; on average, only about 2% of the introduced species have become invasive species. Introduced species are also called *exotic species*.

introgression The transfer of genetic information from one species to another as a result of hybridization and repeated backcrossing.

invasive species An introduced species that damages biodiversity, agriculture, or human health.

isleños Residents of Galápagos who, if they were not born in Galápagos, have lifestyles deeply rooted in Galápagos. *Isleños* are more likely to accept restrictions of life in Galápagos (e.g., restrictions on fishing and interactions with animals).

IUCN International Union for the Conservation of Nature; IUCN's Red List of Threatened Species, which assesses the conservation status of species, includes many species native and endemic to Galápagos.

Jessica Oil tanker that ran aground on January 16, 2001, in Wreck Bay at San Cristóbal. The resulting spill of fuel and subsequent environmental damage highlighted the risks associated with importing fuel to meet the needs of the archipelago's burgeoning population and number of visitors.

Judas goat In Galápagos, a goat used to locate feral goats. Judas goats seek other goats and are equipped with radio transmitters, which enable GNPS rangers to find them and kill their feral companions. Judas goats, which are named for the biblical character Judas Iscariot (who betrayed Christ for 30 pieces of silver), are in hormonally induced estrus, thereby making them especially attractive to other goats.

kleptoparasitism A symbiosis in which desirable resources (food or prey) are stolen from another individual (literally, "parasitism by theft").

La Niña A cooling of surface waters in the equatorial eastern Pacific that is associated with low surface pressures in the western Pacific. Compare with *El Niño*.

lava tube A tube that forms when crust forms on flowing lava, thus insulating the hot core. When the flow of lava stops, a tube is left behind. In Galápagos, some lava tubes extend several kilometers.

leaflet One of the parts of a compound leaf.

leeward The direction downwind from the point of reference; toward the side to which the wind is blowing. In Galápagos, the leeward (i.e., northwestern) sides of high islands such as Santa Cruz are often rain shadows that receive less moisture than the windward (i.e., southeastern) sides of the islands.

magma Molten, underground volcanic material. Cooling magma forms lava and other igneous rock.

mantle plume See *hotspot*.

mid-oceanic ridge An underwater mountain range formed by plate tectonics and typically having a valley known as a rift running along its spine. A mid-oceanic ridge marks the boundary between tectonic plates. The mid-oceanic ridges of the world are connected and form a single global mid-oceanic ridge system that is part of every ocean. The mid-oceanic ridge system is the longest mountain range in the world.

migrant species A species that periodically moves to and from an area, usually at predictable times of the year along well-defined routes.

migration Movement of animals from one region or habitat to another, typically according to seasons.

monoecious Hermaphroditic; producing eggs and sperm on the same individual. Most native and endemic plants of Galápagos are monoecious.

monogamy Having a sexual relationship with only one partner.

monophyletic A group of species that share the same common ancestor, and share it with no other species. *Darwin's finches* are monophyletic; they share the common ancestor that colonized Galápagos, and share it with no other species. Galápagos finches (including the Cocos finch) are more closely related to each other than to any species elsewhere.

mutation A random change in genetic information. Mutations, which can be caused by mistakes during replication or by damage from external agents such as chemicals or radiation, produce the genetic changes that provide new genetic information. Mutations create new alleles. Heritable mutations provide the template of variation upon which natural selection acts.

mutualism An interaction between organisms in which each participant experiences a gain in fitness.

native species A species that arrives and establishes itself in a particular area. Native species in Galápagos live in the archipelago and elsewhere, but were not brought to the islands by humans.

natural selection An evolutionary mechanism that produces differences in survival and reproduction among organisms with different heritable traits. Natural selection is the mechanism for adaptive evolutionary change proposed by Charles Darwin in *On the Origin of Species* (1859). Natural selection produces evolution only if the selected trait is heritable.

niche The part of the environment that is used by organisms.

olivine A translucent, volcanic, olive-green silicate that is rich in iron and magnesium. In Galápagos, large amounts of olivine are found at Punta Cormorán

(Floreana), Punta Pitt (San Cristóbal), and Whale Bay (Santa Cruz).

operational sex ratio The ratio of reproductive males to fertile females within a population, often expressed as a deviation from 1:1.

opposite A type of phyllotaxis in which leaves form in pairs at nodes, across from each other.

orchil A group of lichens (e.g., *Roccella*) from which can be extracted purple dyes. Early settlements of Galápagos supported themselves by harvesting and selling orchils.

ossicle Calcified material forming part of the exoskeleton of invertebrates such as echinoderms.

pahoehoe lava A type of smooth or ropelike lava formed from magma containing relatively little gas. In Galápagos, pahoehoe lava is common on Santiago and on Fernandina at Punta Espinosa. Charles Darwin described pahoehoe lava as "singular ringed & twisted forms, which resemble cables, folds in thick drapery and rugged bark."

panga A small inflatable boat used to go between tour boats and Visitor Sites in Galápagos.

parasitic cone A volcanic cone that forms along the flank of a larger cone. Parasitic cones are abundant on the slopes of Fernandino's Volcán La Cumbre.

parasitism Symbiosis in which one symbiont benefits and the other symbiont is harmed.

pelagic A species that spends all or most of its life at sea.

perennial Living more than one year.

phenotype The physical or physiological traits of an individual; the manifestation of genetic and environmental influences.

phyllotaxis The arrangement of leaves on a stem.

phylogeny The evolutionary history of a group of organisms, often depicted as a branching diagram, or phylogenetic tree; a grouping of populations on the basis of their relationships (usually genetic) that reflect their evolutionary history.

pit crater Crater formed when the roof of a large underground lava tube or magma chamber collapses. In Galápagos, Los Gemelos on Santa Cruz are pit craters.

plagioclase A white form of feldspar common in igneous rocks. Lava at Pinta Island is especially high in plagioclase.

plate tectonics The theory that Earth's crust consists of movable plates that can join or separate over geologic time. The movements of plates explains continental drift, earthquakes, volcanoes, mountain building, and some aspects of biogeography. Although most earthquakes and volcanic eruptions occur near boundaries of tectonic plates, the Galápagos Islands (and Hawaii) were formed by *hotspots* far from the nearest plate boundary.

pneumatophore An upward-growing root modified for aeration. Pneumatophores are typical of black mangroves.

polyandry A reproductive strategy in which one female mates with two or more males.

polygamy A reproductive strategy in which an individual has more than one mate.

polygyny A reproductive strategy in which one male mates with two or more females.

population A group of interbreeding organisms and their offspring.

predation The act of one animal killing another for food.

pumice Porous, lightweight, volcanic rock formed when frothy, gas-rich lava solidifies rapidly. In Galápagos, pumice is found only on Isabela's Volcán Alcedo. However, because it floats, pumice often washes ashore on other islands, especially after the rainy season.

pyroclastic cone A cone formed when basaltic magma contacts seawater, thereby converting the water to steam and the magma fragments into ash particles that react with steam to form palagonite, a lightly colored material that forms the cone. This process is violent; ash is thrown into the atmosphere, and a cone forms as it settles back to the ground around the vent.

radially symmetrical The arrangement of body parts of an organism like pieces of a pie around an imaginary central axis; any cut from the top to the bottom along that axis divides the organism into mirror-image halves.

reproductive isolation The lack or scarcity of interbreeding between populations. Reproductive isolating mechanisms produce speciation.

resident A species that lives year-round in a particular area.

resource competition A relationship in which individuals of different species lose fitness while fighting for control of, or access to, a desirable commodity (e.g., food, territory).

seafloor spreading The process by which new seafloor is created, thereby pushing two tectonic plates apart. Seafloor spreading usually occurs along a mid-oceanic ridge, such as at the Galápagos Spreading Center.

seamount A mountain that rises from the ocean floor but does not reach the water's surface. Because they are volcanic in origin, seamounts are often near mid-oceanic ridges and mantle plumes. Seamounts are abundant between present-day Galápagos and South America.

sequential hermaphrodite An organism that can change from one sex to another. *Protandrous hermaphrodites* begin life as males ("male first"); *protogynous hermaphrodites* begin life as females.

sexual dimorphism Condition in which individuals have different phenotypes because of their sex.

sexual selection Differences in fitness attributable to differences in the ability to obtain mates.

shield volcano A volcano shaped like a shield because of its broad, gently sloping sides formed by slow, repeated flows of lava. Genovesa was formed by a shield volcano; Darwin Bay formed when one side of the caldera collapsed. Other classic examples of shield volcanoes in Galápagos include Santa Cruz, Fernandina, and Sombrero Chino.

siblicide The killing of a sibling or siblings, often as a behavior in various animals. In Galápagos, Nazca boobies exhibit obligate siblicide.

SICGAL Galápagos Inspection and Quarantine System, whose primary goal is to eliminate the introduction of new species in Galápagos. SICGAL is the acronym for Sistema de Inspección y Cuarentena Para Galápagos.

spatter cone Small cone formed by spattered lava which solidifies in colorful, smooth shapes.

speciation The evolution of new species.

subduction The process by which two tectonic plates collide, and one is forced beneath the other, usually in association with mountains and volcanoes. The subduction of the eastward-moving Nazca Plate beneath South America has built the Andes.

subspecies A population of a species that differs from other populations of the same species in appearance and/or size, but these differences do not warrant the rank of species. In Galápagos, subspecies (1) are either more recent arrivals that have had less time to evolve, or (2) have resulted from the subsequent evolution of endemic species.

symbiosis A relationship between two or more interacting organisms. Symbioses occur on a continuum from antagonists (e.g., parasitism, predation, competition) to obligate mutualisms, in which the symbionts cannot function without each other.

sympatric Occupying overlapping geographic range; a geographic region jointly occupied by two or more populations. Compare with *allopatric*.

synapsid A lineage of reptiles characterized by skull morphology; synapsids have a single temporal hole in each side of their skulls.

taxa Taxonomic groups that constitute the ranks of classification, such as family, genus, and species.

trait A genetically determined characteristic.

trophic cascade An ecological phenomenon in which entire food webs can be drastically altered by the addition or removal of a species, usually a top predator.

tuff cone A cone of compacted volcanic ash formed when water enters an erupting volcanic fissure, thereby creating a steam explosion that sends pulverized dust high into the atmosphere. The subsequent consolidation of this ash forms the tuff cone. Tuff cones always occur near the coast. Charles Darwin was the first geologist to realize that tuff cones ("the most striking feature of this Archipelago") are created by underwater mixtures of seawater and rising magma, which produce aerial explosions that fall back to earth as muddy rain. Examples of tuff cones in Galápagos include Daphne Major, Bartolomé's Pinnacle Rock, and Santiago's Cerro Cowan.

UNESCO United Nations Educational, Scientific and Cultural Organization. In 1978, Galápagos was the first site inscribed on the UNESCO World Heritage List.

uplift A landmass created by lava flowing through an underground fault, thereby lifting the overlying land above sea level. In Galápagos, uplifted islands (e.g., Baltra, North Seymour, and Las Playas) have been rising for more than a million years.

upwelling The rise of deep water to the surface.

wet landing An exit from a panga into shallow (e.g., knee-deep) water.

windward The direction upwind from the point of reference. In Galápagos, the windward (i.e., south-eastern) sides of high islands such as Santa Cruz receive more moisture than the leeward (i.e., north-western) sides.

PHOTO CREDITS

Color Images: 8, 11: Sarah Knutie; 20, 25, 26, 37: Daniel Fitter; 22, 32, 35, 36, 38: José Bravo; 27: Julia Ponder; 28, 29, 31: Tom Alfoldi; 30: Jon Degenhardt; 34: Rafael Rubio; Front and back covers: Daniel Fitter

Black & White Images: 2.2: Library of Congress; 2.7: Smithsonian Institution Archives, Record Unit 7231, Box 89, Image #SIA2011-1127; 3.1: Dennis Geist; 5.8: Sarah Knutie; 5.12: Daniel Fitter; 6.1: Julia Ponder; 8.3: Rafael Rubio; 11.1: Tui De Roy/Roving Tortoise Nature Photography; 11.2: C. Josh Donlan; 11.5: Rory Stansbury/Island Conservation

INDEX

Note: Books, boats, quotations, and scientific names are listed in italics, as are page numbers that refer to figures. GPS coordinates for many of the sites listed here are in Appendix 1.